28.40

☀ INSIGHT GUIDES

amazon
WILDLIFe

D1295899

APA PUBLICATIONS   L
Part of the Langenscheidt Publishing Group

INSIGHT GUIDE

amazon
WILDLIFE

Editorial
Project Editor
Maria Lord
Editorial Director
Brian Bell

Distribution

United States
Langenscheidt Publishers, Inc.
46–35 54th Road, Maspeth, NY 11378
Fax: (718) 784 0640

Canada
Thomas Allen & Son Ltd
390 Steelcase Road East
Markham, Ontario L3R 1G2
Fax: (1) 905 475 6747

UK & Ireland
GeoCenter International Ltd
The Viables Centre, Harrow Way
Basingstoke, Hants RG22 4BJ
Fax: (44) 1256 817988

Australia
Universal Press
1 Waterloo Road
Macquarie Park, NSW 2113
Fax: (61) 2 9888 9074

New Zealand
Hema Maps New Zealand Ltd (HNZ)
Unit D, 24 Ra ORA Drive
East Tamaki, Auckland
Fax: (64) 9 273 6479

Worldwide
**Apa Publications GmbH & Co.
Verlag KG (Singapore branch)**
38 Joo Koon Road, Singapore 628990
Tel: (65) 6865 1600. Fax: (65) 6861 6438

Printing

Insight Print Services (Pte) Ltd
38 Joo Koon Road, Singapore 628990
Tel: (65) 6865 1600. Fax: (65) 6861 6438

©2002 Apa Publications GmbH & Co.
Verlag KG (Singapore branch)
All Rights Reserved

First Edition 1990
Fourth Edition 2002

CONTACTING THE EDITORS
We would appreciate it if readers
would alert us to errors or out-
dated information by writing to:
**Insight Guides, P.O. Box 7910,
London SE1 1WE, England.
Fax: (44) 20 7403 0290.**
insight@apaguide.demon.co.uk

www.insightguides.com

ABOUT THIS BOOK

This guidebook combines the
interests and enthusiasms of
two of the world's best-known information providers: Insight Guides,
whose titles have set the standard
for visual travel guides since 1970,
and Discovery Channel, the world's
premier source of nonfiction television programming.

The editors of Insight Guides provide both practical advice and
general understanding about a destination's history, culture, institutions and people. Discovery Channel
and its website, www.discovery.com,
help millions of viewers explore their
world from the comfort of their own
home and encourage them to explore
it firsthand.

Insight Guide: Amazon Wildlife is
structured to convey an understanding of the wildlife and ecology of the
Amazon as well as to guide readers
through its sights and activities:

◆ The **Features** section, covers the
flora and fauna, and indigenous
peoples of the Amazon in a series
of informative essays.

◆ The main **Places** section is a
complete guide to all the sights and
areas worth visiting. Places of special interest are coordinated by
number with the maps.

◆ The **Travel Tips** listings at the
back of the book provide a handy
point of reference for information on
travel, national parks, accommodations, restaurants and more.

The contributors

This revised edition of *Insight Guide:
Amazon Wildlife* has been completely updated by a team of three
experts on the region.

EXPLORE YOUR WORLD®
Discovery
CHANNEL

The chapters on the geography, flora and fauna of the region were checked and updated by **Danny Aeberhard**, a Latin America expert and wildlife enthusiast.

The chapters on the indigenous peoples of Amazonia were provided by **Fiona Watson** of Survival International. Insight Guides are grateful to Survival International (www.survival-international.org) for allowing them to reprint text from their publication *Disinherited – the Indians of Brazil*, which form the "Shamanism", "Land Ownership" and "Genocide" chapters of this new edition.

The task of updating the Places and Travel Tips sections of this book fell to **Claire Antell** of the Latin American Travel Association. She also revised the chapter on "Traveling in the Amazon".

The current edition builds on the solid foundations of previous editions. The book was initially created by **Hans-Ulrich Bernard**, who contributed several chapters, among them the pieces on mammals, insects, traveling in the region and boat excursions from Manaus. **Wilfried Morawetz** wrote about the region's different ecosystems. **Walter Hödl** described the diversity of Amazonia's frogs, Manaus, the ecology of floating meadows, and nature observation in the Guianas. **Martin Henzl** wrote on reptiles, as well as providing photographs, and **Peter Krügel** wrote on the bromeliads.

William Overal contributed text on the Amazon River and Marajó Island. Overal and **João Batista da Silva** wrote on Amazonian flora. **Martin Kelsey** provided the chapter on birds, and also wrote about the illegal trade in parrots. He and his wife **Claudia Camacho** wrote the chapters on Colombia. **George Monbiot** contributed the chapters on fishes, the Manaus fish market, conservation strategies, and on Tomé Acú. **Klaus Jaffe** wrote the chapter on Venezuela, and **William L. Murphy** wrote on Trinidad.

Nigel Dunstone described the national parks of Ecuador and Peru. **Carlos E. Quintela** wrote about the national parks of Bolivia, the Pantanal in Brazil and Brazil's Atlantic forest. **Jeanne Mortimer** described the Rio Trombetas. Much of the outstanding photography is by **Luiz Claudio Marigo**, and **Michael** and **Patricia Fogden**.

This latest edition was proofread by **Emily Hatchwell**, and indexed by **Elizabeth Cook**.

Map Legend

──·─··	International Boundary
────	Province Boundary
⊖	Border Crossing
──•──	National Park/Reserve
────	Ferry Route
✈ ✈	Airport: International/ Regional
🚌	Bus Station
❶	Tourist Information
✉	Post Office
✝ † ⴕ	Church/Ruins
∴	Archaeological Site
∩	Cave
⚊	Statue/Monument
★	Place of Interest

The main places of interest in the Places section are coordinated by number with a full-colour map (e.g. ❶), and a symbol at the top of every right-hand page tells you where to find the map.

amazon WILDLIFE

CONTENTS

A caiman drifting
on a tributary of
the Amazon.

Travel Tips

◆ **Full Travel Tips index
is on page 305**

Features

Places

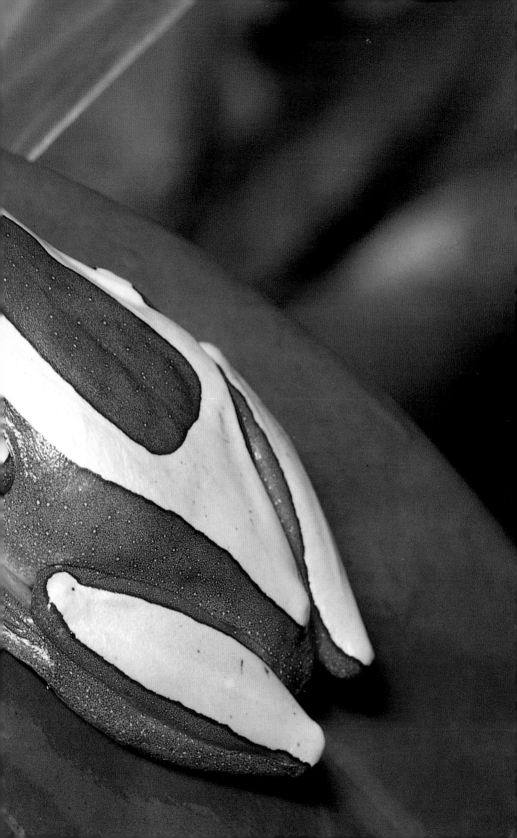

NATURE'S GREATEST SHOW

The forests and rivers of the Amazon are among the most fabulous sights in the world

The word "Amazon" stands for the world's mightiest river as well as for the surrounding biogeographic region, the tropical forests of the Amazon river system. These forests grow in the largest area on earth dominated by moist tropical climate and have evolved our planet's greatest diversity of plant and animal species. The Amazon is one of the largest remaining contiguous tracts of nature on earth, and the only large one in the tropics. More than 80 percent of these forests are still intact, in spite of localized heavy destruction and the justified fear of further losses.

A trip to the Amazon is exhilarating and will also act as a salutory warning as to how much damage humans can inflict on fragile environments. The sheer size of the region and the beauty of your surroundings may, at times, be overwhelming, as can the destruction caused by mining, logging and the spread of settlements. This destruction is not limited in its impact to the flora and fauna of the region; it is equally devastating to the indigenous peoples, and one of the tourist's primary aims should be to ensure that they impact as little as possible on the natural world and local human populations.

The introductory chapters of this book describe the region's ecosystems, the most conspicuous groups of plants and animals, human interaction with the nature of the Amazon, and the human rights abuses faced by indigenous peoples. Subsequent chapters deal with particular geographical areas, with national parks, or, where parks have not yet been gazetted or developed, with nature explorations from the cities of the region. Included is a description of some nature reserves outside the Amazon biogeographical region, which are close to the large cities, and are the gateways to the Amazon.

Information and illustrations were gathered by professional biologists and two of the best nature photographers that work in South America. The result is a comprehensive guidebook to the Amazon for the nature tourist and a documentation of the beauty of the region in this time of increasing environmental concern and conservation movements. With the growing interest in the protection of tropical nature, ecotourism may become an important source of income for the people of the Amazon, and one which is likely to make sustainable use of natural resources. ❑

PRECEDING PAGES: two Jaguars square up; a Common Iguana waits on a branch; a Striped-backed Tree Frog clings to a leaf; a Spiny Devil Katydid in close-up.
LEFT: Venezuela's Angel Falls.

THE GEOGRAPHY

The world's most diverse ecosystem is highly dependent on its geographical environment

The vast lowland basin surrounding the Amazon river and its tributaries – Amazonia – covers 3.7 million sq km (1.4 million sq miles). It is bordered by the Andes in the west, the Guayana shield in the north, the Brazilian shield in the south and the Atlantic Ocean in the east. The area has a generally homogenous moist, warm climate, a prerequisite for the growth of tropical rainforest. This climate changes at Amazonia's borders: to the west, to the cold montane climate of the Andes, and, to the north and south respectively, the dry climate of the *llanos* of Venezuela and the *cerrado, sertão* and *chaco* of Bolivia and Brazil.

Climate and diversity

The extreme diversity of plant and animal life in tropical rainforests may stem from the lack of stress from drought or frost – major selective forces in all other ecosystems. In moist tropical climates, lowland temperatures hover around 28°C (82°F) throughout the year, with daily maxima and minima between 32 and 23°C (89 and 73°F). Annual precipitation normally exceeds 2,000 mm (79 inches), much of which falls throughout the year as showers that most often develop in the afternoon. Prolonged rainfall may occur during the rainy season, which lasts for three or four months, in most parts of the region from November to March.

A moist tropical climate is also found elsewhere in South America. Some areas along the Caribbean coast and in the Guayanas are more or less contiguous with Amazonia and have many common plants and animals. The Pacific slope of the Andes, a region separated by the mountains from the Amazon, has evolved a fairly different flora and fauna, related to that of Central America. To the south, the dry *cerrado* of central Brazil prevents the spread of Amazonian animals and plants, but moist coastal and river corridors have permitted some migration to the Atlantic rainforest in southeastern Brazil, 2,500 km (1,500 miles) to the south of the equator.

Geological influences

One of the most important factors that influenced the Amazon's flora and fauna was South America's long geographic isolation. About 100 million years ago, South America was joined to Africa, Antarctica and Australia as part of the supercontinent Gondwanaland. Subsequently, Gondwanaland broke apart and the four continents drifted apart on separate tectonic plates. For several tens of millions of years, throughout most of the tertiary era, South America existed as an island, without contact with other continents. Many families of animals and plants that became extinct elsewhere were preserved on this island continent. Examples among the mammals include the edentates, which occur only in tropical America, and the marsupials, which, however, have survived in even larger diversity on another isolated continent, Australia.

South and North America fused together at the isthmus of Panama only a few million years ago, during the Pliocene period. Plants and animals could now move with ease between the two continents. Today, some of the South American species like the Nine-banded Armadillo have spread several thousand kilometers into the northern subcontinent, and ungulates and cats, which arrived from the north, are widespread in South America.

Throughout the tertiary era, South America remained close to the equator with its moist and warm climate. However, the climate varied in response to various Ice Ages, when it became drier rather than cooler.

During the last Ice Age, much of the Amazon was covered by a savanna-like vegetation, and the rainforest retreated to small pockets, for example along the eastern foothills of the Andes. These served as refuges for rainforest fauna and flora and as centers of evolution of new species. This temporary isolation is thought to have contributed to the immense biological richness of the Amazon basin, and areas of particular pronounced biodiversity are seen as remnants of previous moist refuges. ❑

LEFT: upland rainforest in Brazil's Pará State.

THE AMAZON RIVER

This immense river system passes through a number of contrasting landscapes, climates and ecosystems. The statistics are startling

The Amazon River contributes almost one-fifth of the total annual amount of fresh-water discharged into the oceans of the world. It has a water flow five times that of the Congo, and 12 times that of the Mississippi. In 24 hours it discharges into the Atlantic as much water as the Thames carries past London in one year. Ocean-going freighters can travel 3,720 km (2,310 miles) inland from the Atlantic to Iquitos, Peru. This makes the Amazon the longest navigable natural waterway in the world. The force of the river reaches far out into the ocean; fresh water can be detected more than 100 km (60 miles) off the coast of South America.

The Amazon River is 6,470 km (4,020 miles) long and has a total water flow of 160,000 to 200,000 cubic meters (5.6-7 million cubic ft) per second, varying seasonally. In the dry season, it flows an average 2.6 km (1.6 miles) per hour. This is remarkable since the main river drops only 70 meters (230ft) in the 3,100 km (1,926 miles) between the Peruvian frontier and its mouth.

The true source of the Amazon was discovered in 1953 to be a small stream, the Huarco, rising near the summit of Cerro Huagra in the Peruvian Andes. The Huarco becomes the Río Toto, then the Río Santiago, then the Río Apurimac, followed by the Río Ene, then the Río Tambo, which flows into a main tributary of the Amazon, the Río Ucayali, still in Peru. The Amazon flows across Peru to Brazil, where it is known as the Rio Solimões until it reaches the confluence with the large tributary, the Rio Negro near Manaus. The last 1,600 km (1,000 miles) from Manaus to the mouth is the Rio Amazonas. In the 990 km (615 miles) from the source of the river to Atalaya, on the Río Tambo, it drops 4,450 meters (14,600ft) in altitude, but from Atalaya to the Atlantic, it drops only another 194 meters (636ft).

The word *Apurimac*, in the local language, means "Great Speaker" because of the roar of its rapids. Due to the sudden drop in altitude, the Río Apurimac is full of falls and rapids.

The steep Andean mountain slopes of the first 990 km (615 miles) of the river are completely different from the lowlands in every feature. However, most people associate the Amazon with the lowland tropical rainforest.

The official source of the Amazon, on Mount Huagra, is a small spring about 15 cm (6 inches) wide seeping out of the spongy high-altitude grassland. The small stream formed from this spring, the Huarco, is soon joined by others filled by the melting snow. Its clear water soon turns a murky yellow with mine tailings and then joins the Río Toro. As the river drops down to the Río Apurimac, the mountain slopes are covered with a moist, rich cloud forest. After leaving the Andes, the present-day Amazon River flows over the tertiary lake bed until below Tefé in Brazil. The last part of it flows through quaternary sediments. The sediments on the bed of the lower Amazon are up to 2,000 meters (6,500ft) deep in places.

The upper part of the river has the cold Andean climate; around its source, snow falls frequently. As it descends, the climate becomes progressively warmer, more humid and tropical. The region of cloud forest is moist and cool. The rainfall in the lowlands near the Andes, at Iquitos in Peru, is high, about 2,600 mm

---

DIFFERENT WATERS

One of the most important features of the Amazon river system is the different types of water which occur within the basin. This can readily be seen near Manaus, where the Amazon and the Rio Negro flow together. The Amazon has a muddy brown color and is full of silt and alluvial matter. In local terminology, this is called a white water river. In contrast, the Rio Negro is the color of strong tea and has very little silt, and is an example of a black water river. Where these two large rivers converge at Manaus, the river water is clearly divided into two colors for 15–25 km (9–15 miles) downstream until they eventually mix together.

LEFT: a small tributary winding its way through lowland forest.

(102 inches). In central Amazonia the rainfall is considerably less – 1,770 mm (70 inches) at Manaus, but it increases again near the coast to 2,277 mm (90 inches) around Belém.

Apart from their different appearances, the two main water types *(see panel, page 19)* differ in many ways that have a profound effect on the flora and fauna of the region, and, as a result, on human settlements. The white water rivers are only slightly acidic, with an average pH of 6.5 to 7.4, and have much less humic matter. In contrast, the Río Negro has a pH of 4.6 to 5.2, making it very acidic, and has a much higher proportion of humic matter. The water properties

Careiro Island near Manaus. Nevertheless, many of its tributaries, such as the Juruá, are meandering snake-like rivers that have changed their course many times.

These winding rivers constantly cut into one bank and build up silt on the opposite margin. They are characterized by a large number of oxbow lakes made by former channels of the river that have become isolated.

Along the meandering rivers one encounters many shortcuts dug through the flooded forest for small boats to use in the flood season to avoid sailing around the long loops of the river. This, and the fact that all the numerous sandy

cause different plant species to grow in areas flooded by the different water types. For example, the famous Amazon water-lily, *Victoria amazonica,* grows only in white water areas; it cannot grow in the acidic black water. In addition to the black and white water types, there is a third classification in the Amazon river system: a few rivers have clear water containing very little humic matter or silt. The biggest clear water river is the Río Tapajós, which flows north into the Amazon, 800 km (500 miles) west of its mouth.

The Amazon River proper cuts a reasonably straight line from where the Ucayali and Napo rivers meet in Peru to its mouth, being interrupted by only a few islands such as the large

beaches are flooded and can be traveled over in boats, makes flood-season traveling much quicker than dry season navigation. There are also numerous lateral channels near the main river. These are called *paranás.* The *paranás* are heavily settled side ducts of the main stream, and are much used as shortcuts.

Seasonally different rainfalls and the level relief of the Amazon bring about regular drastic, but predictable, changes of the river level: in Manaus, high water at the end of the rainy season in March towers 15 meters (50 ft) over the dry season low water mark in October.

The forests that are inundated for several months by the rising water are called *igapó* along black water rivers and *várzea* around white water rivers. Migration between the river bed and these flooded forests is important for many aquatic species. For many fishes, the river bed is poor in food, and the flooding gives them access to special ecological niches, for example, feeding on the fruit that drops from forest trees. For species like manatees, the rise of the water opens access to the floating meadows, which shrink dramatically to small puddles during the dry season, and permits

HIGH WATER

Traveling by boat at high water allows easy observation of wildlife in the flooded forest. The animals are concentrated at the treetops, which are not just accessible but almost at eye level.

to 5 meters (16 ft) in height, whose roar can be heard many kilometers away. Although predictable, local people still head for high river banks for protection.

Geologists have revealed the very peculiar history of the Amazon. Some 250 million years ago, what is today the Amazon River Basin was originaly an oceanic gulf that opened westward to the Pacific Ocean. At that time, all rivers originated on two Archean massifs, the Guayana and the Brazilian shields, and ran towards this western

grazing on a rich diversity of water plants.

The end of the rainy season is the time of optimal food supply for many land animals, and species like monkeys are much more agile during this season than at other times. For nature observation by boat, such as most tours that start in Manaus, the period from the end of the rainy season, March onward, is the best.

Close to the coast the *Pororoca*, a powerful tidal bore, occurs twice per year during the highest tides of the Atlantic. It is a wave of up

LEFT: flooded forest *(igapó)* along the Río Negro.
ABOVE: the floodplain on Marajó Island, at the mouth of the Amazon.

gulf. About 100 million years ago, this gulf was cut off by the uplifted Andes mountains. A huge inland sea grew, containing much of the original oceanic fauna. These animals survived only if they evolved to tolerate fresh water. These geological events are the reason for today's presence in the Amazon river of some species – such as stingrays, which elsewhere occur only in the sea.

Slowly, sediments up to 4,000 meters (13,000ft) deep filled the inland sea and created today's Amazon basin. About 60 million years ago, the sea broke through its eastern escarpment, and the Amazon river system eroded into these sediments. A river which would now drain into the Atlantic was born. ❑

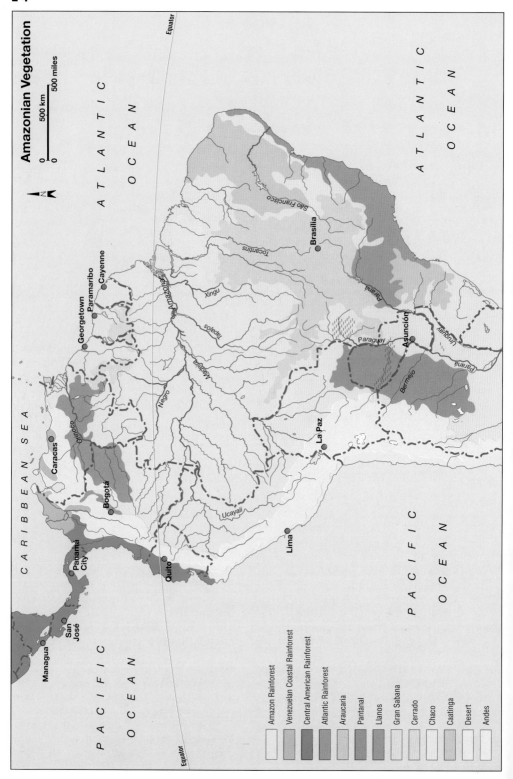

Amazonian Vegetation

Amazon Rainforest
Venezuelan Coastal Rainforest
Central American Rainforest
Atlantic Rainforest
Araucaria
Pantanal
Llanos
Gran Sabana
Cerrado
Chaco
Caatinga
Desert
Andes

HABITATS AND VEGETATIONAL ZONES

The diversity of Amazonian habitats ranges from the mountain cloud forests to the flooded forests of the lower river

Amazonia contains the world's largest closed tropical forest system with an estimated size of about 3.7 million sq km (1.4 million sq miles), roughly the size of the Indian subcontinent.

Strictly speaking, only the lowlands and some isolated mountain ranges are covered by Amazonian plant species, but adjacent or even disjunct vegetation types in South America – such as the Atlantic coastal forest of southeastern Brazil – show strong floral or structural relationships.

Species diversity is high, and at least 20 percent of all existing Angiosperms are found in Amazonia. The age of some plant families (up to 120 million years) and several migration and separation events of the tropical forests during earth's history permitted this tremendously diverse evolution. Migration and separation was triggered on the one side by the frequent flooding of the Amazon basin by the sea or by droughts during the Northern Ice Age.

Species diversity survived climatic changes only in mountain "refuge areas", spreading once more into the Amazon basin when conditions became favorable again. The present floral diversity of most areas is relatively young.

Possibly the most stable habitats are located in the Brazilian coastal mountain ranges, the Central Brazilian shield and the Guayana highlands – regions where many ancient plant families occur in great abundance (like *Winteraceae*, *Monimiaceae* spp.). Plant families which are typical of the neotropics (neotropical endemics) are cacti and bromeliads. Both are thought to originate in rainforest climates, but presently most species are found in dry or mountain areas.

Several phenomena and life forms of Amazonia are also found in the rainforests of Africa and Asia but are unknown in the temperate regions. One example is the extremely high

diversity of plant life. Rainforests are composed mostly of many different tree species (up to 250 species per hectare), in contrast to the uniform fir, maple or beech forests of the northern temperate hemisphere.

Trees are mainly evergreen and broad-leafed and highly sensitive to longer durations of coldness or drought. The active shedding of old and non-functional branches occurs frequently in tropical trees and old leaves are also shed continuously throughout the year. Subsequently, leaf growth may occur abruptly by flushing: young leaves sprout quickly and hang down as red, yellow, purple or even blue bundles of soft foliage; only after some weeks do they turn green and move into their normal position.

The structural diversity of flowers is incredible, ranging from minute and hardly visible green flowers to enormous fleshy ones. The high frequency of bright red flowers (almost

PRECEDING PAGES: rainforest undergrowth and a well-disguised insect.
RIGHT: Euterpe palms.

lacking in non-tropical regions) is related to bird pollination. Unusual in temperate climates, pollination also occurs by beetles, wasps, bats or even mammals. These plant-animal interactions are often highly complex and have evolved over millions of years.

Another example of these interactions are plants like Cecropia and Acacia, which supply peculiar ant species with living space and food (edible bodies; nectar). In return, the plant is protected by the ants, which kill or drive away possible

FLORAL ODDITIES

Oddities are species that begin their life as a liana only to shed their roots afterwards and continue as an epiphyte, while some other shrubs or trees transform after a certain length of time into lianas.

nutrients in the top of the forest, without connection to the soil.

Strangling figs use another strategy. They begin as small epiphytic sprouts on a high branch of a tree, later developing into a liana that eventually strangles the supporting tree and replacing it by developing its own trunk. It is generally believed by botanists that the plant types of the temperate zones are derived from tropical rainforest plants and are impoverished and specialized branches of this complex vegetation.

predators. Some of these ants even kill neighboring plants by bites or "chemical weapons". Anybody who has tried to climb a tree inhabited by ants will surely remember their effective defense mechanisms.

Typical of tropical rainforests are lianas and epiphytes. Lianas can build up a complex network of knots and tendrils, many of them strangling each other or the supporting tree. Epiphytes are plants and shrubs that live in the crowns of other trees to get more light. They represent a special life form which is functionally independent of the supporting tree and not at all parasitic. They have had to evolve mechanisms to store water and get

Vegetation types

Ninety percent of the Amazonian lowlands are covered by "*terra firme*" ("firm ground") forest, a name that refers to the fact that these forests are never flooded during the rainy season. On poorer soils and in drier areas one finds patches of savanna that are small in comparison to the huge forested area, but sometimes still as big as Uruguay or half the size of Germany.

An idiosyncrasy of the Amazon are the flooded forests (*várzea* and *igapó*) which cover two percent of the region. Towards the Atlantic, the seasonally flooded forest is replaced by tidally influenced river mangrove. Structure and flora of the lowland forest do not vary much up to

900 meters (2,950 ft) on the Andean slopes. Higher up grows forest of smaller stature (up to 2,500–3,000 meters or 8,200–9,840 ft), called mountain rainforest, and this is replaced even higher up by elfin or cloud forests. The timber line is formed by a low scrub which turns into "*páramo*" – vegetation in the moister areas (north of 8°S) and into the "*puna*" in the southern drier parts.

Terra firme

Specialists distinguish different forest types, which are never subject to overflooding, but there is still no satisfactory system of well-

canopy 30–40 meters (98–130 ft) in height, with interspersed "emergents" that may reach up to 50 meters (165 ft). Most of the trees found here are evergreen with relatively thin stems. They branch only at the top and form a closed canopy high above the ground. A cross section of the trunk is irregularly shaped, since the tree extends often into buttress or stilt roots. Palms of all sizes and types determine the character of *terra firme* forest, while giant trees with enormous trunks and crowns are rare.

Leaves of different species are similar in shape and texture; this is probably due to an adaption to the constant and peculiar climate. They are most-

defined units. Most of the *terra firme* forests are found on deep, well-drained latosols, the typical reddish-brown, loam-like soil of the tropics, which can be free of rocks and have a depth of more than 50 meters (165 ft). Also significantly different are the forests on white sands. Brazilians also distinguish a "dense forest" *(mata pesada)*, in contrast to the rarer and lighter "liana forest" *(mata do cipó)*.

Primary *terra firme* forest (virgin forest that is undisturbed by humans), usually has a closed

ly medium-sized and elliptic, with a smooth surface and a long pronounced drip tip, which is thought to facilitate the run-off of rainwater.

The understory is dark, and is reached by only 2 to 4 percent of the sunlight. So there are few plants and bushes, and one can easily penetrate without using a machete. The few plants are similar to those that can be grown at home: their adaption to poor light and poor soil allows them to grow even in the dimly lit living rooms of houses in temperate climates.

Most of the other plants found growing at ground level are seedlings from trees. Nevertheless, there are a few shrubs (e.g. *Psychotria*, from the coffee family; or *Rinorea*, from the

LEFT: ground vegetation in primary forest is relatively sparse.
ABOVE: lower mountain forest shrouded in mist.

violet family) which are regular woody members of the rainforest undergrowth. Although hardly any other plants flower there, some trees bear their flowers along the trunks *(cauliflory)* or even at the roots *(rhizoflory)*.

A special case is seen in some species of *Duguetia*, which send long stolons (runners) from the trunk along the ground bearing flowers at the end, sometimes at a distance of more than 6 meters (20 ft) from the mother tree.

Secondary growth

Logged and burnt tropical forests may never properly recuperate. Almost all its nutrients are fixed in living plants and animals. When released from dead, rotting organisms they are immediately absorbed by the surrounding plants and returned into the food chain. The soil is extremely poor, with no humus layer, and when the forest is destroyed the nutrients disappear. It needs many thousands of years to restore them – the denuded earth may remain for years without even grass or plants, becoming a red Amazonian desert.

If the destroyed areas are not too large and some of the biomass remains intact, secondary forest can develop. Along roads or within cultivated land, one finds dense patches of low forest, sometimes shown to visitors as "jungle". These

are recuperating, logged forests (called in Brazil "*capoeira*"). Such forests contain flora different from and poorer than in virgin forest, made up of thin-stemmed lianas and aggressively growing trees and shrubs. Characteristic of these habitats are Cecropias of the fig family (*Moraceae*), with large silvery palmate leaves growing in clusters at the end of the branches, and inhabited by millions of small aggressive ants. In the western Amazon the Cecropias may be replaced by the somewhat similar Balsa wood trees (*Ochroma*).

The giant herbaceous plant *Phenakospermum guyanensis*, of the banana family (*Musaceae*), is related to the travelers' tree from Madagascar. Its banana-like foliage and Strelitzia-like inflorescence can reach more than 10 meters (33 ft) in height. Shrubs with two leaves growing from the same point of the stem but in opposite directions and bright orange latex belong to the genus *Vismia (Clusiaceae)*. Unbranched slender trees with dense clusters of large terminal feather-like leaves are young individuals of *Jacaranda copaia (Bignoniaceae)*. Upon reaching a larger height, they change growth and leaf form and become a typical member of the mature forest.

Mountain rainforests

From a height of about 900 meters (2,950 ft) upwards, the structure of the lowland forest

trees and include hanging cacti, orchids, aroids and bromeliads, which grow not only high up in the crowns but also at eye level. Because of the permanent moisture by fog and the low evaporation rate, small epiphytes like mosses, lichens and ferns cover stems and rocks. Small thin-leaved and transparent filmy ferns *(Hymenophyllaceae)* inhabit slippery rocks and holes in the earth. In valleys, along creeks and waterfalls, a rich herbaceous flora has developed with beautiful *Gesneriads*, *Zingibers* and *Marants* growing under the shelter of large tree ferns.

This forest includes "living fossils" like the

changes gradually: the mean height of the trees becomes lower, the crowns more spherical and the growth of branches denser. Trees don't grow straight but are crooked and less buttressed. The shrub layer is better developed than at lower altitudes. Species diversity decreases and species types change with increasing altitude. *Melastoms*, which are recognized by their opposite leaves with three to five parallel veins and by showy red, blue or white flowers, dominate the forest.

Epiphytes crowd the branches and trunks of

conifer *Podocarpus*, with leaf-like needles; the little known genus *Mollinedia*, a shrub with small greenish grey flowers; or treelets of the genus *Hedyosmum*, known from fossil records more than a hundred million years ago.

At higher altitudes, namely between 1,500 and 2,800 meters (4,900 and 9,200 ft), the forest can turn into elfin or cloud forest, a special type of mountain rainforest. The low trees are even more gnarled and inclined, branching close to the ground and festooned with lianas and nests of epiphytes.

The ground is composed of a layer up to 2 meters (6 ft) of rotten wood and leaves, roots, and branches bent down from the trees. One

LEFT: tree ferns are common in cloud forest.
ABOVE: *Espeletia* plants in Ecuador's high plateau.

walks on a soft, elastic carpet-like surface: any noise is muffled by the ground and the thick moss covering the trunks.

Red and fleshy flowers *(Fuchsia, Ericaceae)* shine out of the thicket and are visited and pollinated by hummingbirds, which are common at this height. Lichens, ferns and thousands of small *Peperomias*, orchids, aroids and bromeliads cling to rocks and trees, or grow on the ground. Another plant characteristic of this forest is a steel-blue fern with simple-shaped leaves.

> ### PÁRAMO FOLIAGE
>
> Azorella forms hard, large green cushions built up by a single plant; its many branches end in minute hard leaves and flowers and the insides of these spherical cushions are filled with dead leaves.

aceae). Espeletia can vary from a stemless rosule plant to a rosule tree covered by old foliage (reminiscent of a monk, from which it derives its name, *frailejón*, in Spanish). *Puya* is best known because of its hapaxanty, which means that the plant flowers after many years of growth with a single huge inflorescence (similar to many *Agave* species), only to die afterwards.

At even higher altitudes, frost is common and, in the morning, snow may cover tropical plants like wild passion flowers

Páramo

At a height of 2,500–3,000 meters (8,200–9,840 ft), the mountain rainforests turn into a closed shrub woodland which eventually gives way to the *páramos*, one of the most typical high-altitude habitats of the neotropics. This vegetation goes up to 4,800 meters (15,750 ft), and ends at the zone of permanent snow. *Páramos* are open grass- and shrublands with small forest pockets at humid and sheltered sites. They can be recognized best by large rosule plants with silvery and hairy leaves, which belong to many different families.

The most frequent and best-known genera are *Espeletia (Compositae)* and *Puya (Bromeli-*

(Passiflora), mountain-bamboo *(Chusquea)* or members of the numerous species of *Melastoms* and *Ericaceae*. Typical of high tropical altitudes is *Weinmannia (Cunoniaceae)*, a shrub or tree with tiny white flowers and small, simply pinnatified leaves. A special feature are *Polylepis* forests *(Rosaceae)*, occurring at over 4,000 meters (13,000 ft) on large boulders.

An interesting feature found here is the occurrence of many different plant genera. Those which are well known from temperate or alpine areas are Plantains *(Plantago)*, Barberry *(Berberis)*, Valerian *(Valeriana)*, Violet *(Viola)* or the many members of the *Compositae* (sunflower family).

In the drier regions further South (Peru), the *páramo* is replaced by *puna* vegetation. This describes open land with tuft grasses, thorny shrubs or extensive cacti, which may be covered with several *Tillandsia* species (bromeliads).

The flooded forests

Typical of the Amazon river system is a forest type which is highly adapted to survive long periods of submergence. Following the rainy season in the Andes, the water level of some rivers (such as the Rio Solimões at Manaus) may rise as much as 15 meters (50 ft) on an

rich white water rivers called *várzea*. Trees in both of these habitats share the same problem: how to stay alive during the high water period when even adult trees can be completely inundated. To prevent saturation, the bark is covered, in many species, by thick cork tissues, and the leaves are protected by water-repelling cuticles. As soon as a part of the crown is submerged, the tree slows its metabolism down to the minimum level required for survival. The leaves are not shed and they function normally as soon as the water levels fall. Parts of the tree that emerge out of the floodwater may then flower or fruit during this season.

annual basis, and the water covers large areas of forests for up to eight months.

Compared to surrounding forests, these flooded areas hold fewer large trees, but they often have well-developed buttress or stilt roots. Herbaceous plants are almost absent, the understory comprises only saplings of larger trees and epiphytic flora is richer than in the *terra firme*.

Flooded forest is called *igapó* in nutrient-poor black water systems. *Igapó* is poorer in species than the flooded forest along nutrient-

An oxygen supply, crucial for survival, is ensured partly by aerial roots. It is suspected that the roots are capable of switching to an anaerobic (without oxygen) metabolism if necessary.

Trees of flooded forest evolved several special mechanisms for seed dispersal. One of the methods of dispersal takes advantage of the fish that enter the forest during the flood. The ripe fruits of some species are swallowed entirely by fish and the undamaged seeds are deposited after passing through the digestive tract. Some fruits float easily and are dispersed by currents. For their first years of life, seedlings have to survive mostly under water, but they usually grow quickly during the dry periods.

LEFT: flooded forest along the Rio Negro.
ABOVE: a Jabiru Stork makes its nest in a flowering Jacaranda tree.

The flora of the flooded forest is much poorer than in the *terra firme*. Some species are confined to one of the two types of flooded forest, like *Pseudobombax munguba* to the *várzea* or the palm *Leopoldina pulchra* to the *igapó*; others like *Annona hypoglauca* occur in both areas. The species composition is often better known than that of the *terra firme* forest, since species collection can be done from a boat.

Quite different to the *várzeas* and *igapós* are mangroves. These grow along the coast or in river mouths in sea- or brackish tidal water. There are about 10 typical species in Amazonia; far fewer than in Asia where they probably

originated. Growing in distinct zones, the outer belt is typically dominated by *Rhizophora*, a species with a tannin-rich bark, often used for tannin; in the inner belt you find species like *Avicennia* and *Laguncularia*.

Like the plant species found in flooded forests, mangrove plants have similarly had to adapt to their extreme habitat. The salt of the sea water can be filtered by the roots or can be excreted by special glands at the leaves. Stilt roots *(Rhizophora)* or pneumatophores *(Avicennia, Laguncularia)* provide the necessary oxygen. Seeds germinate usually on the mother trees and drop as fully developed plantules into the mud, rooting there quickly (vivipary).

Savannas, *caatingas* and *campinas*

The *terra firme* forest is often interrupted by areas of open shrub- and woodland or grass savannas. Like the forests, these open habitats can be further subgrouped.

The "savannas of Humaitá" represent the largest open areas of central Amazonia. They evolved probably on old river beds which offered no appropriate soil for forests. The poorly permeable latosol allowed only grasses and a limited number of forest species to establish.

Furthermore, savanna plants like *Curatella americana*, *Xylopia aromatica* or *Palicourea rigida* may have become newly established on these poor soils or may have remained from the widespread savannas which covered almost the whole continent during dry periods of the Pleistocene era, one to two million years ago. An example of forest species tolerant to these conditions is *Physocalymma scaberrimum*. Most tree species have leaves with silicate incrustations, that give a rough surface and brittle structure.

Completely different are the *campinas* or Amazonian *caatingas*, small groups of dense growth of trees and shrubs in open white sand areas. Signs of early human settlements in *campinas* and the low degree of endemism of plants lead scientists to the conclusion that these open areas are possibly of human origin.

The small, gnarled tree trunks are densely covered with orchids, ferns and bromeliads. There is no special flora: the plant inhabitants are mostly similar to those of the next forests on white sand, which are sometimes connected with the *campinas* by a transitional zone, the *campinarana*. Different to the Amazonian *caatingas* is the *caatinga* proper; a large area of semi-arid thornscrub and dry deciduous woodland in the Sertão region of the northeast of Brazil (Bahia), adjoining the Amazon region to the south.

The savannas of Colombia and Venezuela are called *llanos*. They are usually represented by grasslands interrupted with small forest patches or gallery forests, frequently with Mauritia palms. Their origin is uncertain, although their extension certainly has been influenced by humans. Significantly different are the coastal savannas of the Guianas, which are rich in species and often swampy. ❑

LEFT: palm flowers in the Peruvian Amazon.
RIGHT: the leaves of Cecropia trees have an easily recognizible shape.

BROMELIADS AND THEIR FAUNA

Largely confined to the neotropics,
these plants play a vital role in many Amazon ecosystems

Bromeliads are a specialty of the tropics of the New World, and leap to the eye in all ecosystems. About 2,500 bromeliad species are known from tropical and warm temperate parts of the Americas – only *Pitcairnia feliciana* is recorded from western Africa (Guinea). Several bromeliads as does *Tillandsia usneoides*, "Spanish moss".

Bromeliads which collect water in their leaf axils (due to tightly-fitting, overlapping and inflated leaf bases) are termed tank bromeliads. The arrangement of tanks varies: in some species, water accumulates in a few or many separate leaf axils, yet in others the central

yield edible fruits but only the pineapple *(Ananas comosus)* supports a large commercial industry with a production of about 7 million tons worldwide. All bromeliads are herbaceous perennials and most reproduce both by seeds and vegetatively. Pollination is carried out by hummingbirds, bats and insects such as bees, moths and butterflies.

Specific habitats of bromeliads range from sea level to over 4,000 meters (13,000 ft), from the Peruvian desert to the rainforests of eastern Brazil. The three subfamilies are *Pitcairnioideae*, which are terrestrial; some of the *Bromelioideae*; and most of the *Tillandsioideae*, which are epiphytic (grow upon other plants),

leaves form a large tank surrounded by a few separated tanks provided by other leaf axils.

The tank bromeliads in certain areas constitute a major feature of the landscape. Epiphytic tank bromeliads (called *parásitos* in Latin America) usually occur within or below the forest canopy and so far there is no proof that they absorb nutrients by their roots from their hosts. Bromeliad leaves bear special epidermal cellular structures (trichomes) that absorb water and minerals from the leaf surface. During and after rains, water dripping from trees is enriched with minerals leached from tree leaves. The tanks also tap pollen, dead leaves, twigs, and seeds falling from the trees, which break down in the tanks to form a nutritive soup available to bromeliads and their inhabitants. The total water capacity varies with size and species, from a few milliliters to several liters (up to 45 liters or 80 pints for a *Vriesia imperialis*, a native to Brazil).

Many organisms make use of tank bromeliads: various ants, birds and bats all eat the leaves, flower stalks, nectar, fruit, seeds or pollen. For others, the bromeliad provides a place to hide from humidity or prey, and amphibians seek them out not only to keep moist and avoid heat but also as a convenient place in which to breed.

throughout the year. About 500 species of aquatic organisms have now been recorded from this special ecosystem. There seem to be two extreme types of food chains, one based on algae in sun-exposed tanks, and one on detritus in deep shade, but these are end points of a continuum. Nearly all of the major groups of freshwater invertebrates have been reported from bromeliad tanks including worms, snails, crabs and the aquatic insects such as dragonflies, stoneflies, caddisflies, bugs, beetles, and flies. Larvae of over 200 mosquito species have been found in water in bromeliad tanks, and their numbers usually dominate most commu-

Centipedes, scorpions, roaches, ants, snakes, salamanders, and lizards frequently occupy the older leaf axils outside the central water-collecting portion of the rosette which can no longer hold water. They create a terrarium-like environment, with decomposed organic material resembling peat, able to support animals requiring a moist but well-aerated substrate.

Bromeliad tanks provide relatively stable habitats for aquatic animals because of the long life span of the plant and the retention of water

LEFT: *Aechmea fasciata* in the Brazilian rainforest.
ABOVE: *Bromelia guzmania* from the cloud forest in Ecuador.

nities. Adult female mosquitos of some bromeliad breeding species *(Aedes, Culex, Anopheles)* are capable of transmitting infectious diseases such as yellow fever, filariasis, encephalitis, and malaria. Thus, the term "bromeliad malaria" has been coined, although the bromeliad is neither disease, carrier nor host, only the host of the carrier. In parts of South America, malaria has abated significantly by reducing the local bromeliad population.

While bromeliads usually represent only a small component of the complex tropical forest community, they often play a significant role in the formations and their tanks are focal points for much biological activity. ❑

A SAMPLING OF AMAZONIAN FLORA

The diversity of Amazonian flora is astonishing,
from orchids to plants that have proved immensely useful to people

In the Brazilian forests, legend has it, there exist fierce carnivorous plants capable of swallowing a person whole. Perhaps, but scientists have yet to encounter anything more terrifying than small insect-eaters.

Orchids

In most parts of the Amazon, orchids have taken to the trees in order to compete for light in the forest. This mode of existence is distinct from parasitism because the orchid does not

There are, however, many interesting plants, and equally interesting uses to which many of them are put in the region. Some plants yield food, some medicines, some are made into weapons, others into tools, and there are those, such as the hallucinogenic plants that play a part in the culture of the indigenous peoples. Others, like the rubber tree and rosewood, play an important role in the local economy.

Some plants that originated in the Amazon, like the Brazil nut, have remained commercial specialties of the region. Others, such as the rubber tree, have been cultivated successfully throughout the tropics of the world. The number of such useful plants is incalculable.

harm the host tree on which it is perched. Plants living on trees or other vegetation are called epiphytes. Groups that have adopted this lifestyle in the Amazon include ferns, mosses, lichens, cacti, bromeliads, orchids, and many other flowering plants.

While becoming an epiphyte means more light and probably less competition, certain ecological adaptations are required: the aerial habitat is hotter and drier than conditions on the forest floor, and orchids have special vellum roots for the absorption of water and nutrients. Succulent leaves and pseudobulbs are adaptations for storing water, and the frequent association between orchids and ants probably helps protect the orchid from destructive insects and also to enrich the humus soil collected around the roots. In some cases, the ants themselves carry this soil from the forest floor, while in other cases the soil is deposited by rain and run-off. Symbiosis between orchids and fungi is another strategy for living on reduced resources.

Germination of orchid seeds in nature appears always to depend on an association with ectomycorrhizal fungi, which provide almost all of the necessary nutrients; the seed is very small and has no endosperm as a food reserve. Orchid flowers are highly specialized for polli-

are widely distributed, such as *Catasetum macrocarpum, Encyclia fragans, Epidendrum nocturnum, Oncidium cebolleta, O. nanum, Orleanesia amazonica, Polystachya concreta, Rodriguezia lanceolata,* and *Schomburgkia gloriosa,* all of which can be found throughout the Amazon in diverse types of habitats. Other species are more restricted in their occurrence, both geographically and ecologically: among these orchids are *Acacallis fimbriata* and *Gongora quinquenervis,* both of which are found only in river flood forest. *A. fimbriata* is rare, occurring in low densities in Pará, Amazonas, Rondônia, and Amapá, and the

nation by insects, and the mechanisms by which pollen is transported from one flower to another are often quite complex and bizarre, involving diverse types of mimicry. The orchid bees (subfamily *Euglossinae*) are known to visit orchids which mimic the scent of female bees and which exploit the sexual ardor of their male visitors. Most orchid seeds are wind distributed and a very few lucky ones lodge on tree branches where they can develop.

About 500 species of orchids have been recorded in the Amazon Basin. Many of these

LEFT AND ABOVE: a few of the myriad of orchid species from the Amazon lowland forest.

destruction of its habitat could lead to its extinction. *Eulophia alta,* a terrestrial species growing in open areas, is found widely in the Amazon and has become a roadside flower.

Species like *Zigosepalum labiosum, Koellensteimia graminea, Paphinia cristata, Pleurothallis* sp., and *Maxillaria* sp. display a preference for humid, shady habitats. *Catasetum longifolium,* although widely distributed, is known only from one type of habitat: the crowns of *Morichi* palm trees (*Mauritia* sp.). *Acacallis cyanea* grows in floodplain forest in Pará and Amazonas, but rarely in the open, sandy soil *campinas* where so many orchids come down to ground level. Most orchid species with distinct habitat preferences are found in these *campinas*, in flooded forests, or in the Amazon uplands or Andean foothills.

Some species can be considered endemic to certain Amazon localities. *Brassavola fasciculata*, for example, is known only from one collection made in 1975 on the upper Amazon. *Cattleya eldorado*, likewise, is restricted to white sand *campinas* and black water flooded forests on the Negro and Uatuma rivers near Manaus. *C. araguaiensis* occurs only on the

WIDELY KNOWN

Due to the singular beauty of its flowers, the orchid family is well-known, not only to specialists but also to amateur horticulturalists and collectors, whose knowledge is often impressive.

single flowers to enormous inflorescences. This is the most diversified family of flowering plants, with over 25,000 species mostly concentrated in the tropics and subtropics of both hemispheres.

Within the Amazon, orchids are still the subject of intensive study due to the extensive geographic areas awaiting systematic exploration and the difficulty with which these areas are reached. These same two factors, extensive areas and difficult access, are also saving graces for the preservation of

banks of the Rio Araguaia, and *Catasetum pulchrum*, as far as is known, grows only in restricted localities in Pará. *Catasetum pileatum* is restricted to the upper Rio Negro region.

There are cases of apparently restricted or disjunct geographical distributions among Amazon orchid species which, in reality, reflect a lack of systematic plant exploration in the region. This is probably the case of *Catasetum multifidum*, which was collected only in Rondônia before being discovered at the Serra do Carajás, some 1,000 km (600 miles) away.

The orchid family is among the most evolved of the monocots, showing an impressive range of floral characteristics that vary from microscopic

the Amazon orchids. Two factors threaten these species at present: the destruction of habitats which are sometimes restricted, and the indiscriminate collection of orchids for commercial purposes.

Many large-scale development, agricultural, highway, and colonization projects undertaken in the Amazon since the 1970s have been responsible for the deforesting of huge tracts and have endangered several orchid species. The simple expansion of the urban area of Manaus seems to have extirpated *Cattleya eldorado* from its former habitat. It is likely that several species of orchids, especially micro-orchids that are not sought after by collectors, have disappeared or will disappear from the face of the earth before even becoming known to botanists.

Many regions previously considered unsuitable for agriculture or colonization are now being included in Amazon development projects. The sand from *campinas* is often used in the construction of roads and houses when development encroaches on these patchy environments. This happened to several orchid habitats near the Tucuruí dam site on the lower Rio Tocantins. The Palha *campina* in Vigia, near to Belém, also succumbed, and this former orchid paradise is today completely ruined and its orchid flora on the way to extinction.

Ducke, comprise one of the most valuable forest crops in Amazonia. These trees have been extremely heavily exploited and are considered endangered. In addition, the manner in which they are logged causes a great deal of deforestation of the surrounding trees. Although other lauraceous trees have a similar scent, these two are the commercial sources of linalool, the major ingredient of the rosewood oil used in making perfumes.

About 50 factories exist in Amazonia for the extraction of this oil. Many move from place to place, even in the least exploited areas. To extract rosewood oil, first a factory with a still

Only by preserving whole plant and animal communities in national parks, biological reserves and the like will we ensure that the joy of seeing orchids in their native habitat will not be lost to future generations. However, habitats are being destroyed at a far greater rate than we are able to gather the data needed to establish effective, sustainable reserves.

Rosewood, *Pau Rosa*

The two species of the *Lauraceae* family, *Aniba duckei Kosterm.* and *A. rosaeodora*

LEFT AND ABOVE: orchid species from the elfin forest of the Andes.

is set up in an area of virgin forest. Field men select trees above the legal limit of 30 cm (12 inches) in diameter (about 15 years old). These trees are then felled, cut into transportable logs and taken to the factory, where they are fed into a large chipper, and the resultant chips are carried in wheelbarrows to the stills. It takes two and a half hours for 500 kg (1,100 lb) of chips to be distilled. Each ton of wood yields 7½–9kg (17 to 20 lb) of oil. Nearly all the oil is exported. Seventy percent goes to the United States, 14 percent to Great Britain, 3 percent to France, and 10 percent remains in Brazil for its own cosmetic industry and for later exportation in refined forms.

The rosewood was once a very common tree, but exploitation is making it more scarce. There are laws controlling its use, but in areas near factories trees soon become scarce. The factories, therefore, must move often in order to keep up with the supply.

A positive step is that research is now being carried out to find a practical way to use the linalool in the leaves instead of the wood, so that entire trees need not be destroyed. Some planting of young trees is also being done, but meanwhile the pleasure of finding a sweet-scented rosewood tree while collecting in the forest is growing rarer.

its collection is a time-consuming job. It is extracted in the dry season, so the same people can work as rubber gatherers and Brazil nut harvesters.

There have been attempts at establishing rubber plantations in Amazonia, but these have not yet proven very successful. Fordlândia, for example, on the Rio Tapajós, was a failure because the locality had unsuitable soil and topography. Another venture from Malaysia also failed due to an attack of the fungus *Microcylus ulei*. Some fungus-resistant strains have now been developed in Brazil, and the plantations are yielding more rubber. The rubber

RUBBER GATHERING

Rubber gatherers begin a circuit at about 5am along a trail devised to include 100 to 120 rubber trees. They make a diagonal incision on the trunk of each tree with a special sharp knife, and attach a small tin below this slit to collect the latex. On a large tree, they may make several slits and collect several tins of latex. This first round takes four to five hours.

At midday they begin the circuit again, this time to collect the fresh milky latex from each tin. This must be got back before it congeals. By three or four o'clock, they are ready to process their latex into a large ball of crude rubber for sale.

To do this, they build a smoky fire and erect a large wooden spindle over the smoke. They then pour a little latex onto this spindle and rotate it to form a smoke-darkened ball, gradually adding the day's collection until the ball weighs about 20 to 40 kg (45 to 90 lbs). Then they cut or slip it off the spindle, and it is ready to sell or trade.

Rubber Tree, *Séringa* (Seringa)

The economic wealth of the Amazon region has depended to a great extent on the rubber industry. The great rubber boom that lasted from about 1880 to the First World War led to the fabulous excess of the city of Manaus and to prosperity throughout the whole region. But when the rubber plant, *Hevea brasiliensis (Euphorbiaceae)*, was exported to Asia and was cultivated in vast plantations there, the rubber boom in Amazonia ended. Yet, today, rubber still ranks as one of the major industries of Amazonia, and one of its most important exports. Most of the rubber grown is wild, and

grown on plantations has a higher yield than that of wild rubber, and requires less labor to harvest. This benefits multinational companies, not small-scale, independent rubber collectors, who sell (very cheaply or by barter) individual balls of rubber to traders, who then sell the rubber on for export.

Guaraná

Guaraná (*Paulinia cupana var. sorbilis*, from the *Sapindaceae* family) has been cultivated for centuries in Amazonia, and is still an important crop. Humboldt and Bonpland first collected it in 1800 from the upper Rio Negro, where the indigenous groups used it as a stimulant and as a medicinal plant. When planted in the shade it takes the form of a liana, and in the open it is a shrub. Guaraná bears fruit three years after planting, but the yield increases greatly after five years. The fruit is usually ground into a powder by the natives to make a drink, although there are other uses as well.

In folk medicine, the drink is an important stimulant, comparable to the African cola bean. It is cited as being good for the heart, liver and kidneys, and as an aphrodisiac and analgesic. Today, it is used as a flavoring in "Guaraná", a popular carbonated soft drink in the region. The straw platter and toast them in a primitive sort of oven. They then pound them vigorously to separate the seeds from the seed coats. The separated seeds are transferred to a wooden pestle and ground to fine powder. This is left out for one night to be moistened by the dew, and the following day water and cassava flour are added to form a paste.

Some indigenous groups have traditionally made models out of Guaraná paste, which are then baked and hardened. In modern times, this has been updated by local artisans, who use the paste to make models of animals, which they sell as souvenirs.

preparation of today's commercial drink is basically the same as the age-old method, though there is a reliance on machinery rather than hands to do the job. The original brew of the indigenous peoples is much stronger and stored in gourds, whereas in the factories, sugar is added and the liquid is kept in sterile bottles.

The indigenous people at Maués first remove the red outer pericarp of the fruit and the white aril, leaving the black seed. They dry these seeds rapidly without fermentation on a large

LEFT: latex being tapped and collected.
ABOVE: the Guaraná fruit is used in the manufacture of soft drinks.

Brazil nut, *Castanha-do-Pará*

Although Brazil nuts, *Bertholletia excelsa (Lecythidacease)*, are popular the world over, few people are aware that they grow like the segments of an orange, arranged 12 to 24 within a large wood cup *(pyxidium)*. This cup is the fruit of the Brazil nut tree, which grows to 30 meters (100 ft) tall and begins bearing fruit when it is 10 to 15 years old. One adult tree may produce 500 kg (1,100 lb) of nuts each harvest. The fruit of the Brazil nut tree matures in January, about 4 months after the large, attractive yellow flowers appear.

Harvesting Brazil nuts is a hazardous job. It takes place at the height of the rainy season, so workers have to walk and camp in heavy rain. In addition, the fruits are heavy, trees are tall, and harvesting is done on the ground after the fruits have fallen; it is not unusual to meet a worker who has been injured by falling fruit. Workers gather the fruits, chop off the top third with machetes and shake out the nuts. These are then washed in a nearby stream, dried, and shipped.

The annual commercial yield of Brazil nuts is about 50,000 tons, a large percentage of which comes from the area around the Rio Purus. Most of the nuts come from wild trees in the forest, but there are also a few experimental plantations that contribute to the annual crop.

lamellated trunk is actually made up of long, thin, twisted buttresses which radiate from a central core. Each buttress is itself divided into irregular, plant like sections which may be no more than an inch thick.

The resistance of the wood and its unusual shapes dictate its uses – it is a common source in Amazonia and the Guianas for paddles and axe handles, which can be produced with little labor from such thin wood. One frequently encounters a *carapanuaba* tree in the forest with a paddle-shaped piece of buttress missing where someone has taken the wood to make a paddle.

Although the vast majority of the crop is exported, the natives also use the nuts both for food and oil. The nuts are very rich in oil (60–70 percent), similar to olive oil, and can be burnt for cooking or to generate light. The cups, too, are useful. One tribe even used them as containers for their arrow poison.

Paddlewood, *Carapanuaba*

The Amazonian name for this tree comes from the mosquitos *(carapanã)* that breed and live in the dark damp recesses of its fluted trunk. Aside form its function as a mosquito hatchery, *Carapanuaba* is valued as a highly resistant wood which is not attacked by termites. The

Water vine, *Cipó d'água*

This is a valuable plant to expeditions in the remote forest areas, since it yields drinking water. The water vine is a large liana common on high, non-flooded ground away from rivers. When a piece of the stem, a few feet long, is cut and held up, a good quantity of fresh water runs out freely. The liquid also contains some minerals. The natives contend that the bulk has medicinal value. The *cipó d'água* plants belong to the family *Dilleniaceae,* and can be of several species such as *Doliocarpus rolandri,* and *D. coriaceus.*

Water can also be drunk from plants of some other families, such as from the surface roots of a species of Cecropia, a common tree in Amazonia, but be sure to make a positive identification first. Some poisonous vines yield a small quantity of water, too.

Monkey ladder vine

The name *escada-de-jabuti*, or "turtle ladder", as it is called in Brazil, is given to a number of species of *Bauhinia (Leguminosae, Caesalpinioideae)* with an unusual type of stem growth.
It is a common plant in all the tropical forests of South America, and various areas have different

COLONISTS

Many ornamental plants and park trees, spread by humans, originate in Amazonia. Visitors to other tropical regions may identify more American than endemic plants in local parks and gardens.

Examples of plants endemic to Amazonia that have successfully spread across the world include: the yellow flowering *Allamanda*, various species of *Plumeria* or *Frangipani*, the raintree *Samanea saman*, the palm *Royastenia*, or *Heliconias*, often considered equal in beauty to orchids. No plant, however, has been as successful as the water hyacinth, although the kind of hold it took was certainly not on the mind of the horticulturists who exported it.

vernacular names for it. In Venezuela it is called *bejuco de cadena*, or "chain liana", and in Guyana it is known as "monkey ladder". In fact, these names all refer to the irregular growth of the liana trunk in a ribbon-like "S" or scalariform pattern. The leaf is shaped rather like a cloven hoof, from which is derived another of its Venezuelan names, *pato de venado*, or "deer's foot". The liana climbs to the top of the tallest trees and flowers in the crown of the forest.

LEFT: the fruit of the Brazil nut tree is of great importance to the Amazon economy.
ABOVE: heliconia originated in the Amazon region.

Water hyacinth, *Aguapé-purua*

Unfortunately, the water hyacinth's beautiful blossom, which lasts only one day and wilts after sunset, is not its most notable characteristic. The plant's over-efficient system of growth and propagation has made it a problem in waterways throughout the tropical world.

A mother-plant, once established, quickly sends stolons branching in all directions, with new plants sprouting from their tips to send out stolons of their own. The stolons usually break away or decay when the young plants are established, but often they remain connected. The result, after only a few generations (which is only a matter of months), is a huge mat of interconnected plants, its petioles swollen into spherical bladders which keep it afloat. The water hyacinth propagates rapidly by seed as well.

The water hyacinth, *Eichornia crassipes (Pontederiaceae)*, is a native plant of Amazonia, where it has never been a serious problem because the balance of nature keeps it under control. However, it has found its way, either by accident or as the result of importation as an ornamental plant, to other parts of the world where it *has* become a serious problem, due to a population explosion of monumental proportions. It has clogged drainage trenches in Guyana, caused concern in Queensland, Aus-

tralia, and has become a real concern in several regions of Louisiana (having gone to New Orleans as a horticultural exhibit in 1884). It was reported in the Nile for the first time in 1958, and six years later impeded over 1,600 km (1,000 miles) of the water and dam system. In Suriname, just two years after the Brokopondo Dam was opened, water-hyacinths were reported to be covering 53 percent of the surface of the large lake behind it.

A plant whose seed can remain dormant for 20 years and then give rise to 65,000 new plants in a single season is obviously a challenge to weed control experts. The most widely used herbicide is *2,4-d*; however, it needs frequent

reapplication. There are experiments in biological control as well. In Louisiana, a small moth caterpillar which feeds on the plant has been introduced. Although this has proved effective, it is not enough. Meanwhile, water-hyacinths keep appearing in new areas, where they continue reproducing at their phenomenal rate.

Just as no perfect control has been found, no large-scale use for the water hyacinth has yet been devised. Its low protein content makes it useless as food or forage. In India, it has been used as mulch for young tea plants and has been found to have value as a plant fertilizer. But the demand for tea plant mulch hardly makes much more than a tiny dent in the vast supply of water hyacinths, and therefore that supply multiplies geometrically every day.

Crops

Most crops that grow close to the equator are today distributed throughout the tropics. Most of them originated, however, only in one continent or even in a very restricted area. Examples are the coconut, which was once endemic to Southeast Asia, or the oil palm, which once evolved in West Africa. Many of these widespread crops are of Amazonic origin, like cacao, papaya, or manioc. Manioc *(Manihot utilisima)*, also known as yuca or cassava, is a starchy root crop that is the staple carbohydrate crop of Amazonia. There are two types: bitter and sweet. Bitter manioc is poisonous unless prepared correctly to remove the cyanide. To do this, the natives soak the roots, then grind and roast them to prepare a flour which is eaten with all meals. Sweet manioc, called *macexeira* in Brazil, is not poisonous and is boiled and eaten like potato. Manioc grows well in poor soil, and in recent times has been used to produce alcohol for powering motor vehicles. It is of particular importance since no other cereal crops, not even rice, grow well in Amazonia.

Cacao, or the cocoa bean, is a native crop of Amazonia. The main source of cocoa used to be Bahia in eastern Brazil; however, now it is grown extensively in Amazonia – especially in Rondônia. Coffee is mainly grown in southern Brazil. *Robusta* coffee is being grown locally within Amazonia, but yields a poorer quality coffee than *Arabica*.

Jute is an extremely important crop of the floodplain. The plant is cultivated alongside rivers. Upon harvesting, the cane is soaked and then beaten into separate the fibers. The fiber is then sent to factories in Manaus for processing.

Fruit trees are grown mainly in house gardens and around small farms. The most common fruits include bananas, limes, papayas, mangoes, cashew nuts, passion fruit, bread fruit, jackfruit and many lesser known local fruits seen in the markets in Iquitos, Manaus and Belém. Although previously there were no large fruit orchards, citrus orchards now exist and tropical fruit production is being promoted for export. Cashew is now being cultivated in large quantities in the Salinópolis region of Pará, near the estuary.

LEFT: the over-successful water hyacinth *(far left)* and a fruiting papaya tree.
ABOVE: palm flowers and fruit.

The shallow, sloping banks and beaches of the floodplain constitute a fertile, useful natural area for cultivation of quick-growing crops. This area is mostly used to grow bananas, corn and more recently soya. The soil is naturally re-fertilized each year by the floods. Sugarcane is found in many gardens. Along the Trans-amazon Highway it is being grown commercially to produce alcohol as a motor fuel.

A relatively recent introduction to Amazonia is black pepper. This is grown mainly by Japanese farmers near Belém and Manaus. This crop requires fertilizer, but because of its high selling price, it is economical even with the additional cost. ❏

MAMMALS

Amazonia's long period of geographical isolation has resulted in the evolution of a fascinating variety of mammals

However astounding the plant life or ecology of tropical forests can be, the vast majority of visitors to the rainforest yearn, first and foremost, to spot flagship species of birds and mammals during their stay, from macaws to Jaguars. Some come with a rather unrealistic expectation of just how many species they see, underestimating how much depends on patience, keen vision, a good guide, the time of year, or just plain luck, and some species of mammals can be particularly hard to track down. Whereas many bird species are colorful and fairly vocal, making them relatively easy to locate, most mammals are of subdued color, fairly silent, shy and often nocturnal.

You stand a healthy chance of seeing several types of monkey, but count yourself exceptionally privileged if you sight a Jaguar (South America's largest predator), Giant Anteater, or one of the many mammals that have become rare and elusive due to habitat loss and extreme hunting pressures of the last century.

To avoid disappointment, it is best to approach the Amazon with the attitude that nature comes as a profusion of plant, insect and bird life. Mammals are an occasional bonus rather than the principal objective of nature observation. On a typical trip of one or two weeks, it is realistic to expect to see about a dozen mammal species, mostly monkeys, peccaries, coatis, the two species of dolphins, bats, and, with luck, sloths and otters.

In order to best understand South American mammalian fauna, you need to have an idea of the ancestry of the widely divergent species found here. These fall into three broad categories. Firstly, orders, such as the marsupials and the edentates, that are found in no – or few – other places in the world. Secondly, there are mammals which have relatives on other continents, but differ very much from them; examples are the New World monkeys or certain rodents. A third group of mammals differs only slightly from related forms in North America or

even Asia, for example many cats or ungulates.

Scientists believe that marsupials and edentates (order of animals possessing no or few teeth) were present in South America when the southern supercontinent, Gondwanaland, broke up about 100 million years ago. Protected from competition with mammals that evolved elsewhere, these two orders thrived in isolation for tens of millions of years, evolving unusual forms like the Giant Ground Sloth or predatory marsupials, which became extinct thousands of years ago. On other, less isolated continents, the edentates could not compete with the success of higher mammals, and survived as an idiosyncrasy of the South American fauna.

The marsupials were somewhat luckier than the edentates, since from South America they reached an area of Gondwanaland that later broke off to become Australia. Protected completely from competition by "higher" mammals, they proliferated – until the Europeans arrived – surviving in a wider variety than in South America, where competition wiped out all marsupials bar the opossums.

The combination of evolution in isolation or immigration resulted in today's fauna of some 600 neotropical mammals. The exact number depends on the inclusion or exclusion of montane and savanna species, and on the species or subspecies status of some animals.

NORTHERN INVADERS

Ancestors of mammals such as the jaguar and puma, reached South America from North America a few million years ago, when the two landmasses became connected at the isthmus of Panama. Unclear, however, is the exact origin and the means of arrival of intermediately divergent mammals, such as the New World monkeys and rodents like the Capybara. Primates and rodents evolved only subsequent to the separation of South America and were initially not present on this continent. However, they are far too diverse in South America from their relatives elsewhere to have arrived from the north only a few million years ago.

PRECEDING PAGES: Brown-throated Three-toed Sloth.
LEFT: a Monk Saki.

Marsupials

Marsupials, or pouched mammals, are considered primitive on grounds that they split away from the common stock of placental mammals at a very early moment in mammalian evolution, more than 100 million years ago. In addition, they normally became extinct once they has to compete for survival with "higher" mammals. They probably evolved in the region which is now North America, and spread from here through South America and the connected and ice-free Antarctica to Australia. Just like in Australia today, they flourished in South America and filled many ecological niches, even range. The Western- and the Bare-tailed Woolly Opossum are two of many species with the prehensile tail of specialized tree dwellers. There are also several species of Mouse Opossums, and the nocturnal Water Opossum – the only marsupial that lives in water. With the help of its webbed hind feet, it has adopted an otter-like lifestyle and diet. It also has just about the shortest gestation period of any mammal – usually 12–13 days, but sometimes as few as eight.

Interestingly, the insectivores, placental mammals like shrews, hedgehogs and moles, do not occur alongside marsupials. They occur worldwide but not in Australia and most of South

evolving into hyena-like and sabre-toothed, cat-like animals. Many of these forms became extinct upon the arrival of placental mammals across the Panamanian landbridge, in the Pliocene era. Nevertheless, 81 marsupial species exist today in the New World, and about a dozen of them occur in Amazonia.

America's modern marsupials nearly all belong to the opossum family. They normally have a rat- or shrew-like appearance, with head-body-tail lengths ranging from 12–110 cm (5–43 inches). Although of similar appearance, they occupy quite diverse ecological niches, and a variety of tree- or ground-dwelling species can coexist in the same geographical America. They passed the Panamanian land-bridge in the Pliocene era, but have reached only as far as the slopes of the Andes, and no species occur in the Amazon or Orinoco river basins.

Edentates

The edentates are an order of toothless or nearly toothless mammals comprising three quite diverse looking groups – the anteaters, the sloths and armadillos – and 16 of the 29 living species occur in the Amazon. The edentates are confined to tropical America, but a related order, the Pholidota or pangolins, is found in parts of Africa and Asia. In earlier geological eras, many more species of edentates inhabited South

America. The most impressive representative of this group was the Giant Ground Sloth, of the Pleistocene era, which died out some 9–5,300 years ago.

While out-competed or preyed upon by modern mammals, the group still holds on strongly to some niches in today's ecosystems. Their low metabolic rate has helped them in this respect, since it allows them to specialize in nutrient-poor diets, namely ants in the case of the anteaters and armadillos, and a wide range of leaves in the case of sloths.

GIANT SLOTHS

The size of a small elephant, the Giant Ground Sloth was still around when early humans colonized the continent, and it may be the inspiration behind a number of indigenous legends.

against horny papillae of the mouth. This anatomy doesn't completely protect anteaters from bites, and so they avoid aggressive types, like leafcutting or army ants.

The Giant Anteater, a beautiful animal with a diagonal black and white shoulder stripe and a bushy tail, can measure more than 2 meters (6 ft) in length, and weigh up to 40 kg (88 lb). It occurs in forest and savanna alike, and is adapted to a life on the ground. In contrast, the Southern Tamandua, or Lesser Anteater, which is less than half the size

This low metabolic rate protects armadillos from overheating in their subterranean lifestyle.

Though edentate means "toothless," only the anteaters have no teeth at all. Their snout is extended to form a rigid trunk. They have a long tongue, which can be extended as far out as 60 cm (24 inches) in the case of the Giant Anteater. Ants are trapped between backward pointing spines by rapid movements of the tongue – more than one hundred per minute – and then they are masticated by being crushed

of the Giant Anteater, is adapted to foraging in trees, where it makes full use of its prehensile tail, as does the rarely-observed, nocturnal Pigmy or Silky Anteater, which is less than 50 cm (20 inches) long. While anteaters are rarely hunted for food, their numbers have been reduced, in many areas drastically, by the pet trade and for trophy hunting, and they face regional extinction.

Sloths are either two-toed or three-toed. However, both subgroups have three claws on their hind legs: the count rather refers to the number of claws on their forearms/forelegs. They have round heads with flat faces, and their long arms and legs are extended into long curved claws, which are used to anchor their body while climbing upside

LEFT: a Grey Slender Mouse Opossum.
ABOVE: the smallest of the three anteater species, the Silky Anteater.

down along branches. The sometimes greenish color of their long fur is due to the growth of blue-green algae that works as an extremely effective camouflage. This and the slowness of their movements are effective against detection by predators and tourists alike.

In many forests, the long-haired, two-toed sloths of the genus *choloepus* – Hoffmann's two-toed and the Southern two-toed – are the most abundant larger mammal. In contrast to the choosy monkeys, these animals feed on a variety of leaves, which take weeks to digest.

The brown-dappled three-toed sloths, of which there are three species, are also common through-

The largest of the species, the Giant Armadillo, can reach a length of 1.5 meters (5 ft) and weigh up to 60 kg (132 lb), but the Glyptodon, a huge extinct Pleistocene species which died out 10,000 years ago had a 3-meter (10-ft) long body shell used by aboriginal peoples as roofs for huts or burial sites. The Giant Armadillo, which is heavily hunted for food, is considered endangered. But most of the other four species in the Amazon, such as the Nine-banded Armadillo, which has spread all the way into the southern United States, are faring well, despite being considered suitable for human consumption. Another species, the Yellow Armadillo, can be

out Amazonia, though more frequently encountered at the edges of the forest and on riverbanks. They feed exclusively on the leaves and fruits of the Ymbahuba tree. The Brown-throated Sloth is the most widespread of the three species; the pale-throated is found in the far northeast of Amazonia; while the maned Three-toed Sloth is confined to patches of remaining Atlantic forests. The Three-toed Sloth is the world's slowest mammal on land (2–3 meters/6–8 ft per minute).

The armadillos evolved an armor of bony plates which makes them look, at first glance, more like a reptile than a mammal. When attacked, they either crouch to the ground or roll up into a ball to protect their unarmored belly.

found further south in the Pantanal. Armadillos are often observed crossing the road at night.

Bats

The nocturnal lifestyles and similar general appearance of bats discourage most people from making detailed observations of these interesting mammals. Bats are the only mammals capable of active flight, and this ability has facilitated their distribution throughout the world. There are nearly 1,000 bat species known worldwide, about 100 of which occur in Amazonia. Taxonomists divide bats into 19 different families. Seven of them are restricted to the New World, and three more occur both in the New and the Old World.

Missing from the New World is unfortunately the attractive group of flying foxes and fruit bats, which is so typical in the tropics of Asia and Africa. Their niche is filled by the Spear-nosed bats, which are smaller than flying foxes.

The prototype bat is an insect eater, catching them with the teeth or with the tail membrane. They use echo-location to navigate at night, and to localize their prey. Despite sharing many characteristics, bats have nevertheless become anatomically quite diverse evolving to take advantage of a variety of ecological niches. Some bat species have become carnivores and prey on small mammals, lizards, frogs or even other bats.

Primates

Two families of monkey occur in South America's rainforests, namely the small marmosets and tamarins, and the larger capuchin monkeys. Advanced primates like great apes and gibbons, or primitive forms like lemurs don't exist in America. Higher primates evolved fairly late, namely about 35 million years ago. Since it is believed that the New World species are related to Old World Monkeys, consequently they must have reached the isolated South American continent by some unknown means, possibly on natural rafts from Africa. This event was much more probable in the geological past, during

A specialist, the Fishing or Bulldog bat is able to locate fish underwater and to grab them with the sharp claws of their huge feet. Many bats have important functions in the ecosystem as pollinators or dispersers of seeds. The Common Vampire is a small bat that is able to land and run on the ground to approach large mammals. It makes a small incision with its sharp teeth, unnoticed by the sleeping mammal, and laps up the small trickle of blood. It's commonest in cattle country, but is rare in the rainforest proper.

LEFT: a Three-banded Armadillo from Brazil's Pantanal region.
ABOVE: a Vampire Bat.

the period when the Atlantic Ocean was still fairly narrow. Despite a distant common heritage, New World monkeys differ from primate relatives in Africa and tropical Asia. A long and, in the case of the capuchins, often prehensile tail, and a broad nose with nostrils pointing sidewards, rather than downwards, are typical for New World species. There are more than 50 monkey species in America, of which about 30 are associated with Amazonian forest. Related, but often different species occur west of the Andes or in southern Brazil's Atlantic forests.

Marmosets and tamarins are small, normally of about 300g (10 ounces) in weight. Many species are quite handsome: they have manes and mustaches which, just like their limbs can be colored differently from the rest of the body. In contrast to other monkeys, they have sharp claws, an adaption to their permanently arboreal lifestyle. Unusual for most monkeys, they are monogamous, and a founder pair heads a family group. They eat fruits and flowers and also forage for small animals. Some species have a predilection for plant saps and gums. Only a single marmoset species can inhabit any particular area, but sometimes several tamarin species fit into

EVOLUTION STRATEGY

The Night Monkeys may have evolved their nocturnal habits as a protection against diurnal predators, and in order to use food sources for which there is competition during the day.

The eight tamarin species of the region have evolved numerous subspecies with quite divergent colors. Rivers form frequently impassible barriers to particular species and subspecies. Some species are mostly black, like the Black-mantle, the Saddle-back, and the Red-handed Tamarin, which can be distinguished by its reddish legs, or reddish arms and legs, or just bright red hands and feet. The Emperor Tamarin has gray fur, a reddish tail and long white mustache.

The marmosets and tamarins of the Amazon

divergent ecological niches of the same habitat. The center of diversity is upper Amazonia, and no species occurs in northern South America.

Three species of marmosets are found in the region. The brownish Pigmy Marmoset found in the western Amazon is, with a weight of no more than 190g (7 ounces), the smallest of the world's monkeys and close to being the world's smallest primate. The Tassel-ear Marmoset occurs between the Madeira and the Tapajós rivers, while the Bare-ear or Silvery Marmoset, a nearly white species with a pink face, lives only south of the Amazon. Related to the marmosets is the blackish Goeldi's Monkey, with a western distribution.

fair well as long as extensive tracts of their habitat exist. The limited distribution of some species between particular river systems makes them very vulnerable, however. Because of reduced habitat, some relatives west of the Andes or in the Atlantic forests of southern Brazil, such as the Cotton-top Tamarin or the Golden Lion Tamarin, may face extinction.

Another family of monkey, the capuchins, include not just their namesakes, but also sakis, Squirrel Monkeys, woolly monkeys, titis, spider monkeys, uakaris and howlers. These are all bigger animals than marmosets and tamarins. They are also more diverse in their appearance and their ecological adaption. Of the 30 capuchins,

24 are associated with the Amazonian rainforest, but they are more widespread than the marmosets and tamarins and occur in many parts of Colombia and Venezuela. Some have extended their range from the lowland into the upper montane forest above 2,000 meters (6,500 ft).

Most capuchins combine a white face with a brown or black upper head, a pattern that reminded early naturalists of the hood and habit of monks, and gave rise to these monkeys' vernacular name. The Brown Capuchin is widespread, while Weeping Capuchin only inhabit areas to the east of the Rio Negro and north of the Amazon. The White-fronted Capuchin is to

Among the many related species is the Squirrel Monkey, which is probably the most common primate of the region and which successfully drives out other monkeys in search of food by sheer group size. With its black snout, black upper head and white areas around the eyes, it has a somewhat Mickey Mouse-like face, although the big ears are light rather than black.

The six species of sakis and two species of titis are attractive animals. The male and female of the Pale-faced Saki have different coloration – the male has black fur with a white, somewhat owl-like face and a black snout, while the female is uniformly brown. The Monk Saki,

be found to the west and south of the river.

The capuchins have evolved the world's only night-active higher primates, the Night Monkeys. These have been classified into some five to seven similar-looking species – all are fairly small monkeys with a gray or brown back and a light or orange underside, but no prehensile tail. The eyes are surrounded by white patches, which are separated and surrounded by black stripes. To make contact with mates, the male calls in an owl-like manner on moonlit nights.

LEFT: a wide-eyed Night Monkey, a night-active higher primate.
ABOVE: an Emperor Tamarin.

with long gray and white-mottled fur, is encountered south of the Rio Negro and Amazon. In contrast to the long-haired sakis and titis, woolly monkeys have a very dense and short-haired fur, as well as a prehensile tail. The Common or Humboldt's Woolly Monkey is widespread, while the Yellow-tailed Woolly Monkey is restricted to Peru.

Rare, and confined to the flooded forest, are the Red and the Black Uakari, the only New World monkeys without a long tail. The former has white fur, a red and wrinkled face, and a bald head. It occurs on the islands of the upper Amazon, and many riverboat tours make a special effort to find this unusual looking primate.

Amazonia's three howler monkeys are unrelated to Southeast Asia's gibbons, but their inspiring – though rather frightening – vocalizations play a similar part in the nature experience. Like sloths, they are successful through the ability to feed on the amply available leaves of the forest. The Red Howler Monkey is distributed north of the Amazon, the Red-handed Howler Monkey to the south, and the Black Howler Monkey further south still, with a range that includes northern Argentina. In a different way reminiscent of gibbons are the Black and the Long-haired Spider monkeys, particularly in the ease with which they swing hand over hand through the canopy.

mouse-like rodents, the big spiny rats and tree rats are indigenous to the forest.

In contrast to the mouse-like rodents, squirrels, of which there are seven species, are a favorite among visitors, with their appealing bushy tail and their lively lifestyles. The Fire-vented Tree Squirrel, with its bright red head, underside and limbs and a head-to-tail length up to 60 cm (2 ft), is one of the most attractive representative. In general, squirrels are rarer here than in the Old World or in North America.

Mice, rats and squirrels were found in South America only during the last few million years, and don't differ much from North American

Rodents

With over 100 species, the rodents are the biggest mammalian order in Amazonia. Many rodents are of mouse- and rat-like stature, and hence don't attract too much human interest. Their handicap is to be secretive and small, and considered as pests, a prejudice which is appropriate only to a few species like the House Mouse or the Black Rat, which were introduced from Europe, inadvertently, with the first sailors. They stay close to human settlements and do not penetrate into the forest. Since many cities of the region, like Manaus, are quite dirty, it is difficult to avoid this kind of wildlife observation. In contrast, numerous native species of

forms. But South America's first rodents colonized the region nearly 35 million years ago. Some of these ancient rodents such as the New World porcupines and the Capybara, have, in the absence of ungulates, evolved differently from the main rodent groups in the rest of the world. The Capybara, for example, falls into an ecological niche filled elsewhere by pigs and hippopotamuses. This is the world's largest rodent, weighing up to 66 kg (145 lb). Some extinct relatives of the Capybara could reach the size of small horses. Sometimes capybaras are referred to as water pigs, since they spend much of their lives close to water. They are, of course, unrelated to pigs, but are indeed reminiscent of them with their compact, barrel-shaped bodies and similar habits. They live in groups of up to 40 animals. The male has a large scent gland, called a *morrillo*, on top of the snout, and both sexes have two scent glands on each side of the anus. These glands produce in each individual a different mixture of chemicals. The odor provides an important means of individual and sexual recognition. They are considered attractive game just like Agoutis, Acouchis, and the Paca. The latter has a shape similar to Capybaras, but reddish brown fur with white lateral stripes and lines of dots. The orange or brownish Agoutis and Acouchis have the appearance of a short-

Carnivores

About 22 species of the order Carnivora occur in the region, belonging to five families: cats, dogs, raccoons, weasels and bears. The only representative of the latter group, the Spectacled Bear, is a rarely seen denizen of upper Andean forests and Páramos, and never enters lowland forest.

The Jaguar, with a weight that can exceed 100 kg (220 lb), is the only big cat of the New World. Its fur is golden brown and spotted with black dots and rosettes. It resembles the leopard of the Old World, but is decidedly the bigger animal, looking more impressive also with its broader forehead. Just like the Leopard, the Jaguar has a

eared hare. An example is the Red-rumped Agouti of the eastern Amazon.

In contrast to their Old World relatives, New World porcupines spend much of their time in trees. In an example of convergent evolution, several species, such as some monkeys, opossums and anteaters, have evolved prehensile tails to function as a fifth limb to facilitate their arboreal lifestyle. Hares, distantly related to rodents, occur only with a single species, the Brazilian Rabbit, also called the Tapiti.

LEFT: a young Red Uakari, a species facing extinction.
ABOVE: a Red-rumped Agouti.

completely black variety. These are individuals which differ only in some genes, and are not a separate subspecies. The Jaguar lives in dense forest, often close to water, and preys on all large mammals. It rarely attacks humans, but preys on domestic stock. This adversary relationship with humans, and the value of its fur, has led to severe hunting and a dramatic reduction of its numbers.

Seven species of smaller cats include the puma, a highly adaptable species, which occurs throughout the Americas in habitats as diverse as desert, steppe, northern or tropical forests. The Puma is uniformly sandy-brown, gray-brown, beige or sandy colored, while the margay and the Ocelot are spotted, like the much bigger Jaguar. Like all large cats, they are endangered by the fur and pet trades. Similarly to the Jaguar, they are very secretive and can rarely be observed, and the best way to monitor their presence is to identify their tracks in the sand and mud of river banks.

Six species of dog include four of fox-like appearance such as the Short-eared Dog. The attractive Maned Wolf is reddish brown with black, unusually long legs, which give it a somewhat antelope-like gait. Occurring at the southern rim of Amazonia, the wolf's long legs are adapted to moving easily over the tall grass of savannas.

One species of raccoon, the Crab-eating Raccoon, is related to three strangely looking

tion in Peru's Manu National Park, where it is one of the sanctuary's prime tourist attractions.

The grison is of gray color with a black face, breast and limbs, and the tayra looks similar to a black marten with a pale head.

Ungulates

Hoofed mammals are poorly represented in the region, particularly since many groups, such as wild cattle, antelopes and goats, never made it from North into South America. Four camel species such as the llama occur on the alpine grasslands of the Andes, while the rainforest holds only the tapir, peccary and deer family.

denizens of the forest, the South American Coati, the Kinkajou and the Olingo. Although carnivores by relationship, these three mammals live mostly on fruits and insects. The Coati is a lovely animal, curious, playful, inquistive; with a long, ringed tail and a long, upturned snout, while the Kinkajou and Olingo look a little like a cross between a monkey and a weasel.

To the weasel family belong the Amazonian Skunk and two otters, the South American River Otter and the Giant Otter. The head-body-tail length of the Giant Otter can reach 2 meters (6 ft), and exceeds the size of the Sea Otter of the North American Pacific Coast. The fur trade has dramatically decimated this species. It has protec-

There are 11 deer species in South America, and six of these occur in Amazonia. All species are similar to deer elsewhere. The White-tailed Deer has extended its range from North America into the north of the region, and occurs in most parts of Colombia and Venezuela. South of it occurs the Marsh Deer, which can be observed in the Pantanal. The Red and the Gray Brocket Deer have only half the shoulder height of the previous two species, and very small antlers.

The Collared and the White-lipped Peccary are the closest relatives of the Old World's wild pigs and are of similar stature. Living in herds of up to 100 animals, they are one of the Jaguar's principle prey. Peccaries are famous for their unusual altruistic behavior: if a group with young cannot evade a predator, a single adult individual will heroically confront it, often with fatal results.

Tapirs are the only South American odd-toed ungulates. The Brazilian Tapir is widespread, while the Mountain Tapir is restricted to Andean forests and Baird's Tapir to the northwest. These tapirs are brownish animals, in contrast to the black-and-white Malayan Tapir. The stout body helps them push through the forest underbush. Their snout is a short trunk, which

ECHO-SOUNDING

The dolphins have, in the turbid river water, lost some eyesight, and for navigation and hunting, they depend on echolocation. This sense is so precise they can even tell differently shaped fish apart.

America's largest non-marine mammal. Unrelated to seals, they have a similar body shape, but no hind limbs, and the tail is a whale-like horizontal fluke. They are strictly herbivorous, and feed during high water seasons on the rich plant life of floating meadows. They fast after retreating into the river during the low water season. The manatee was once abundant in the Amazon river system, but has not reached the Orinoco basin. With the arrival of the Europeans, it became heavily hunted for food, and is now rare,

can be used to grasp food such as leaves, in a similar way to the elephant.

Aquatic mammals

Three mammalian groups have adapted to live more or less permanently in the water. While one of the groups, the seals, is absent from the Amazon, the river is home to two species of cetaceans and two species of manatee. Amazonian Manatees are, with a length up to 3 meters (9 ft) and a weight of up to 500 kg (1,100 lb), South

endangered, and still inefficiently protected. A related species is the West Indian Manatee, which occurs in coastal waters.

The Amazonian River Dolphin, or Boto, occurs throughout the Amazon and Orinoco systems. It is a long-snouted river dolphin, and its only relatives today occur in the Indus, Ganges and a lake in China. A second species, the Tucuxi or Gray Dolphin, is related to marine dolphins, the coastal forms sometimes considered a separate species. The Boto, with a length up to 2.6 meters (8.5 ft), is the larger animal; the Tucuxi is only half this size. As the objects of local myth, they are not often hunted and are hence relatively widespread, even in very small tributaries. ❑

LEFT: a pair of Bush Dogs.
ABOVE: the Puma, or Mountain Lion, is found from Canada to Argentina.

BIRDS

The world's richest bird life is to be found in the
vast expanse of the lowland rainforests of Amazonia

Estimates of the number of bird species present in lowland Amazonia depend on the geographical criteria employed, but it certainly has well in excess of 1,000 species. Hundreds more species can be added by including the Amazon catchment, including the alpine *páramos* of the Andes and the subtropical and temperate forests of the foothills. Within western Amazonia (the richest area of all) studies at more than one site have shown that within just a few square kilometres of lowland forest it is possible to record more than 500 species.

It is a matter of considerable biological debate why the lowland Amazon rainforest is so rich in avifauna. It may be partially explained by the inherent stability of the tropical forest, which has, in evolutionary terms, brought about a high degree of specialization with a very intricate level of animal-animal and animal-plant interactions. It is therefore possible for a large number of species to co-exist in the same area, as each has very narrow and specialized requirements. This habitat also encourages certain traits to evolve which would be difficult in temperate or strongly seasonal environments: for example, diets almost wholly consisting of fruit.

It is, however, a misconception to think of this as a vast, uniform, unchanging environment. The forest is a very dynamic place, and vegetation is dictated precisely by soil or moisture levels, so that within a small area one can encounter distinct types: areas dominated by particular palms, or by patches of bamboo, sandy areas, or seasonally flooded areas. Change constantly occurs, varying from the movement of the flow and course of rivers to natural clearings caused by treefalls. All this brings corresponding changes to vegetation and thus to the bird life. There are species that may not only just be restricted to zones of forest which flood seasonally (*várzea* forest), but to particular types of vegetation within those zones. A

visitor who explores the forest's various parts, will see more species of birds than one who concentrates on a narrow range of habitats.

Bird-watching

As in tropical forests everywhere, bird-watching in Amazonia can be a frustrating business. Birds of the forest floor often move stealthily, more often heard than seen. Even when birds are glimpsed, the poor light conditions can make the recording of plumage details difficult. Most birds are in the under-story, canopy or at mid-levels in the forest profile. Observing them will cause an aching neck and you cannot expect to identify every small bird briefly seen between gaps in the foliage over 40 meters (130 ft) above.

Many insectivorous birds travel around in mixed species flocks. This creates clumps of birds, so walking through the forest you may not see anything for a long time. Once a flock is encountered, however, there can be great confusion as perhaps 20 or 30 species pass by rapidly. Each species is normally represented by pairs or small family parties, so that a flock may contain almost a hundred individuals. As flocks occupy distinct territories, after more than a few days in an area it is usually possible to identify the territories and so relocate different flocks.

Fruit-eating birds of many different species will often visit the same type of tree or shrub. If a fruiting tree is found to attract birds it is worth sitting nearby in a concealed place to wait for a range of fruit-eating birds to arrive. The most difficult of all birds to see are those which habitually fly above the canopy: these include birds of prey and swifts. Seeing these requires getting above or close to the canopy, or using areas with a good view of the sky, such as the banks of lakes or rivers, or clearings.

Throughout the neotropics, there are 85 families of birds of which 30 are endemic, or nearly so, to the region; partly because South America as a continent remained isolated until just three million years ago. The majority of the bird families are represented in the Amazon basin and each of the important ones is described below.

PRECEDING PAGES: the Black-necked Red Cotinga.
LEFT: a Straight-billed Hermit, a species of hummingbird, attaches its mud-nest to the tip of a palm leaf.

Water birds

The changing pattern of the watercourses in Amazonia creates oxbow lakes, which in turn produce weed-choked pools and swamps. This variety of wetlands supports herons, egrets and ibises. Some of these are forest-dwelling, feeding along small creeks, and are not often spotted out in the open. They include the tiny and rarely seen Zigzag Heron, so called for its intricate pattern of barred plumage. The elegant Agami Heron, with a very long, dagger-like bill, also feeds close to cover, whereas birds like the Cocoi Heron are easy to spot along river banks. However, the population of these herons, and the Scarlet Ibis, has

often seen perched atop bushes or trees. Its far-carrying call is an extraordinary loud donkey-like braying sound. Despite its bulk, once airborne the bird flies powerfully and can soar and glide like a vulture. Other fowl-like birds are the curious trumpeters. They are terrestrial, long-legged birds, dark-plumaged with lax, pale plumes on the back. They occur in small parties, feeding on reptiles and amphibians and are much prized by indigenous peoples as pets because they produce loud calls on the approach of intruders.

In swamps and lakesides occur Purple and Azure Gallinules, and Wattled Jacanas. The latter species is polyandrous, the female laying more

plummeted in many regions of Amazonia since the 1970s, perhaps due to overfishing by humans. The Green Ibis feeds quietly in damp areas on the forest floor, but at dawn and dusk it is extremely vocal, flying across rivers on stiffly-held wings, giving out a loud rollicking call. More than 14 species of herons and egrets may be found within a single locality.

Of all tropical areas, the Amazon basin has a surprisingly small number of species of waterfowl. They include the Orinoco Goose and the Muscovy Duck, the ancestor of the familiar farmyard variety. One distinctive species is the Horned Screamer, a huge, heavily-built bird which feeds on waterside vegetation and is most

than two clutches, with different males tending each one and the female playing no role in parental care. A small number of crakes and rails occur in grassy fringes of the forest. The Gray-necked Wood-Rail is particularly well known, producing a loud dueting chorus at sunset.

Along shady creeks, close to overhanging vegetation, the shy Sungrebe may be seen. This is a member of the Finfoot family, a pantropical group represented by a single species in each of the three great tropical zones. A family unique though to the New World and comprising a single species is the Sunbittern. This elegant, dainty bird is widespread but difficult to see. It picks its way slowly along forest streams, sometimes climbing into the branches to call, giving a slightly wavering whistle. Its plumage is beautifully patterned with intricate barring of golds, browns and blacks, but when its wings are outstretched, revealing brilliant orange patches, the bird's full splendor can be seen.

The strongest seasonal effect throughout most of Amazonia is the changing water levels of the rivers, caused by rainfall patterns far away in the Andean catchments. This translates to river level fluctuation of over 10 meters (33 ft) in many areas. When

ACUTE SMELL

It is likely that the Turkey Vulture detects food by an acute sense of smell. The large King Vulture soars at great heights and may watch for the Turkey Vultures descending to a carcass.

of prey form a large part of the avifauna, and over 30 species can be recorded in a single locality. The globally distributed Osprey is familiar throughout the Amazon. It does not breed there, but non-breeding individuals are present throughout the year. The Harpy Eagle is one of the most powerful raptors in the world. It is remarkably agile, twisting and turning through the canopy to seize large monkeys, its principal prey. It is, however, difficult to see, since it very rarely soars above the canopy.

the water level is low, great sand and mud banks are revealed. These provide important nesting areas for Large-billed and Yellow-billed Terns, Black Skimmers and some waders, such as Collared Plovers. Migrant waders from North America may also congregate to feed on such sites.

Scavengers and birds of prey

Four species of New World vultures can be seen over the rainforest. They feed on carrion such as dead sloths and other mammals. Diurnal birds

LEFT: the Boat-billed Heron can be seen near forested rivers *(far left)*, and a Rufous-necked Puffbird.
ABOVE: a Greater Yellow-headed Vulture.

Other sub-canopy hunters include the Forest-Falcons which feed mainly on small birds. At dusk Bat Falcons appear. These are small elegant raptors which, as their name suggests, feed mainly on bats, although they also take large flying insects. A very common bird along the water's edge is the Yellow-headed Caracara, which is mainly a scavenger.

Tinamous and cracids

Tinamous are ground-dwelling birds which feed mainly on fallen fruits and seeds, although some insect prey may also be taken. They are rather furtive birds, often "freezing" when approached, although sometimes taking flight in a loud and clumsy way when flushed. Their plumage is dull gray or brown, some species having barred or spotted upper parts or flanks. The calls are most distinctive, having a haunting quality, usually consisiting of tremulous whistles. These may be single notes or in a rising or falling series, depending on the species. They are most often heard at, or just before, dawn and during the late afternoon. Some sites in Western Amazonia may support about five or six species.

The cracids are made up of the chacalacas, guans and curassows. They are all fruit- and seed-eaters, spending most of their time in trees. Chacalacas are noisy birds, often seen in small

long-tailed birds which creep through tangles of vegetation and hop along branches in a fashion resembling squirrels. They often join mixed species flocks of insectivorous birds, preying on large caterpillars. The Greater Ani frequently occurs in flocks of over a hundred strong, often beside watercourses and in apparent association with troops of monkeys. The Red-billed Ground Cuckoo is a timid terrestrial species which eats lizards, frogs and large arthropods. It sometimes preys on the animals flushed out by raiding swarms of army ants.

Night birds include the nocturnal curassows, owls, nightjars, nighthawks and potoos. The

parties, and appear fairly tolerant of areas in which humans are active. In contrast, the curassows are extremely vulnerable to hunting pressure and require forest with little or no disturbance. Therefore, they are excellent indicators of the conservation status of a forest area. Curassows are large, fowl-like birds which have characteristic whistling or humming songs.

Cuckoos and night birds

Sites in western Amazonia may support nearly a dozen species of cuckoo, including the migrant Yellow-billed Cuckoo from North America. Typical birds of the forest are the Squirrel Cuckoo and Black-billed Cuckoo,

PARROTS

Parrots are well-represented and over 20 species can be found in a single locality. These range from the spectacular large macaws, such as the Blue-and-Yellow, the Scarlet and the Red-and-Green to sparrow-sized Parrotlets. The parrots are some of the most obvious of the bird groups, often seen in the late afternoon flying in flocks to roosting sites on river islands. They are attracted to exposed earth banks in certain areas where they consume various minerals. The most spectacular sites are in the Manu National Park and Tambopata in Peru, where large numbers of several parrot species congregate, but which are most famous for Scarlet and Red-and-Green Macaws.

latter are similar in appearance to the frog-mouths of Asia and Australia. Cryptically colored, they spend the day perched at the end of branch stumps. At night they undertake sallying flights to catch moths, usually returning to the same perch after each flight. The Common Potoo has an eerie series of mournful whistles, an unforgettable sound, mainly heard on moon-lit nights, which local people traditionally believe is given not by the potoo but by a sloth.

The hoatzin

Related to rail and gallinaceous birds is the extraordinary prehistoric looking Hoatzin.

Hummingbirds and trogons

Over 15 species of hummingbird may occur in the same area, although each tends to occupy a specific niche. The hermits are large, forest-dwelling hummingbirds, traveling sometimes a kilometer or more close to the forest floor, feeding on the hanging flowers of *Passifloras* or *Heliconias*. Other species, such as the Glittering-throated Emerald, are more pugnacious, defending flowering shrubs on the forest edge. Jewelfronts or Gould's Jewelfronts and Black-eared Fairies are also forest birds but feed mainly in the canopy or subcanopy, collecting small insects from the tips of leaves of flowering trees.

Living in groups beside lakes and swamps, its crop resembles the digestive system of a rumi-nating animal. Pairs are aided by helpers at the nest, and the young, if alarmed by a poten-tial predator, will leap from the nest into the water. There they will swim back to the bushes, clambering back through the branch-es using claws on their wings, an evolutionary throw-back, as seen on Archaeopteryx, and unique among birds of the world. It is found only on the young birds.

LEFT: a Sunbittern.
ABOVE: a Wattled Jacana foraging at the edge of an oxbow lake.

A pantropical group which shows its greatest diversity in the neotropics are the trogons. The Amazon forest supports the largest number of all. Seven species may coexist, including the Pavo-nine Quetzal. Trogons have hooting calls, a char-acteristic sound of the Amazon forest. Normally perched at mid-story in the forest, they peer from side to side before taking flight to pluck a large caterpillar or cricket from nearby green foliage. All are brightly colored, yet their unobstrusive behavior can make them difficult to spot.

Kingfishers and jacamars

Five species of kingfishers are widespread. Three of them, the Ringed, Amazon and Green, are found mainly along rivers, whereas the Green-and-Rufous and the Pygmy birds frequent smaller water-courses, usually within the forest itself.

The jacamars are a neotropical family which fill much the same niche as Bee-eaters in the Old World. They normally occur in pairs, perched beside small clearings and openings in the forest cover, darting from their perches to catch flying insects in their long bills. Up to seven species may be recorded within the same area.

Puffbirds and barbets

Puffbirds are also restricted to the neotropics. These include the Black-fronted Nunbirds. Read-

leaves in which animals like cockroaches seek refuge. Though found in their greatest abundance in Asia, they have been found throughout the tropics. In most areas of Amazonia, usually no more than three species will occur together.

Toucans and woodpeckers

Toucans, characteristic of the Amazon, are also mainly fruit-eaters. No fewer than seven members of the toucan family may coexist, including the smaller aracaris and toucanets. Large toucans, such as the Toco Toucan and the Yellow-ridged Toucan, have croaking calls and are usually seen in pairs or small parties, flying from tree to tree,

ily observed in semi-open areas around small settlements, three or more of these smart black birds with their long, slightly decurved red bills, will often gather and produce excited choruses.

They are unusual among Amazon birds in nesting in burrows on the ground, a habit shared by some of the other puffbird species, including the Swallow-wing, which nests in sand banks and is common in open country. Forest puffbirds include the Spotted Puffbird, which nests in termite colonies.

Barbets are a group of birds that subsist mainly on fruit, but which at times also take arthropods. These are often captured acrobatically, with the bird hanging upside down to tear open dead

making a great "whoosh" sound with their wings.

Woodpeckers range from the large crimson-crested and red-necked to the tiny piculets. They share the tree trunks with woodcreepers, a group found only in the neotropics.

Unlike the woodpeckers, which are generally boldly patterned, the woodcreepers are well camouflaged. With 15 or more similar species present in the same area, they can cause considerable identification problems. Whereas woodpeckers have powerful bills which are used like chisels, the woodcreepers have decurved bills, used to prise into crevices to pick out small animals. The most extreme of all are the Scythebills, which have extraordinarily long, decurved bills.

Oven- and antbirds

A very diverse group are the ovenbirds or furnariids, of which more than 25 may be found within one site. These include the Pale-legged Hornero which does indeed produce a nest made of mud that resembles an old brick oven. The group also includes spinetails, which are generally long-tailed, dull brownish birds that skulk in long grass, although there are also forest spinetails and some which creep along branches. Ovenbirds often carry names which describe their behavior – thus Foliage-Gleaners collect prey by searching through leaves; Leaftossers feed on the ground by overturning

antshrikes, habitually enter mixed species flocks of insectivores; indeed *Thamnomanes* antshrikes, are key flock species present in most mid-story flocks. Antbirds occur from the forest floor (antpittas are essentially terrestrial) to the canopy (some of the antwrens).

Most species are habitat-specific, restricted to certain types of forest in preference to others. Some occur only in certain vegetation such as bamboo clumps, others exist only on river islands and rarely cross to the river banks. Of all the forest species, the antbirds are perhaps the most vulnerable to forest fragmentation. The ant-following species in particular seem to be

leaf litter; and Palmcreepers creep in palms.

The antbirds, like the furnariids, are exclusively found in the neotropics. Fifty species of this group can be seen in some prime western Amazonian sites. Their name derives from the habit of some to follow swarms of army ants, collecting large arthropods as prey items, flushed by the raiding swarm. However, the group as a whole has diverged to fill a wide range of niches. They usually occur in pairs or family groups, although antpittas are normally solitary. Some, like many of the antwrens and

the first to disappear from isolated forest patches. This may be due, in part, to the spatial requirements of the army ant colonies.

Manakins and cotingas

Some of the most attractive of the small forest birds are the manakins. In almost all species the males are gaudily plumaged and assemble in courtship leks (a loosely gathered group of males assembled for courting females), where they perform elaborate dances. These may take place, in the example of White-bearded Manakins, in a special "court" where the birds have taken care to clear the area of small leaves and twigs. Others, such as the Golden-headed

LEFT: a pair of Hoatzins.
ABOVE: a Yellow-ridged Toucan.

Manakin, perform their displays in the canopy. Probably the most curious display of all is that of the Wire-tailed Manakin. Both sexes have long filaments in their tails, those of the male being particularly long. During the dance, the male strikes the female across the sides of her head with these projections. One of the most complex of the dances belongs to the Blue-backed Manakin, in which two or even three males perform a tightly synchronised bouncing dance.

PIGEONS AND DOVES

Represented by fewer than ten species at any site, these include canopy birds like Ruddy Pigeons, ground-dwelling quail-doves, and forest-edge and open-site birds like Ruddy Ground-Doves.

Flycatchers, wrens and thrushes

Probably the biggest group of all is the Tyrant Flycatchers. Over 60 species have been recorded in a single locality. They vary in size from wren-sized, pygmy-tyrants and spadebills to the thrush-sized Boat-billed Flycatcher. The longest species is the Fork-tailed Flycatcher, an elegant bird which can sometimes be seen in large roosting flocks beside lakes. The kiskadees are a readily recognizable group which perch in open situations, most commonly the Great Kiskadee,

Related to the manakins, and also essentially fruit-eaters, are the cotingas, another family endemic to the neotropics whose greatest diversity is in the lowland rainforests of the Amazon. Most are polygynous, with brightly-colored males. The Screaming Piha, however, is a very drab grayish bird, and males form widely dispersed leks in which they display with a loud, explosive call. This is one of the most characteristic sounds of the Amazon forest and is quite unforgettable. Bare-necked Fruit-Crows are one of the largest cotingas and can often be observed flying across rivers in small flocks. The Amazonian Umbrellabird has an umbrella-like crest and a long, pendant wattle.

which incessantly calls its name and is known locally as "Victor Díaz". The smaller tyrannulets give visitors the biggest headaches. Many are canopy-dwellers and can be identified only with experience and knowledge of their calls.

Very few North American passerine birds winter in the Amazon lowland forest. Exceptions are the Barn Swallows and Bank Martins, which arrive in August. The former, in particular, can be a common sight along the rivers. Migrants also arrive from the southern part of South America, such as Southern Martins, which can form huge roosts in towns like Leticia and Iquitos.

The loud duets of the Thrush-like Wren are a distinctive sound throughout the region. This is a large wren, spending most of its time close to clumps of bromeliads and other epiphytic plants in the canopy. Probably related is the Black-capped Donacobius, a very smart dark brown and cream bird inhabiting the tall grasses growing beside rivers and lakes. Pairs will frequently perch high on the grass stems, producing loud whooping calls. In the forest, the small crake-like Nightingale Wren is a denizen of the forest floor and produces a captivating series of notes, delivered singly with ever-increasing pauses between them. The Musician Wren also keeps mainly to the ground and produces a song remarkably similar to that of a human whistling.

The Yellow-rumped Cacique is commonly found beside water. It is a colonial nester, often selecting isolated trees surrounded by water or close to human settlements, presumably to reduce the risk of raids by monkeys and even toucans. The species is polygynous. The male performs a richly mimetic song, as it shakes and bobs its plumage. This display is even more ornate among some of the larger oropendolas. The Russet-backed Oropendola produces a song akin to the sound of water pouring from a bottle.

The honeycreepers and tanagers sometimes number more than 40 species in a single locality. Many of these species are gems of birds which

Thrushes are poorly represented, but may include one or more migrant species from North America. One of the resident species, Lawrence's Thrush, is an accomplished mimic, imitating the calls of a range of other species, including manakins and antbirds.

Icterids, tanagers and finches

Another important group of birds are the icterids. These are quite diverse, including the attractive, orange-and-black colored troupial (*Icteris jamacaii*) and orioles, caciques and oropendolas.

occur in twittering, mixed-species canopy flocks. These may include the Paradise Tanager, Opal-rumped Tanager or the Green-and-Gold Tanager. Others are common in open or secondary habitats such as the Silver-beaked Tanager or Blue-grey Tanager. Some, like the Ant-Tanagers, are found in the forest shrub layer and understory.

Finches build their nests mainly by river banks, forest edges and cultivated areas. Of these, the Red-capped Cardinal is an attractive riverside bird. Grassy patches support small seed-eaters such as Chestnut-bellied Seed-eaters. One species encountered in the forest, for example, is the Slate-colored Grosbeak, with its fluty, thrush-like song, delivered from high, canopy perches. ❑

LEFT: a Wire-tailed Manakin.
ABOVE: a Blue-napped Chlorophonia.

REPTILES

From Anacondas that can reach 10 meters (33 ft) in length to caimans and huge river turtles, the Amazon plays host to some strikingly impressive reptiles

It is not just the giants of the reptile world that play a prominent role in the Amazonian ecosystem. Equally interesting are the many smaller species of lizards and snakes that have developed an incredible diversity of lifestyles and have colonized all available habitats ranging from subterranean and aquatic to the forest canopy. Of the 1,200 South American species of reptile, 550 occur in rainforests, and about 270 in the lowlands of the Amazon basin. In lowland forests, most species are widespread, whereas the eastern slopes of the Andes contain a high number of endemics. The Andean foothills also support the most species-rich reptile fauna in the world, with more than 90 species of reptiles found within a few square kilometers. Some reptile species suffer tremendously from exploitation by humans, hunted for their skins, their flesh, or simply because they are considered harmful.

Caimans and turtles

In Amazonia, many vertebrates are difficult to find, as they occur in low densities and live a secretive solitary life. However, some reptiles, such as the large aquatic turtles and caimans, are an exception. Caimans are readily observed from a boat at night by the reflection of their eyes in a strong torch light. Additionally, the Spectacled Caiman *(Caiman crocodilus)* and the Black Caiman *(Melanosuchus niger)* sunbathe for long periods on beaches.

Both species construct nests of rotting debris, which provide a constant temperature and humidity for their eggs. On hatching, the females, and occasionally the males, help their offspring out of the nest and guard them for some weeks. Caimans feed on a variety of fish, mammals, and birds (attacks on humans are exceedingly rare), which are generally captured and eaten in the water. Spectacled Caimans grow up to 3 meters (10 ft) long, whereas Black Caimans may reach more than 5 meters (16 ft).

PRECEDING PAGES: a Black Caiman with its catch of a Piranha.
LEFT: the beautiful Emerald Tree Boa.

Two small (1–1.5 meters/ 3–5 ft) species of caimans, the Musky Caiman *(Paleosuchus palpebrosus)* and the Smooth-fronted Caiman *(Paleosuchus trigonatus)*, inhabit small forest streams and ponds. Unusually for crocodilians, the Smooth-fronted Caiman is semi-terrestrial and feeds largely on terrestrial vertebrates. Unlike neotropical caimans (which, with alligators, form the family *Alligatoridae)*, true crocodiles *(Crocodylidae)* are absent from Amazonia, though two species occur in northern South America.

Some visitors to the Amazon are lucky enough to witness nesting colonies/groups of the Arrau Sideneck Turtle *(Podocnemis expansa)*, one of the most fascinating wildlife experiences in South America. While females of this species attain a carapace length of 89 cm (35 inches) and a weight of more than 90 kg (200 lb), its fossil relative, *Stupendemys geographicus*, from the Miocene of Venezuela, was the largest turtle that ever lived – the Museo Nacional de Ciencias Naturales in Caracas owns a carapace that is 230 cm (7½ft) long.

The Arrau inhabits the main rivers of the Amazon and Orinoco river systems during the dry season, and enters floodplains during the wet season, where it feeds on fruits that fall into the water and other plant material. Nesting occurs at night during the height of the dry season when the sandy river banks and islands are exposed.

TORTOISES

A single land tortoise, the Yellow-footed Tortoise *(Geochelone denticulata)*, inhabits the Amazonian forests. The length of this tortoise varies from 30 to 40 cm (12–16 inches). During the early morning and late afternoon, the tortoise searches for food, which consists of fruit and other plant matter, mushrooms and carrion. Local people collect tortoises around certain fruit trees or attract them with carrion, and keep them in enclosures as live meat preserves. In fact, tortoises are regularly offered at Amazonian markets and their flesh is highly appreciated. For this reason, they have become exceedingly rare around all populated areas.

Females generally bury 80 to 90 eggs. After an incubation of about 45 days, 5-cm (2-inch) long hatchlings emerge and race for the water.

Since the European colonization of Amazonia, a drastic decline of the Arrau has taken place. In 1814, Alexander von Humboldt estimated that 33 million eggs were harvested annually on the Orinoco, and in 1863, Henry W. Bates reported the annual export of turtle oil from the upper Amazon amounted to the destruction of 48 million eggs. This heavy exploitation continued until the turtles and their eggs were finally protected by laws in the 20th century. Today, it remains doubtful whether the populations of the Arrau are

(scales) and its broad head with fleshy appendices. Typically resting in quiet shallow water, a snorkel-like snout allows this species to breathe unnoticed by stretching its long neck towards the surface. Small fish which pass close to the head are sucked in by the sudden opening of the mouth and expansion of the throat.

Anacondas and iguanas

Some of the most distinct lizards and snakes are also found in or near water. Once heavily persecuted for its skin and still hunted for meat, the illegal skin trade or out of ignorant suspicion, the magnificent anaconda *(Eunectes murinus)* is now

stabilizing. Presently, the only remaining nesting concentrations are on the Río Trombetas, the Rio Tapajós, and a few places along the Río Orinoco in Edo, Apure, Venezuela.

A smaller relative of the Arrau, the Terecay *(Podocnemis unifilis)*, is still abundant in most tributaries of the Amazon. With a length of up to 46 cm (18 inches) and 10 kg (22 lb) in weight, it is the major source of turtle meat in the area. The Terecay basks on beaches or logs in the river during low water. Single females nest on sand or clay beaches along the inhabited rivers. Probably the most curious turtle, the Matamata *(Chelus fimbriatus)*, is recognized easily by its flat carapace with rough, conically raised scutes

frequently displayed at tourist lodges. Despite their gigantic dimensions of up to 10 meters (33 ft) long and 250 kg (550 lb) in weight, the largest snakes in the world will not normally attack humans unless severely provoked. They kill their prey, large vertebrates, through the fearsomely powerful constriction of their coils.

Like all boas, anacondas give birth to live young, with a litter containing 30 to 80 baby snakes of about 70 cm (28 inches). The only venomous aquatic snake, the Surinam Coral Snake *(Micrurus surinamensis)*, inhabits all of Amazonia and feeds on eel-like fish. Caiman Lizards *(Dracaena guianensis)* are well adapted to life in water and somewhat resemble their namesake.

Another lizard living along the rivers is the Common Iguana *(Iguana iguana)*, a large (2.2-meter/7-ft) green lizard with a prominent dorsal crest, spiny head scales and a huge dewlap. Iguanas perch on limbs of trees and jump into the water when threatened. In some places, local people tame iguanas by feeding them fruit, in other cases they are hunted for their meat.

Common Iguanas belong to the largely neotropical lizard family *Iguanidae*, represented by 40 species in Amazonia. Most rainforest iguanids are arboreal and each species is specialized for inhabiting certain parts of trees. One group of slender iguanids, the anoles *(Anolis)*, are abun-

Teiids and geckos

The second large neotropical lizard family are the teiids *(Teiidae)*. Although, most Amazonian teiids are small, secretive inhabitants of leaf litter, the large Northern Tegu *(Tupinambis nigropunctatus)* prefers open, sunlit patches in the forest. The omnivorous tegus run extremely fast and seek shelter in ground holes. The related smaller, partially green Ameiva *(Ameiva ameiva)*, is abundant near housing and plantations.

Tropical Geckos *(Hemidactylus mabouia)* are often the first reptiles encountered by tourists. Following the colonization of Amazonia, the Tropical Gecko has spread along all the major

dant on leaf litter, trees and even huts near forest. Males display their huge colorful dewlaps and rapidly move their heads up and down during courtship and combative encounters. Climbing on thin twigs of trees and bushes, the iguanid, *Polychrus marmoratus,* resembles an Old World chameleon. Another iguanid, *Plica plica,* dwells only on the lower parts of big trunks and feeds on tree ants. Tucanos of Colombia worship this lizard as the "*Vaimahse*" (Lord of animals), for its distinct, large penis with outward curled tip, which it probably displays during courtship.

rivers and roads. The only indigenous nocturnal gecko, the Smooth Gecko *(Thecadactylus rapicauda)*, occurs on trunks inside primary forest, but may be found in palm huts close to the forest. Geckos are harmless and help inside houses by feeding upon roaches. They are adapted to nocturnal life with their cat-like vertical pupils, and have broad toe-pads enabling them to climb even smooth surfaces like glass.

Boas and colubrids

Most lizards are habitat specialists and diet generalists, whereas most snakes are food specialists able to survive months without nutrition, but with a broad range of habitats. Snakes are rarely

LEFT: a Vine Snake *(Oxybelis argenteus).*
ABOVE: a Parrot Snake *(Leptophis aheatula).*

encountered in rainforests, although their highest diversity occurs there, and even snake experts (herpetologists) need months to find a substantial fraction of the sometimes 60-plus species at a single site. Promising times to look are after heavy rains or storms.

The boa constrictor is the second largest snake in Amazonia, with a maximum length of more than 5 meters (16 ft). Boa constrictors can be aggressive and will emit loud hisses when harrassed. Probably the most common boa in Amazonia, the beautiful

USEFUL COLUBRIDS

Two bulky colubrids, the Mussurana *(Clelia clelia)* and Cribo *(Drymarchon corais)*, may reach 2.5 meters (8 ft) long, but are considered useful as they feed on rats and even venomous snakes.

Green Whip Snake *(Oxybelis fulgidus)*, a slender arboreal snake feeding on frogs and lizards. Also green and arboreal is the non-venomous Parrot Snake *(Leptophis ahaetulla)* but it will open its mouth threateningly when molested. Remarkably thin Blunt-headed Snakes *(Imantodes cenchoa* and *Imantodes lentiferus)*, with extremely large eyes, search trees and bushes at night for small lizards and frogs. Snail-eating Snakes *(Dipsas)* are similar in appearance, but specialize on snails as prey.

Rainbow Boa *(Epicrates cenchria)*, is hunted by local farmers since it frequently preys on chicken. Rainbow Boas are rivaled in beauty by Emerald Tree Boas *(Corallus caninus)*, which have a bright green dorsum with white vertebral marks and a yellow venter, providing a perfect camouflage in the foliage. Their brownish relative, the Garden Boa *(Corallus enydris)*, often lives in palm-thatched roofs, where it hunts mainly bats.

Colubridae are the most species-rich family of snakes; more than 40 different species may occur at a single site. Most colubrids are non-venomous, but some bear enlarged grooved teeth in the back of the jaw and may cause slight poisoning. Among the slightly venomous forms is the

Coral snakes

Several species of colubrid mimic the venomous coral snakes. Imitation supposedly protects the non-venomous or slightly venomous species from predators. The False Coral Snake *(Erythrolamprus aesculapii)* is one of the most stunning examples, as its variable patterns resemble a specific species of coral snake that it overlaps with in distribution. The false and true coral snakes of the United States are different species.

Although they are not aggressive and rarely bite, look before stepping over logs; don't turn over dead logs (or stones) and be careful near woodpiles. Real coral snakes *(Micrurus)* resemble colubrids in body shape, but have very short tails and tiny eyes. The most common coral snakes typically have a pattern of distinct rings with alternating colors, usually black or with two bright colors and may reach over 1 meter (3 ft) in length. Their comparatively short fangs inject a potent neural venom that causes paralysis of the skeletal muscles. The earliest symptom of envenomization is double vision, and death may occur from respiratory arrest.

Pit vipers

The snakes most feared throughout the New World are pit vipers *(Crotalinae)*, which account for the majority of deaths caused by snake-bites.

Pit viper venom affects primarily the blood, and first symptoms include local pain, vomiting, sweating, headache and swelling. Without treatment, mortality is about 7 percent, caused by hypotension, renal failure, or intra-cranial haemorrhage. Responsible for the majority of envenomizations in Amazonia are Common Lanceheads *(Bothrops atrox)*, which occupy cultivated areas as well as rainforest. This predominantly terrestrial snake is about 1.5 meters (5 ft) long and feeds on mammals, frogs, and birds. Exclusively arboreal, the slender Two-Striped Forest-Pit vipers *(Bothriopsis bilineata)* and Speckled Forest-Pit vipers *(Bothriopsis taeniata)*

Pit vipers have large sensory holes (pits) between the eyes and nostrils that serve as heat receptors. The membrane in these holes registers infrared radiation and is sensitive to temperature changes of only 0.003°C, thus helping to localize prey and potential enemies by their body temperatures.

Pit vipers are also characterized by large triangular-shaped heads and very short slender tails. All vipers possess highly developed teeth (fangs) for venom injection, which are erected from their horizontal resting position during a strike.

LEFT: color is often useful for determining the sex of reptiles, as in this male iguana.
ABOVE: and the female of the species.

are usually found coiled around twigs of bushes and trees. Neotropical Rattlesnakes *(Crotalus durissus)*, the only representatives of their kind in South America, avoid forests, but occur in some savanna enclaves within Amazonia. All pit vipers mentioned here give birth to live young, whereas the giant Bushmaster *(Lachesis muta)* lays eggs.

This, the largest viper in the world (up to 3.7 meters/12 ft long) is reported anecdotally to be extremely aggressive, but, in fact, few bites occur. Bushmasters have short activity periods at night in which they hunt for mammalian prey. Mostly, they hide under roots and logs, and after having fed, will remain there for the time of digestion which may take two to four weeks. ❑

AMAZONIAN FROGS

With new species being discovered every year,
Amazonia possesses the richest frog fauna in the world

The Amazon basin, with its high humidity and warm temperatures throughout the year, provides excellent conditions for amphibians, whose body temperature is dependent upon that of the immediate surroundings and whose skin is not well protected against water loss. Two groups of amphibians – salamanders *(Urodela)* and worm-like caecilians *(Gymnophiona)* – are rarely encountered. The limbless caecilians are either terrestrial burrowers or aquatic. Due to their slender appearance, they can be mistaken for snakes or large earthworms. Salamanders are surprisingly scarce (in species as well as in individuals) and represent an exception to the general tendency for taxonomic units to show a high diversity in the tropics.

Frogs *(Anura)* are by far the most abundant amphibians. Over 300 species of these jumping amphibians occur in Amazonia, and new species are being described every year. With a density of more than 80 species within a few square kilometers, the western Amazonian lowlands host the richest frog fauna in the world.

Over 75 percent of the Amazonian frog species are nocturnal. Nightly choruses heard along river banks, oxbow lakes, flooded forests or forest ponds are dominated by frogs. Calls emitted only by males guide the mute females to their mating partners. These species-specific vocalizations sound like low rumbles [Cane Toad *(Bufo marinus)*], low metallic beats [Gladiator Frog *(Hyla boans)*], whistles [Whistling Frogs *(Leptodactylus* sp., *Adenomera* sp.), some Poison-dart Frogs *(Dendrobatidae)*], mooing cattle [Boatman Frog *(Phrynohyas resinifictrix)*] and rasping barks (several treefrogs of the genus *Hyla)* or high-pitched "clicks" (small *Hyla* sp.).

These calls are often mistaken for those of birds, insects or even mammals. Calling males reveal where frogs live (or at least breed): they are found virtually everywhere since they have become effective predators on the ubiquituous arthropods. These amphibians occur in burrows dug by rodents, in cavities beneath the superficial root system, on ground level or within leaf litter, or in the canopy. Even though most species prefer *terra firme* primary forest, frogs are also abundant in large water bodies, floating meadows, flooded forests, in fields and human settlements.

Reproduction in Amazonian frogs does not always follow the usual pattern involving aquatic eggs and larvae of most temperate species. In fact, less than 45 percent of the Amazonian lowland species are known to lay their eggs in water. About one-third lay eggs outside of water bodies but still undergo an aquatic larval phase. Larvae reach water either through flooding, dropping from arboreal egg-laying sites to the water below, or are even carried to the water by the male parent.

Nearly 20 percent of the Amazonian frog species are independent of water bodies throughout their developmental phases: eggs are laid either in unflooded foam nests or undergo direct development within the egg capsule. Rather than depositing their eggs, a few species carry them in dorsal brood chambers and pouches until metamorphosis, or at least until the larvae hatch.

Cane toads and glass frogs

The Cane Toad *(Bufo marinus)* is regularly observed near huts and houses, where it leaves its day-time retreats soon after dawn and starts to feed on insects attracted to artificial light. This toad has a thick, glandular skin and grows to a body size of 20 cm (8 inches). It is a frequent guest at dog- and cat-food bowls, where it consumes any kind of food offered. Thus, the Cane Toad is the only Amazonian frog known to be an occasional vegetarian and occasionally to feed on inanimate objects.

Even though considered to be highly venomous by some locals, this large Amazonian toad is innocuous to humans; however, it is best left alone, since secretions from large ear glands may cause irritations or brief, excruciating pain when coming into contact with the eye or open wounds. It may cause serious problems for dogs

PRECEDING PAGES: the Reticulated Poison-dart Frog *(Dendrobates reticulatus)* carries its tadpoles to water.
LEFT: a tree toad *(Phyllomedusa tarsius).*

and cats who pick them up with their mouths. The most common toad in forested areas is the Leaf Litter Dweller *(Bufo typhonius)*, whose back resembles the shape and color of a dead leaf. Due to its cryptic coloration, individuals are usually detected only while moving. This species is a so-called explosive breeder: breeding assemblies and chorusing males can be found only on a few days of the year. Members of the diurnal Harlequin Frogs *(Atelopus* sp.) are small colorful toads which establish territories near forest streams. When ready to mate, the female is dorsally embraced by a male; both partners then move to the stream where the eggs are laid.

Tree dwellers

In an (originally) almost completely forested ecosystem such as the Amazon basin, the dominance of tree-dwelling species is not surprising. Out of the 130 known lowland frogs, more than 80 show climbing ability. The most species-rich group of Amazonian frogs are the Tree Toads *(Hylidae)*. As with all frogs with climbing capabilities, Tree Toads possess enlarged finger and toe discs, which help them attach to stems and leaves. The large eyes of these nocturnal frogs allow good vision.

Tree Toads show a great variety in life history. The medium- to large-sized genera, *Osteo-*

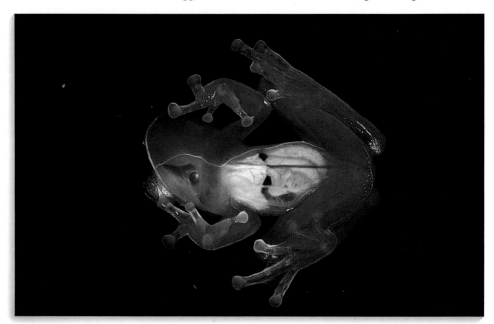

GLASS FROGS

Fast-flowing streams – common at the flanks of the Amazonian basin but scarce in its lowlands – are the breeding habitat of the nocturnal Glass Frogs *(Centrolenidae)*, so-called because their transparent skin reveals the intestinal system, the beating heart, and green bones. Eggs are attached to the underside of leaves overhanging running water. The clutch is attended by a male until the larvae hatch and drop into the stream below. The tadpoles are adapted to bury themselves in the detritus or gravel of the stream bottom; this minimizes predation by fish and aquatic insects and prevents them from drifting away.

cephalus, Phrynohyas and *Phyllomedusa*, live primarily high up in the canopy and may descend to the ground only to breed; some even breed in water-filled leaf axils or tree holes at varying heights. A canopy species frequently heard in the Amazonian *terra firme* primary forest but hardly ever seen is the Boatman Frog *(Phrynohyas resinifictrix)*. This frog breeds exclusively in tree cavities at great heights. Male frogs are spaced at large distances (100 meters/ 330 ft) and call from water-filled treeholes. These function as acoustic resonators and allow long-distance communication. The frog's loud "queng-queng" call is repeated three or four times. This explains its common name Boatman Frog: the croaks imitate the tapping of oars against the side of the canoes, which are used by the Indians to maintain the rhythm of the stroke when rowing. Morphologically similar to the Boatman Frog, yet behaviorally and ecologically different, is its relative, the Poison Tree Toad *(Phrynohyas venulosa)*, which breeds in water bodies at ground level. This medium-sized frog extrudes a sticky substance from dorsal glands when caught.

Maki Frogs *(Phyllomedusa* sp.) are easily identified by their bright green color, vertical cat-like pupils and their somewhat sluggish behavior. These frogs possess a wide array of biologically active peptides in their skin; these

members of the genus *Hyla* lay their eggs directly into water. The large Gladiator Frog *(Hyla boans)* heaps up a barricade of mud to enclose a pool of water. Here the male and female mate, about 3,000 eggs are laid and early larval development occurs. These pools measure less than a meter wide and are created at the border of a body of water during the wet season. Subsequent flooding allows the tadpoles to leave their confined home. A close relative, the Wavrin Frog *(H. wavrini)*, is one of the few frogs heard in the *igapós* in central Amazonia (Manaus and the Anavilhanas) during high water level. Widely-spaced males perch on leaf-

effectively thwart attacks by predators. Some of these peptides are related to mammalian hormones or neurotransmitters and play an important role in pharmacological studies. Amazonian Maki Frogs deposit non-pigmented eggs on leaves or in folded leaves, which at least partly conceal the eggs.

Some Tree Toads, such as the members of the small brightly-colored *Hyla leucophyllata* group deposit their eggs on leaves or mosses above water without forming specific nests. Yet most

less branches up to four meters (13 ft) above water level and alternately produce low, continual croaks of one-second duration.

The floodplains of white water rivers are the breeding ground of many colorful tree toads, which compose the main part of the spectacular nightly chorus at the floating meadows. *Hyla leucophyllata*, a yellow frog with orange webbings and chocolate brown dorsal markings, frequently calls in the vegetation dominated by waterhyacinths. Greenish frogs of the semi-aquatic tree toad genus, *Sphaenorhynchus* are also found here or on floating carpets of aquatic ferns.

Water ferns are the habitat of *Lysapsus limellus* of the family *Pseudidae*, a small, silvery-

LEFT: a species of Glass Frog.
ABOVE: the Narrow-mouthed Toad *(Synapturanus mirandaribeiroi)*.

lined, green semi-aquatic frog frequently used by fishermen as bait. The enormous tadpoles of this medium-sized frog measure some 250 mm (10 inches) – up to four times the length of their parents – making them the largest in the world.

A common guest in bathrooms – even in large cities – is the Brown Tree Toad *(Ololygon rubra)*; it runs along vertical structures, similar to its frequent coinhabitant, the Tropical Gecko.

Open country species

Open formation habitats such as roadsides, grasslands and human settlements in the Amazonian lowlands are frequently inhabited by the water. Although protected from aquatic predation, eggs and the long-tailed tadpoles are frequently destroyed by parasitic maggots or taken by wasps *(Angiopolybia* sp.), which pluck the eggs or even tadpoles out of the viscous foam. By excavating small burrows, males of *Leptodactylus fuscus* protect offspring from parasitic flies and predators. At the entrance of its burrow, a male attracts a female to the nesting site by continuous whistle-like calls. At the onset of the rainy season these calls are frequently heard during evening hours along street and roadsides. Once a female approaches, the male guides her into the chamber and embraces her

foam-nesting members of the nocturnal Whistling Frogs *(Leptodactylidae)*. In *Physalaemus* sp., foam nest construction takes place on the surface of small puddles. Floating in pairs close to the shoreline, the male beats oviductal secretions of the female into a frothy mass. During the beating motions, unpigmented eggs are fertilized and distributed within the foam, which protects the eggs and early larval stages from aquatic predators. After a few days, the foam nest dissolves and the rapidly developing larvae are released into the water.

The large South American Bull Frog *(Leptodactylus knudseni)* produces a foam nest lasting three weeks in a depression on land next to from behind. The mating pair remain within the burrow until the foam nest is finished.

The Broad-mouthed Horned Frog *(Ceratophrys cornuta)*, with its terrifying appearance, does not resemble a typical frog. This large, green- or brown-colored leaf litter frog is a typical sit-and-wait predator and capable of devouring large amphibians as well as small mammals.

Members of the genus *Eleutherodactylus* possess enlarged toe and fingertips and are found in forests from the ground floor up to the canopy. This species-rich genus (more than 400 species and the richest genus among the vertebrates) is characterized by direct development (small replicas of the adult frog leave the egg capsule).

Poison-dart frogs

The best-known Amazon frogs are the Poison-dart Frogs *(Dendrobatidae)*. Small and colorful, they are active only during the day, feeding on termites and small insects. Most dendrobatids live in leaf litter. Males establish territories and continuously call for mates from elevated sites such as fallen tree trunks. The Penemkwitsi *(Allobates femoralis)*, a brown frog with golden lateral stripes, may occupy breeding territories of up to

> ### DEADLY SKINS
>
> The Colombian Chocós use frog-skin alkaloids as hunting-dart poison. The Chocos-region species Kokoi *(Phyllobates latinasus)*, has a powerful venom which causes paralysis and respiratory failure.

out conspicuous markings belong to the non-poisonous genus *Colostethus*. Individuals with vivid warning colors belong to genera *Phobobates*, *Dendrobates* containing more or less dangerous skin toxins. The largest Amazonian dendrobatid, the Striped Poison-dart Frog *(Phobobates trivittatus)*, uses call sites up to 1.5 meters (5 ft) above ground; its "tinc-tinc-tinc…" call is heard throughout the day, most frequently in the early morning. *Dendrobates quinquevittatus*, with bright golden stripes on a dark background, releases tadpoles in bromeliads or small

200 sq meters (2,190 sq ft) for up to four months. The Secoya of Peru named this frog after its four-note call "Pe-nem-kwi-tsi". Their frequency-modulated (whistle-like) calls prevent other males from entering their territories; the mute females, attracted to the calling males, follow them to a nesting site. Eggs are laid on land sheltered between leaves. The male frog stays with the fertilized eggs for two or more weeks and carries the tadpoles on his back to water after they hatch. Small brown dendrobatid frogs with-

water-filled tree holes. It lives higher up trees than all other Poison-dart frogs. Liana paths are frequently used to reach breeding sites.

Not uncommon, but rarely seen are the Narrow-mouthed Toads *(Microhylidae)*. Most of these squat, small-headed termite- and ant-eating species are explosive breeders, and hundreds of males call at forest ponds a few days after the wet season heavy rains.

During the complex courtship of Tongueless Toads *(Pipidae)*, eggs are embedded in the dorsal skin of the female which develop directly into froglets in the dorsal brood chambers. The large, bizarrely shaped Surinam Toad *(Pipa pipa)* is frequently found in open standing water. ❑

LEFT: mating Tree Toads *(Hyla brevifons)* leave their eggs on a leaf tip.
ABOVE: the Horned Frog *(Hemiphractus* sp.).

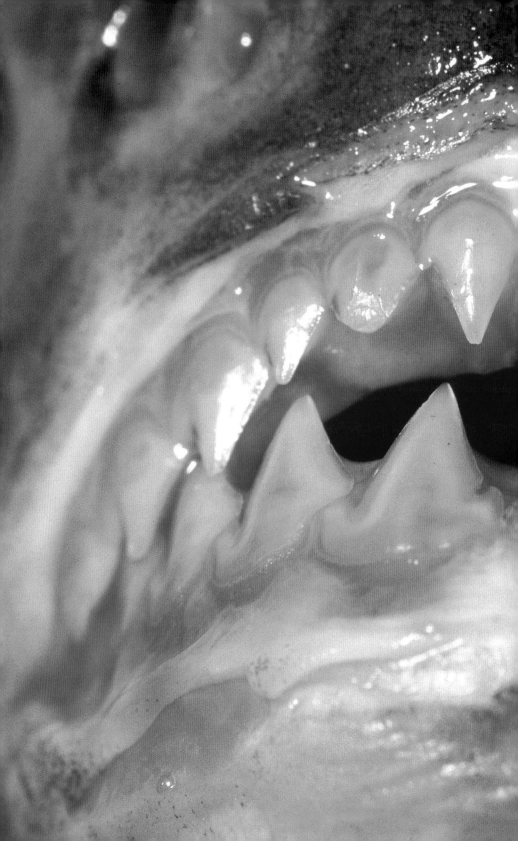

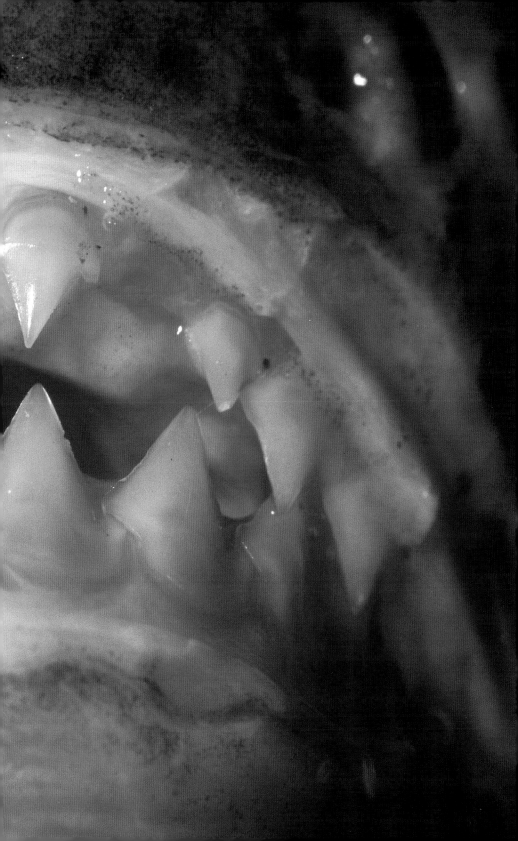

FISH

The only systems on earth with a fish diversity
comparable to that of the Amazon Basin are the oceans

The rate at which new fish species continue to be found suggests that there may be as many as 3,000 in the Amazon rivers and lakes. Among them are physical forms and ways of life unique to the fauna of Amazonia, many of which evolved in response to the poverty of their habitats and the flooding cycles in the Basin. In the white water and black water rivers of the Amazon's catchment, the mineral content of the water is so low that, during the dry season, many fish species have to rely chiefly upon their fat reserves for survival since aquatic plants are almost non-existent.

The fish find most of their food during floods, when the waters spread many kilometers into the surrounding forests. Moving out of the main channels, the fish shoal among the branches of the trees. Among these fish, the fruit-eaters have a niche peculiar to the Amazon.

Characins

The most striking of these fish are the fruit-eating characins. Like many species in the Amazon, their strange dentition reveals much about their diet. The largest characin, the Tambaqui (*Colossoma macropomum*), resembles a thick-set, bull-nosed European bronze bream, until it opens its mouth. It reveals a set of teeth similar to the molars of a sheep, which, driven by the impressive muscles of the head, can crack the seeds of the rubber tree and the jauari palm.

The Tambaqui and other characins, such as the Pacus (*Mylossoma* and *Myleus*) wait beneath certain trees in the flooded forest for the fruit to fall. They destroy much of the crop produced by the larger-seeded trees; but for those with small seeds they are important agents of dispersal, as the seeds pass through their digestive tracts unharmed. While the survival of much of the fish fauna depends on the forests, the fish in turn are an indispensable part of the forest's ecology.

PRECEDING PAGES: the jaws of a piranha (*Serrasalmus* sp.).
LEFT: Tambaqui feeding on seeds in flooded forest.

The flooded forest's contribution to the aquatic ecosystems is such that if it is removed the freshwater foodwebs are likely to collapse. Regrettably, the Brazilian government has been promoting the clearance of the floodplains for agriculture, believing that this will increase food production. As many of the riverside soils are infertile, clearance is likely to produce just the opposite effect, as the critical protein supplies provided by the region's fisheries are replaced by unproductive cattle pastures.

Many of the fish species feeding in the forest migrate to take advantage of the seasonal fluctuations. Some of the fish in the Amazon system migrate downstream to spawn, laying their eggs in or around the white water channels before returning to the forests surrounding the black water rivers. Many of the species making use of the flooded forest appear to be highly specialized feeders, and concentrate upon the food that they are best equipped to eat.

One of the strangest phenomena involving the fish of the Amazon system is the *friagem*, which takes place some years during June. Cold winds from the south of the continent chill the surface of the rivers, and the cold water sinks. This displaces the water close to the riverbed which, in the white water systems, has been depleted of its oxygen by bacteria. This water rises through the river, and the fish coming into contact with it float gasping to the surface, a bonanza for fishermen. This does not occur on the black water rivers, as there is insufficient organic material on the riverbeds for intense bacterial activity to take place.

The characin family has diversified in Amazonia to a remarkable extent. Besides the fruit-eaters, its members include the Hatchet Fish (*Gasteropelecidae*), surface-living characins which have deep keels and greatly extended pectoral fins that enable them to glide above the water when pursued, similar to the marine flying fish. The tetras, among which are the iridescent Neon and Cardinal Tetras, are heavily exploited in the Amazon for the tropical fish trade. They are also an important part of the subsistence of many river dwellers. But undoubtedly the most famous and perhaps the most bizarre of the known characins are the piranhas.

The ferocity of the piranha has, like so many of the hazards of the natural world, been greatly exaggerated. There is no doubt that some species, under certain conditions, have killed people and other large mammals; but most of them are usually harmless. Only a handful of the 20 species of piranha are dangerous. The behavior of the predatory species seems to depend on the state of their habitat. In the main river channels and the large lakes of the Amazon, especially during the flooding seasons, they appear to leave swimmers unmolested. But when the shoals are confined to limited bodies of water with scarce food supplies, such as the pools left behind by a falling river, they can become dangerous. The Black Piranha *(Serrasalmus rhombeus)*, which reaches a length of 40 cm (16 inches), has a particularly evil reputation, and has been blamed for several deaths.

But several of the piranha species are harmless to people. Some, like the Opossum Piranha *(Serrasalmus elongatus)*, have adapted to eat the fins and scales of other fish, which is why so many of the fish on sale in the markets look worn. Others eat fruit, seeds and even leaves, while several species are partly insectivorous.

'MARINE' FISH, LUNGFISH AND CICHLIDS

The Amazon is remarkable for the number of species belonging to families normally considered to be marine. Bullsharks have been found 3,000 km (1,800 miles) inland, and sawfish frequent some turbid tributaries. Stingrays, a possible "hangover" species from an earlier oceanic gulf, are common in the Amazon. Their spines are sold in some riverside medicine markets.

As in Africa, there are lungfish in the Amazon – cocooning themselves in mud during the dry season – and cichlids, well-known to aquarists. The Tucanare or Rainbow Perch *(Cichla ocellaris)*, which can reach 10 kg (22 lb), is one of the most important food fish of the Amazon.

Catfish

The catfish of the Amazon are as diverse as the characins, having evolved to fill some extraordinary niches. The largest are predatory, and grow to more than 2 meters (6 ft) and 150 kg (330lb). Some very small species, by contrast, are parasites feeding upon larger fish, either eating the mucous on their skins, or sucking their blood, penetrating the skin by means of long fine teeth. The blood-sucking Candirus *(Trichomycteridae)* have been known to target bathers, entering the nose, ears, anus or urethra, from which they are notoriously hard to extract. Other catfish are armored with the most baroque adornments. The Bacu Pedra *(Lithodorus dorsalis)*, whose entire skin surface is covered with heavy plates rather like those of the prehistoric ankylosaurs, and whose fins are protected by thick spines, resembling in shape and texture weathered rock. Quite why these catfish should arm themselves so heavily has yet to be determined.

Osteoglottids and eels

One of the largest freshwater fish in the world is the Pirarucu *(Arapaima gigas)*, a member of the primitive air-breathing osteoglottid family. While the giants reported by some of the earlier explorers in the Amazon are now rare, specimens of over 2 meters (6 ft) in length, weighing 125 kg

impulses, but the only one which can kill or stun large prey. It can produce up to 650 volts and regularly produces 400 volts with a 1-amp current.

Leaf fish

Some of the most striking work by fish biologists in the Amazon has involved the leaf litter banks of black water streams. Some surveys have revealed as many as 46 species within 100 meters (330 ft) of leaf litter. Most of them are strange looking and dwarfed: the average adult length of one species, the second smallest in the world, is 1 cm (0.4 inches). Among them are some of the most outlandish of all the Am-

(275 lb) or more, are still caught. The other well-known Amazonian osteoglottid is the Arawana *(Osteoglossum bicirrhosum vandelli)*, known as the "water-monkey". It can leap up to 2 meters (6 ft) from the water – twice its maximum body length – to snatch invertebrates, birds, bats and reptiles from the overhanging vegetation.

Besides the piranha, the Amazonian fish best known to foreigners is probably the Electric Eel *(Electrophorus electricus)*, only one of several species which can generate and detect electrical

azon's species. The Leaf Fishes imitate flattened, blotched and tattered dead leaves with remarkable fidelity, their extended lower jaw mimicking a stem. Even when disturbed, they are hard to distinguish from the substrate, sinking with a motion similar to that of the other settling leaves.

The Wormfish, a catfish which has yet to be classified, lives at the head of the litter bank. It is blind, scaleless and bright scarlet, and breathes through its skin, sharing their habitat with earthworms. Such a discovery hints at what there may still be left to find, as scientists begin to investigate the smaller species of the world's most diverse freshwater fauna. ❏

LEFT: a Leaf Fish *(Monocirrhus polyacanthus)* from the Brazilian Amazon.
ABOVE: an Armored Catfish.

INSECTS

Some experts believe there may be several tens of millions of insect species on earth, and the Amazon is the richest habitat of all

During the evolution of life on earth, the clear winner of the competition for species diversity was the humble insect. There are so many species that even the quest to estimate their correct number is a topic for scientific controversy. To date, several hundred thousand species have been identified, about 10 times more than all mammals, birds, reptiles, amphibians and fishes put together.

While nearly all vertebrates are known to science, we may know as little as one percent of all insects. Because of the huge numbers involved, science relies not on the recording of individual species, but on more indirect calculations. Some are based on the catching of insects in particular habitats: the number of species that is caught (without necessarily becoming individually identified) in a particular habitat within a set time frame, multiplied by the estimated number of different insect habitats, leads to approximations of several million insect species. Some experts assume that a major proportion of all insects is specific to one particular plant species. The number of insect species collected after the fumigation of a tropical tree with insecticide, multiplied by the number of tree species, leads to the high estimates of species diversity.

Taxonomists help us to simplify the diversity by placing each of the millions of species into one of about 30 orders. Of these, about a dozen have conspicuous species of interest to the general tourist.

In contrast to mammals and birds, insects had already evolved to a large degree of diversity 100 million years ago, before Gondwanaland broke up to create the isolated continent of South America. This is one of the reasons that no higher insect taxanomic groupings – like the edentates among the mammals – are restricted to this continent. But some lower taxa, like the Morpho butterflies and the leafcutting ants, evolved in and are typical of the neotropics.

Termites and ants

Termites and ants are the insect groups with the largest number of individuals and they represent the largest insect biomass in the rainforest. It is even believed that their combined biomass in a tropical rainforest exceeds that of all vertebrates. Termites, although often referred to as white ants, are a fairly primitive order of insects in contrast to the advanced ants, and they are actually completely unrelated to ants.

Termites are nearly colorless due to their life spent underground or in rotting wood. They live mostly on the cellulose component of plant material. They cannot digest cellulose with their own enzymes, but do so with the help of bacteria that live in their intestines, a similar dependence as that of ruminants. Termites are social insects and some species live in societies of millions of individuals, all being descendants of a single royal couple.

Ants live in similarly large societies and show a baffling variety of social structures and cooperative and antagonistic interactions between different species, including the dependence of some species on slaves captured from another ant species. Possibly the most conspicuous ants of the neotropics, and endemic to them, are the leafcutting ants. Their large columns are easily visible, as each individual carries a piece of leaf several times its size to its underground

ARMY ANTS

Most ants have a permanent dwelling, but army ants move about carrying their eggs and pupae with them. Colonies can number over 1 million. They advance in military-style formations, sometimes several meters in breadth and a hundred meters (330 ft) in length, and devour every living thing unfortunate enough to be in their way. Birds often follow these army ants on the march and prey upon insects that try to escape. Ants contribute to the overwhelming success of Cecropia trees, the most abundant vegetation in neotropical secondary growths. Cavities in the branches of these trees are colonized by ants that attack every animal – including humans – that touches the Cecropia.

PRECEDING PAGES: a Waxy-tailed Planthopper.
LEFT: Transparent-winged Cicadas.

dwelling. There, the leaves are chewed into a paste, which forms the substrate to grow a fungus, on which the ants live. The dependence of fungus and ants is mutual: the fungus will only grow well in the ants' dwelling, and the queen, when founding a new colony, assures the transfer of the right fungus by carrying a little piece of the fungal mycelium with her.

Ants belong to the order of the *hymenoptera,* which they share with wasps and bees, to whom they are related. Taxonomists normally list the *hymenoptera* as the last order of insects, meaning that they were the most recent insect species to appear in the evolution process.

are thought to be the most complex among insects, and involve a resonating apparatus of air-sacs, membranes, and covers that can be used to modulate the sound output. Grasshoppers mostly chew plant food, while cicadas suck plant sap.

Planthoppers belong to the same order as cicadas and can boast some of the most bizarre-looking specimens of the insect world. Perhaps the most well-known is the Lantern Fly, which, in spite of its name, is neither a fly nor a luminescent insect. The head of this insect resembles that of a crocodile: dark and light marks seem to simulate eyes, nostrils, and teeth, and

Grasshoppers and cicadas

Without doubt, two unrelated insect orders, grasshoppers and crickets, and cicadas, have the biggest impact on the tropical nature experience, as they provide most of the acoustic backdrop of rainforest landscapes. The combination of trilling, thumping, groaning, or outright roaring sounds roll through the forest like a bizarre outdoor concert. Only the males call in both insect orders, but with quite different musical "instruments". Grasshoppers follow the principle of playing a violin, using a scraper which is rubbed across a file; cicadas, meanwhile, have exploited the principle of drumming. The noise-making instruments of cicadas

these colors may be part of its natural "disguise" to scare away hungry birds. The shape of the jutting frontal protuberance has also given this insect its alternative name: the peanut-headed bug. Another peculiarity of some planthoppers is their ability to modify their appearance with wax excretions: a typical representative, the Waxy-tailed Lantern Bug, is easily found in the flooded forest. The ribbons of wax that resemble a trailing tail protrude from the ducts of the insect's wax glands, and may protect the animal against enemies.

Cicadas and planthoppers are closely related to bugs, and often unified with them in the same order. Many members of this large insect group are very colorful, and they are some of the most attractive insects to be found. Most feed on plant juices or aphids, although some, such as the bed bug, are a bane to humans. More serious, however, are bugs of the group *Triatominae,* which are responsible for spreading Chagas' disease, a tropical malady caused by the protozoan *Trypanosoma cruzi,* which is widespread in South America. This bug, known in Spanish as the *vinchuca,* in Portuguese as the *barbeiro,* and in English as the Reduvid or Kissing bug, lives

CAMOUFLAGE

Some stick insects are brownish or grayish drawn-out creatures that may exceed 30 cm (12 inches), while others are flat and green resembling leaves, like those of the plants on which they reside.

Mantids and leaf insects

Two orders of insects, the mantids and the walkingsticks or leaf insects (commonly known as "stick insects") are restricted to warm countries, though not strictly to the tropics. The Praying Mantis gets its name from the way it sits motionless, or slowly swaying with the wind, with its fore legs held up as if folded in prayer. Quite in contrast to this seemingly harmless posture, the mantis, virtually invisible to other insects, is predatory, and the fore legs trans-

in cracks in walls and ceiling thatch. The disease is spread when an infected bug bites you and defecates near the wound. You then scratch the bite and transfer the parasite to your blood. If you get bitten, wash with soap and water and get a blood test as soon as you get home. Capture the bug if you can: some local places can test for infection. Not all vinchucas are carriers, but it is a serious disease if you catch it and must be treated early. Vinchucas are not found at altitudes greater than 2,500 meters (8,200 ft).

LEFT: the grasshopper *Maskia nystrix.*
ABOVE: a Lantern Fly *(Fulgora lanternaria),* also known as the peanut-headed bug.

form into a deadly trap once potential prey wanders within reach.

In contrast to the mantids, walkingsticks and leaf insects are vegetarians with a nocturnal lifestyle. During daylight, they do what they do best, namely mould into their surroundings. They move very slowly and, if detected by a predator, have to rely on a foul-smelling liquid deterrent that they squirt out. There are many species of grasshoppers that compete with leaf insects in their astonishing abilities to disguise themselves, and the wings of these insects may not only imitate the color and shape of particular plant leaves, but even contain specks and fringes reminiscent of fungal growth on a decaying leaf.

Other members of these orders have spiny-like protrusions and other bizarre-looking features that make it difficult for a predator to swallow.

Butterflies and moths

Without doubt, butterflies and moths are the single most important insect attraction for visitors. In particular, amateur entomologists aspire to collect members of the Morpho group, large butterflies with shining blue wings, that are restricted to the neotropics. Among them, *Morpho hecuba* has a wingspan of 18 cm (7 inches), making it the biggest butterfly of the Amazon. Only the male Morphos have this

Nature presented the insect world with the gift of disguise to help a species outlive the dangers of the rainforest. And while leaf insects and grasshoppers are the masters of camouflage, butterflies are the champions of mimicry. One strategy is to look like members of other arthropode groups that are avoided by birds: some moths take on the appearance and movement of bristly spiders, and some butterflies have body shapes where details of the wings, waist, and mouth resemble certain wasp species, even to the point of mimicking wasp flight patterns and seemingly growing false organs, such as ovipositors, which only occur in wasps.

vivid coloration, the females being rather dull brown. Morphos are frequently sold as souvenirs, and fortunately are now mostly bred commercially. However, trying to take them out of the country may cause you difficulties at customs. Three more butterfly families – among the more than 100 that belong to this order – are restricted to the neotropics: the *Ithomiidae,* the *Heliconiidae,* and the *Brassolidae.* The latter are called owl butterflies, since they have a large eyespot on each wing, probably a means of protection against birds. Among the moths, the Great Owlet *(Thysania agrippina),* with a wingspan of 30 cm (12 inches), is probably the moth with the largest wingspan on earth.

Butterflies and their caterpillars have developed brightly colored patterns to signal inedibility, which can, for example, result from the caterpillar's diet of poisonous plants. Based on identifying these colors, the insect's enemies eventually learn to leave particular creatures alone. This phenomenon has triggered chains of convergent evolutionary processes: "edible" insects imitate colors and shapes of their inedible counterparts, and even fly together with their look-alikes. And since a predatory bird's ability to identify an unsavory item is limited, insects seem to agree on certain limits, such that even unrelated species adopt similar warning colors. For example, more than a dozen species of unrelated but similar-looking butterflies often join together to form something called a mimicry ring.

Beetles and mosquitos

With about 300,000 known species, beetles are the biggest order of all living beings. Beetles cannot compete with butterflies in terms of their tourist profile, or with ants in complexity of social structure and their contribution to the biomass, or with cicadas for noisemaking, but they are probably the insect group with the greatest diversity in exploiting various ecological niches.

palm stood naturally in small stands or as isolated individuals in the forest, but it can be very harmful in today's extensive stands of the coconut palm, which is not native to the region.

Together with flies, mosquitos form the order *Diptera,* the "two-winged" insects whose name originated from the change of their hind wings into halteres, an organ for balancing. The word "mosquito" comes from the Spanish language, and means "little fly", but Spanish-speaking South Americans also use the term *zancudos.*

Mosquitos are a serious problem along white water rivers, but they are virtually absent from the black water rivers such as the

What's more, some are extraordinarily beautiful and, because of this, beetles of the *Buprestidae* are sometimes used by Amazonian Indians for ornamental purposes. The biggest beetle of the neotropics, the Hercules beetle *(Dynastes hercules)*, has a maximum length of 18 cm (7 inches). Its home, however, is Central America and the West Indies, rather than the Amazon. The South American Palm Weevil, with an average length of 4 cm (2 inches), is unusually large. This insect species did not do much damage as long as palms like the local Mauritius

Rio Negro, whose waters are too acid for larval development. Only female mosquitos bite, since they need a blood meal as a nutrient source to complete their egg production. Male mosquitos indulge in drinking only the nectar of flowers.

Mosquitos are justifiably feared as transmitters of various dangerous diseases. For instance, Aedes mosquitos transmit the viral diseases yellow fever and dengue fever, Anopheles mosquitos the Plasmodium protozoa that cause malaria. However, only a very small fraction of all mosquitos carries any pathogen, and their numbers decrease the further away one gets from human habitats. ❑

LEFT: a Bush Cricket *(Cyclopetra speculata).*
ABOVE: a leafcutting ant returning to its colony.

INDIGENOUS PEOPLES

*Over the centuries, Amazonian peoples have evolved a close
relationship with the environment and spiritual ties with the land*

It is generally believed that 30,000–40,000 years ago hunter gatherer peoples crossed the strip of land which linked Asia to Alaska and spread throughout North and South America, eventually reaching Brazil by about 10,000 BC. However, recent discoveries of rock paintings and remains of settlements in the arid interior of Brazil suggest that the region may have been populated as long ago as 50,000 years, possibly by peoples who had arrived by sea.

Before 1492, indigenous peoples numbered about 5–6 million in Brazil alone and were extremely diverse. Some lived in large settlements of up to 10,000 inhabitants, on the coast and along the flood plains of the major rivers. Archaeologists have found enormous earthwork mounds containing pottery, stone tools and burial chambers on Marajó Island and near Santarém. Flood plains were intensively cultivated; manioc and corn were grown and ditches, canals and trails built to irrigate and transport produce. Turtles and fish were caught and some crops dried and stored.

Tribes such as the Omagua and Tapajós were capable of mobilising powerful armies and there were extensive trade networks. Accounts by the early European chroniclers expressed admiration for the abundance of food and social organisation of these societies.

Other tribes lived as hunter gatherers in uplands and forest far from the rivers. They lived in small, mobile groups and were probably egalitarian and nomadic, very similar to today's Amazonian hunter gatherers.

Arrival of the Europeans

The five centuries after the arrival of the Europeans in Brazil were cataclysmic and the indigenous population fell by over 93 percent. Initial friendly contact rapidly turned into open hostility as the Europeans began to exploit and abuse the locals. They annihilated whole tribes through murder, slavery, and the transmission of diseases against which the local people had no immunity. Some tribes resisted bravely but were unable to withstand the onslaught. Many, like the Manoa, after whom the Amazon city Manaus is named, became extinct. Only those who fled or who lived deep in the Amazon survived. By the end of the 19th century there were barely 300,000 indigenous people in Brazil.

The local population plummeted still further in the 20th century as governments decided to open up and exploit the mineral and timber wealth of the Amazon region. By the 1950s there were little more than 100,000 indigenous people left in Brazil. Although the Indian Protection Service (SPI) was founded by the sympathetic Cândido Rondon in 1910 to protect them, its ultimate aim was to integrate or assimilate them into national society and divest them of their culture and land.

The SPI became increasingly corrupt as officials sold off land and human rights abuses became rife. Darcy Ribeiro, a Brazilian anthropologist and senator, calculated that between 1900 and 1957, 87 tribes disappeared. After a criminal investigation which catalogued thousands of crimes of mass murder, torture, and theft, the SPI was disbanded and replaced by a new department, FUNAI, which still exists. The Villas Bôas brothers created a "safe haven" for 16 tribes in the Xingu park. Yet even this came

---

RUBBER SLAVES

The 19th century rubber boom bought appalling hardship to indigenous peoples, captured by slave traders and forced to gather rubber until they literally dropped dead from exhaustion and ill treatment. In the Putumayo area of Peru alone, 40,000 people died. The Andoke of Colombia were reduced from 10,000 to a mere two dozen within 40 years. Today, having brought themselves out of debt bondage to local rubber merchants, their population has grown, their culture is flourishing and they are building longhouses again. In remote parts of the Amazon, there are still cases of debt bondage where traders exploit indigenous people who are left permanently in debt for goods they have bought.

PRECEDING PAGES: swimming in the Amazon;
a local girl relaxes in a hammock.
LEFT: the decorated face of a Secoya man.

under threat as a road was bulldozed through it. In the 1980s, Brazil embarked on huge development projects, such as the Carajás and Polonoroeste schemes, devastating many tribes by forcing them off their land and exposing them to outside diseases.

Today, there are some 215 tribes in Brazil, with a population of 350,000, of which nearly two thirds live in the Amazon. There is great diversity, and each tribe has its own customs and most speak their own language. Some, like the Kanoê and Korubo, have almost no contact with national society and their tiny populations of a few dozen or less are extremely vulnerable

and the Awá in the eastern Brazilian Amazon, are nomadic. They live in small, mobile family groups, which constantly move through the forest in search of game and fruits. They have few possessions and sleep in shelters made of palm leaves. The men are superb hunters, skilled at tracking and mimicking animals.

Some tribes, like the Yanomami in Brazil and Venezuela, live in large communal houses known generally throughout the Amazon as *malocas*. A typical Yanomami *maloca* measures up to 40 meters (131 ft) across, with a thatched roof open at the top to let in the light as well as acting as a funnel for smoke from the hearth fires. Up to 200

to attacks and disease from loggers and settlers encroaching on their land. Others, such as the Tikuna, who number over 30,000, are increasing in size and have successfully campaigned for their land to be officially recognized. The Tikuna have their own schools with instruction in their own language. But they too face the problems of illegal logging and drug trafficking on their land. In 1988, 14 Tikuna were murdered by gunmen acting for a timber merchant.

Traditional lifestyles

Most Amazonian peoples are settled and live along the rivers. A few tribes, such as the Maku living in the border area of Colombia and Brazil

or 300 people may live under one roof, each family with its own hearth opening on to the central area where children play and community feasts and ceremonies are held. Like many Amazonian peoples, the Yanomami are semi-nomadic horticulturalists as well as hunter gatherers.

The women tend gardens where they grow manioc (an important staple from which they make flour), corn, sweet potato, plantains, pineapples and papaya. Brazil nuts and peach palm are harvested seasonally. Because of the low fertility of the soil, gardens are carefully rotated and are abandoned after a few years. New gardens are opened in the forest while the old ones are left to regenerate. *Malocas* sometimes split when the population is too big for the land to sustain it, and are abandoned every 5 to 10 years, allowing the game and soil to recuperate.

Natural resources

Over generations, Amazonian peoples have developed many successful natural poisons using plants, vines and tree barks. Most hunt with bows and arrows, often tipped with curare, a powerful muscle relaxant that kills the prey swiftly without harming the meat. Some tribes, such as the Matis, hunt with blow pipes up to 3 meters (10 ft) long. They are extremely accurate and are used to hunt birds and monkeys.

poisons in the Amazon. They are pulverized and beaten, and the resulting pulp placed in a stream. The poison stuns the fish which float up to the surface and are easily caught in baskets. They are fully edible as the poison is not absorbed. The whole community participates in the event.

The Amazon is a hugely diverse environment and its indigenous peoples have a profound and sophisticated knowledge of the natural world. Different parts of hundreds of plants are used for food and medicines, building canoes and houses, as well as for weaving hammocks, slings and baskets and making blow guns, darts, bows and arrows. Plants are used for soap, deodorants,

Fish is an important food for many Amazonian tribes, reflected in ceremonies such as the Waiãpi's dance to mark the fish spawning cycle. The Enawene Nawe camp for several months by selected rivers, which they dam to catch large quantities of fish. Men and boys dive into the pool created by the dam to recover the fish trapped in large baskets. The catch is then smoked and kept in the village as a valuable food source over several months. More than 30 plants and lianas are used as fish

LEFT AND ABOVE: body decoration forms an important role in the social and cultural practices of many Amazonian indigenous peoples.

contraceptives and perfume, as well as for body painting and ritual purposes. The Yanomami use 500 different plants for food, medicines and building houses and the Ka'apor use 112 plant species for medicinal purposes alone. Amazonian plants and indigenous knowledge have given the world quinine and curare, both used in medicines throughout the world. Guaraná, a popular drink in Brazil, has been made by peoples like the Sateré Mawé for hundreds of years. It is made by grinding and roasting the guaraná nut and is a natural stimulant used for hunting and ritual purposes. Today, guaraná is an important source of income for the Sateré Mawé. Amazonian peoples also use plants for ritual and religious purposes.

Uncontacted peoples

The Amazon is home to the greatest number of "uncontacted" peoples anywhere in the world. There are about 50 groups in Brazil, mainly in the Amazon region, with more in Peru, Colombia and Bolivia. Some are probably the remnants of tribes decimated by the forces of colonialism a century ago. Although they have no contact with national society, they often secretly observe outsiders, including other tribes, but have chosen to remain in isolation since history has shown them the dangers of mixing with outsiders. Most live in remote and infertile terrain up the headwaters of small

The riverside *caboclos*

Today there are many settlers or "*caboclos*" living beside the major rivers. In some cases, they live and depend entirely on the river and *várzea* (floodplain), but many cultivate some crops on *terra firme*, or non-flooded areas. The *caboclos* have adopted many techniques of the indigenous population and indeed some have local ancestry. Their life is strongly influenced by the annual cycle of the river. In the dry season, beans or corn are cultivated on the exposed river margins. In some places, especially near Manaus, there are cattle which must be moved during floods to *terra firme* or kept on rafts and fed grass.

Amazon tributaries, where it takes considerable skill and knowledge to survive.

During the 1970s and '80s, the Brazilian government, believing that indigenous peoples were "primitive" and an "obstacle to progress", had a policy of integration. Expeditions were mounted to make contact with them. Usually these were disastrous and frequently over half a tribe would die from common diseases like flu within the first year of contact. Those tribes who survived often took years to recover psychologically and physically, but the loss of wisdom through the death of shamans and old people was devastating and irreplaceable. Today, Brazil carries out contact missions only if the tribe is under serious threat.

Lives of the *caboclos*

The *caboclos'* palm-thatched houses are built on stilts or float. Floating houses, built on a number of large tree trunks, can still be seen along the major rivers. A common feature of such houses is a raised garden, usually in an old canoe shell. This garden is protected from the floods, as are chickens and other domestic animals. The people depend heavily on fishing for their livelihood and, in the dry season, on catching turtle.

Life on a white water river is quite different from that on a black water river. In the black water river floodplain, there is no rich alluvial soil beside the river. The flooded areas are usually on white sand. There are fewer fish than in white water areas, and so far fewer settlers along the black water rivers. Most settle the Amazon and its major white water tributaries.

Their diet generally consists of cassava flour mixed with fish. They also grow some fruit and keep a few chickens. Many beliefs and legends originate from indigenous culture, including medicine lore. Where there are groups of people primary schools may be found. However, the vast majority of *caboclos* live too far from such schools to receive any formal education.

Many riverside dwellers live where they do, usually in the *várzea*, because there is a sufficient number of rubber trees worth tapping. The an ideal crop for *várzea* land when the river level is low. Many other fibers are known and used, especially by indigenous peoples; for example, fiber from the trunk of the *piaçaba* palm tree *(Leopoldina piassaba)*, used mainly for brushes.

Although most settlers use small dugout canoes to get around in, a large fleet of vessels of all sizes moves up and down the river system, bringing people to buy rubber or Brazil nuts from the the *caboclos* in return for supplies. Often the price charged is exorbitant and settlers become indebted to the boat owners. In more frequently traveled sections, the rivers are plied by small launches which act as a regular "bus" system.

National Council of Rubber Tappers represents many people working in small-scale projects such as those collecting latex, Brazil nuts and and *babaçu* coconuts. Fulfilling the vision of the rubber tapper Chico Mendes, the council has created 40 reserves with more than 5 million hectares of forest. Here, over 30,000 families extract and sell rubber and other forest products.

Jute is also an important crop and one of the most important sources of income for the *caboclos*. Since it takes just four months to grow, it is

LEFT: local trade carried on from canoes near Boras, Peru.
ABOVE: a Yanomami man and child.

Quilombos

The *quilombos* are settlements founded by runaway African slaves, brought over in the 17th century to work on sugar plantations. Those who managed to escape often hid in the forest where many lived undetected. Slavery was abolished in 1888, yet an estimated several hundred *quilombos* remain. Today, their inhabitants, known as *quilombolas*, are increasingly organised. The constitution recognizes their right to own their land and some communities have secured title to their land, although many more still await legal recognition. They live much as their ancestors did: hunting, fishing, planting crops and harvesting Brazil nuts. ❑

SHAMANISM

*Healers and communicators with the spirit world,
shamans are a vital part of traditional Amazon societies*

Shamans communicate with the natural world and its spirits. Generally healers, they use both natural medicines and a belief in the spirit world in their cures. Shamans may represent the spirits, and are respected for their powers. All tribal peoples in Amazonia have – or at least had – individuals who act as shamans. Different peoples have their own chant, and mind-altering plants to enter trances and communicate with spirits. Most interpret dreams and the meanings of daily events, and are experts in the mythic cycles of their people. Becoming a shaman can require years of demanding apprenticeship, often involving dietary and sexual restrictions.

shamans. Different peoples have their own names for shamans, and they take different roles in the community: some, such as the Guaranikarais, are seers; some are singers and poets; others are regarded as tricksters or entertainers. The Tukano believe shamans can turn into jaguars, the most powerful and feared animal in the forest.

Among some peoples, only men can become shamans, in others, women also. Some peoples believe you must be born into the tradition, while the Araweté, for instance, believe everyone has the quality or the ability to be a shaman. For the Waiãpi, you can have this quality, but you can also lose it. Shamans use dance,

Yanomami shamans

Davi Yanomami, shaman in his village Watoriki-Theri, gave the following description of the *shapiri* [spirits] and shamanic practices: "Whoever is a shaman has to accept them, to know them. You have to leave everything: you can't eat food or drink water, you can't be near women or the smell of burning, or children playing or making a noise – because the *shapiri* want to live in silence. They are other people and they live differently. Some live in the sky, some underground, and others live in the mountains which are covered with forests and flowers. Some live in the rivers, in the sea and others in the stars, or in the moon and the sun. Omame [the creator] chose them because they were good for working – not in the gardens, but for working in shamanism, for curing people. They are beautiful but difficult to see. The *shapiri* look after everything. The *shapiri* are looking after the world."

"We Yanomami learn with the great *shapiri*. We learn how to know the *shapiri*, how to see them and listen to them. Only those who know the *shapiri* can see them, because the *shapiri* are very small and bright like lights. There are many, many *shapiri* – not just a few, but lots, thousands like stars. They are beautiful and decorated with parrot feathers and painted with *urucum* [redberry paste]. Others have earrings

Hallucinogens

Like many peoples, Brazilian Indians use plants to alter their mental state, to enter the world of spirits and religion. Tribes in the north of the country take a hallucinogenic snuff called *yopo* or *yakoana*. They roast the inner bark of certain trees, mix the ash with powdered leaves and blow it into their nostrils with a hollow cane.

Other tribes make a drink called *caapi* (also called *ayahuasca*) by boiling particular species of creepers and vines. This is taken during special festivals where the Indians re-enact their mythology, and seek support from good forces and protection from malevolent ones. Amazon-

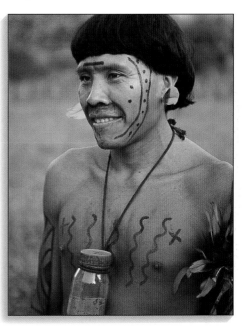

and use black dye and they dance very beautifully and sing differently."

"The whites think that when we Indians do shamanism we are singing. But we are not singing, we are accompanying the music and the songs. There are different songs: the song of the macaw, of the parrot, of the tapir, of the tortoise, of the eagle, of all the birds which sing differently. So that's what the *shapiri* are like. They are difficult to see."

LEFT: a shaman inhales hallucenogenic snuff.
ABOVE: a Yanomami shaman heals a feverish child
RIGHT: Levy, a Yanomami shaman carrying his jar of hallucenogenic snuff *(right)*.

ian Indian designs, such as paintings on houses, are influenced by the visions. Tribes in northwest Amazonia grow and use *ipadú*, or coca. The leaves are roasted, pounded and mixed with ash. The resulting fine green powder is placed in the mouth where the active ingredients (including cocaine) slowly induce a stimulating effect. All such plants are themselves considered to be powerful and sacred, and potentially dangerous if mistreated. They are only used in ritual contexts under strict conditions, and never casually or for entertainment. ❑

© *Survival. This is an extract from the book* "Disinherited – the Indians of Brazil" *published by Survival International.*

LAND OWNERSHIP

The most effective way of ensuring the survival of
indigenous peoples is to grant them land rights

Under international law tribal peoples own the lands they live on and use. This has been the case since the 1957 convention (no.107) of the United Nations' International Labour Organisation, a law which Brazil signed up to as long ago as 1965. In spite of this, Brazil and Suriname remain the only countries in

(using physical markers set upon the ground).

These steps usually take years to carry out and are never even begun without vigorous pressure from Indian supporters within and outside government. The process invariably comes up against powerful lobbies, often miners or loggers, or local politicians who are seeking votes or

South America where Indians are deemed to have no ownership rights whatsoever over their own lands, and it places all Indian land there in an extraordinarily vulnerable position.

In Brazil, Indians are legally minors and their communities can own no land at all – they can simply live on and use certain areas of government-owned land which have been recognized as "Indian areas" or "parks". This recognition, usually by presidential decree, can easily be modified or annulled by subsequent presidential decisions: all too often this is exactly what happens. For land to be set aside for India "use", it must first be "delimited" (its boundaries outlined on a map) and then "demarcated"

a share of their profits, or both. Most senior army officers have also proved staunchly opposed to the creation of Indian lands near the country's borders, wanting to maintain their power both within the country and over what they see as strategically sensitive areas.

The effect of all this is that there is constant pressure on the government not to create new Indian areas, to reduce the size of existing ones, and actually to annul those already demarcated. In1996, the minister of justice introduced a decree giving third parties such as loggers or settlers the right to challenge the limits of demarcations. Eight areas were to be "revised" – in other words, reduced. Cases such as these can drag on unresolved for many years. The same minister signed an act to reduce the demarcated land of the 12,000 Indians living in Raposa/Serrado Sol – he wanted the Indians to live in a few small enclaves, as was proposed in other cases, whilst freeing up at least a fifth of their land for mining and ranching.

Such reductions of Indian land are invariably presented as being "in the national interest" – yet they are motivated rather by the economic interests of a few powerful individuals. Certainly, they are disastrous for the Indian peoples concerned and of little or no benefit to the Brazilian population as a whole.

Maria Cleonice Servino, a young Wapixana woman, describes a typical incident when military police, called in by the local rancher, attacked her village in 1987: "I was three months pregnant at the time. Twelve trucks full of police rolled up. They went about breaking everything and hitting people. They broke the ribs of one of my brothers, and threw women down onto the ground – their children were crying and hiding under the tables. I remained standing, so a soldier came up to me and ordered me to lie down. I said I would not. 'I am not a dog that you can order about. I am in my home.' By now it was raining and all the

Cleonice's story

The Makuxi, Wapixana, Ingarikó and Tau-repang peoples live in the northern Amazon, and face some of the worst violence against Indians anywhere in Brazil as they struggle to have their land, Raposa/Serra do Sol, legally recognised. Over a dozen Indians have been killed, and hundreds more have been beaten up and had their homes and livestock destroyed by the local police and by ranchers and settlers who oppose the Indians' campaign.

LEFT: Ava-Canoeiro children playing.
ABOVE: an Awá hunting group with their catch of tortoises, agouti and birds.

children were covered in mud. They threw a table at some of the men and people were thrown on top of each other. Everyone was crying except for me –I don't know why.

"The soldier hit me in the stomach with the butt of his gun. 'Why don't you kill me? I am three months pregnant and if my baby dies it will be your fault. You may be the boss in the barracks, but the bosses here are us.' I tried to push the gun away with my hand and he said, 'You've escaped this time, but you won't next time.'"   ❏

© *Survival. Extract from the book* "Disinherited – the Indians of Brazil" *published by Survival International.*

GENOCIDE

Indigenous populations are frequently the victims of mass murder,
whether by loggers, miners or the forces of the state

During August 1993 a scrappy note arrived at the FUNAI office in the city of Boa Vista in the northern Amazon. It read: "The Indians [from near Haximú] are all here... they don't want to go back because the goldminers went to a *maloca* [communal house] nearby and killed seven children, five women and two men and

fire and was chopped up with a machete. After it was over and the miners had left, some survivors crept back and cremated the dead.

The final total was 16 Yanomami dead. After seemingly interminable delays, a case was eventually brought to court at the end of 1996 and a judge found five miners guilty of geno-

destroyed the *maloca*." It had taken a month for news of the killings to reach the outside world.

It had started several months earlier, when miners had killed other Yanomami whose relatives later retaliated, killing two miners. It was then that a group of heavily armed miners went to the Yanomami community of Haximú on the Venezuela-Brazil border. On arrival the miners opened fire on the *maloca*, in which mainly women and children were at home, before moving in to burn it down. Those who could, fled for their lives, and a handful of survivors took refuge in the forest. An old and blind woman was left behind: the miners kicked her to death. A baby lying in a hammock survived the gun-

cide. Although 19- to 20-year sentences were handed down, only two men actually ended up in prison, the others having fled.

This was Brazil's second conviction for genocide. The first, two years previously, was levied on a rubber tapper following the murder of eight Oro Uim Indians in 1963. He organized an attack on the group, and after the massacre the survivors were taken to his plantation where they were enslaved. By the 1990s, the Oro Uim numbered only 55 individuals.

The Brazilian courts' recognition of these killings as genocide is an important acknowledgement of the seriousness of the crime. However, pointing the finger at a handful of miners and a rubber tapper could be described as missing the point – if their actions are described as genocide, to what extent is the Brazilian state's appalling treatment of Indians also genocidal?

Government culpability

In the Yanomami case, the Brazilian government must bear some of the responsibility: for over four years they failed to expel miners working illegally in the Yanomami area, allowing disease and violence to spread. As the mining invasion and health crisis accelerated, the government actually threw out all health teams working with the tribe. The Yanomami popula-

Other policies are openly racist, and even more clearly genocidal in effect – in 1999 one politician called for an amnesty for miners who have committed crimes in indigenous or protected areas, and several are pressuring the government to open up all Indian areas to mining. Faced with this, many predicted the demise of the entire Brazilian Indian population. This is no longer a danger: the Indian population is now increasing overall, but small, isolated tribes are still being put at risk, and their people killed. ❏

© *Survival. Extract from the book* "Disinherited – the Indians of Brazil" *published by Survival International.*

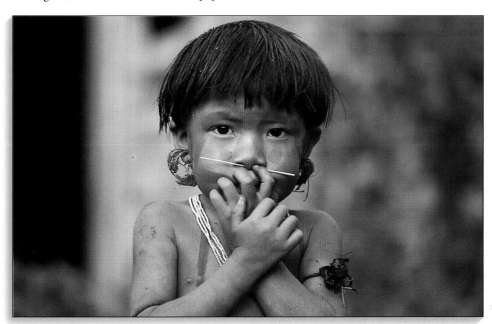

tion fell – through disease and attacks – by nearly one-fifth in the space of seven years.

The Yanomami are not an isolated case, and many others are even more extreme: government policies of integration, settlement or development have been directly responsible for the demise of many Indian tribes. The deliberate state neglect of Indians (whether because of corruption, underfunding or political expediency) has destroyed many more.

FAR LEFT: a Nambiquara woman stands by the highway that has laid waste to her environment. **LEFT:** an Awá woman, one of three survivors of a massacre. **ABOVE:** a Yanomami child.

RACIST ANTHROPOLOGY

Portrayals of Indians as violent savages remain common. Perhaps the worst recent example is that given of the Yanomami by the US anthropologist Napoleon Chagnon, whose studies have become standard references. He fabricated a sensationalist, racist image of the Yanomami, calling them "sly, aggressive, and intimidating" and falsely claiming they "live in a state of chronic warfare".

Chagnon's work has been severely criticized by others with extensive experience of the Yanomami and has undoubtedly been detrimental for the Indians. It has been used to justify a proposed fragmentation of their lands and the withholding of education development funding.

PEOPLE AND THE AMAZON TODAY

The rape of the Amazon's natural resources may leave the region
not only devoid of its biodiversity but also of no sustainable economic use

Over-exploitation of the Amazon began with a Spanish expedition in 1541 that started from Quito and worked its way eastward into the lowlands and down the Amazon in search of gold. The quest for quick profits became a continuing theme, and today's mining, logging, damming of rivers, cattle ranching, development disastrous Transamazon Highway across the south of the Basin, or the catastrophic BR364 road in Rondônia in the 1980s, have also contributed to massive forest loss and degradation. However, large-scale mining and industrial logging are now increasingly responsible as the expanding road networks built by loggers attract

of inappropriate forms of agriculture, and indiscriminate settlement projects may destroy one of the last great wildernesses on earth.

Brazilian Amazonia contains about 40 percent of the world's rainforests, nearly 2 million hectares (5 million acres) of which disappear every year, and there are alarming signs that the rate of deforestation may be increasing. In 1970, only 1 percent of the Brazilian Amazon had been deforested, but by 2000 almost 15 percent had gone. Subsidized clearance of huge, unproductive cattle ranches, often used for speculation, was, until recently, responsible for nearly all this deforestation. Gigantic colonization programs of the sort that accompanied the construction of the colonists and ranchers, and facilitate access to the forest which is difficult to control. Migrants from the northeast and south of Brazil have flocked into the Amazon to seek a living. Some were encouraged by government incentives in the 1980s to settle on colonization schemes like the Polonoroeste project in Rondônia.

Land issues

Most colonists are not drawn to the Amazon by the myth of fertile lands and higher standards of living. Many are internal refugees, often expelled from their lands by landlords working with hired assassins, attempting to prevent the peasants from gaining political power. Others are dislodged by government policies favoring agro-industry over traditional subsistence farming, or are ruined by the region's economic problems. Plantations of cash crops and fast-growing timber, and increased mechanization of agriculture have also forced people off their small holdings. Dispossessed of their land and livelihoods, they have no choice but to move either to city shanty towns, or the Amazon. For most settlers, life remains hard, faced with soils which cannot support crops, diseases, a lack of infra-

DEVASTATION

During the 1998 drought caused by El Niño, two months of fire, started by subsistence farmers, destroyed almost 25 percent of Roraima. This recurs, as burnt forest is 10 times more likely to catch fire again.

many have been killed by gunmen hired by rural landowners, and by police agents. An estimated 100 million hectares (250 million acres) of Brazilian forest have been illegally "privatised" by farmers, loggers, and corrupt registration officials using fake papers. In Bolivia, Peru and Colombia, many settlers have no economic option but to grow coca (used to make cocaine). Fire, accidental and intentional, also destroys forest as farmers in the Amazon traditionally use slash and burn methods to clear and fertilize land.

structure, and ranchers trying to take over their lands. New arrivals buy territory from earlier settlers, squat on the property of big landowners, or subscribe to the private colonization schemes which are now a significant cause of deforestation in states such as Mato Grosso.

Brazil has one of the most unequal land ownership distributions in the world – nearly 50 percent of the land is owned by 1 percent of the population. The Landless Movement, representing about 4.5 million people, is successfully lobbying for land reform, but progress is slow and

PRECEEDING PAGES, LEFT AND ABOVE: 18th-century prints portraying Amazonia as an idealized Eden.

Mining and logging

Many move to the Amazon to mine. Freelance goldminers – or *garimpeiros* – have been working in the rainforests for several centuries, but the gold rush truly began in 1980, when *garimpeiros* found an extraordinarily rich deposit in Serra Pelada, near the Rio Tapajós. There are now around 300,000 people in the Brazilian Amazon directly dependent on freelance mining. As they are highly mobile and the government's efforts to restrain them are feeble, they represent a significant threat to the survival of the region's ecosystems and its indigenous peoples.

In 1987, goldminers moved into the territory of the 10,000 Yanomami in northern Brazil, who had had little contact with the outside world. By 1989, 40,000 miners had arrived, and in the next seven years, 2,000 Yanomami died from malaria, tuberculosis, and common colds introduced by the miners. In the Yanomami area, the gold is found in riverbeds. Excavations of the river valleys make the water so turbid that the fish the Indians depend on cannot survive. Though most of the miners have now left the Yanomami lands and moved into tribal territories elsewhere, pockets still remain and the effects of their presence persist. Many Yanomami now have malaria, and mercury used in the extraction of gold is likely to become a serious health risk to them as it accu-

est. Tribes such as the nomadic Awá, many of whom are uncontacted, are at great risk. Their lands have been heavily invaded by loggers, ranchers and settlers attracted by the Carajás railway and roads, and many people have been killed by invaders stealing their land and resources.

In Ecuador, European and North American corporations are drilling for oil in some of the most diverse Amazonian rainforests. Inadequate safeguards have led to 30 major spills from one pipeline alone, and chemicals are discharged directly into the rivers. There is oil exploration in the Cuyabeno Wildlife Reserve (Siona territory) and the Yasuni National Park (Waorani

mulates in the food chain. Uncontrolled mining poses similar serious problems for indigenous peoples in neighboring Venezuela.

Mining interests also threaten the Amazonian forest. The Grande Carajás Program to tap the world's largest deposit of iron ore could transform 18 percent of the Brazilian Amazon. The mine, backed by the World Bank, the EU and the Japanese government, has been carefully managed; but associated developments, such as ore smelting, cattle ranching, timber cutting and colonization, are proving destructive. New industries in the Carajás region use charcoal made from forest trees, resulting in massive deforestation: Maranhão state has lost one third of its for-

territory). The Waorani have lost vital hunting grounds, and the Siona, Secoya and Cofan tribes have been devastated by the impact of 40 years of oil extraction and subsequent colonization. In Peru, the uncontacted Yora, Mashco-Piro and Amahuaca, having seen Mobil prospect for oil in their forest headwaters, now face an invasion by mahogany loggers. The Brazil-Bolivia gas pipeline will pass through several tribal lands.

Timber cutting is fast becoming the greatest threat to the forests. Logging was previously relatively unimportant as the great diversity of the trees made exploitation difficult. Now, however, importers accept a wider range of woods, and sawmills have moved in from other parts of Latin

America. The money that colonists and ranches make by selling the timber on their lands helps people reach new frontiers. Roads that the timber cutters build bring development to parts of the Amazon which would otherwise have remained untouched. The export industry causes the most destructive cutting, even though it handles only a small proportion of the wood felled. As high-value trees like mahogany are widely dispersed through the forest, timber cutters build roads to reach them, destroying far more trees than they harvest. Asian timber TNCs are investing heavily in the region and control at least 13 million hectares (32 million acres) of Brazilian forest.

taries in Mato Grosso. Oil palm planting in countries such as Ecuador is destroying the territories of several indigenous communities, and the international trade in pets and skins is now a serious threat to the survival of many species.

Brazil's national power company had planned to build hydroelectric dams in most of the major river valleys. While some produce reasonable quantities of electricity, others are environmental and economic disasters. The Balbina dam, north of Manaus, has flooded at least 2,400 square km (925 square miles), while producing heavily subsidized electricity. The decision to build was political rather than economic, as the

Dangerous development

Timber is not the only product whose sale abroad is helping to destroy Amazonia. In the *cerrado* scrublands to the south of the Basin, soyabeans are grown for European cattle farmers. To help exporters, Brazil is investing in two huge waterways and asphalting roads, projects strongly opposed by local indigenous peoples who fear their lands will be exposed to invasions and settlement. Agrochemicals used for soyabeans are contaminating Amazon tribu-

lucrative construction contracts and jobs could be used as political patronage. In the near-stagnant reservoir, acids released by the decomposing forest destroy the turbines every few years. The upstream migration of turtles and several fish species has been stopped, depriving the Waimiri Atroari people of a food source. Opposition by indigenous peoples and other groups and economic reasons led the national power company to postpone many of the dams. However, the government intends to build a series of dams along the Xingu river to alleviate Brazil's energy crisis. Many believe these will not provide sufficient electricity and that impacts on local populations far outweighs any benefit.

LEFT: the impact of agriculture is clear from this photograph of the Peruvian Andes.
ABOVE: road-building wreaks environmental havoc.

Perhaps the most dangerous government initiatives in the Amazon are the Calha Norte and Avança Brasil projects. Calha Norte, administered by the Brazilian army, started in the 1980s to establish a military presence by opening a new development frontier along the entire northern border of Brazil, a region inhabited by many indigenous peoples with some of the least disturbed forests in Amazonia. The army intends to support large projects such as road building and hopes to attract mining companies, timber cutters, ranchers and settlers to populate the region. It is planning to build several barracks in indigenous territories, a move bitterly resisted by the Makuxi and Yanomami in Roraima, who say past experience of military barracks in indigenous areas has been very negative.

Calha Norte has been hindered by bad planning and the resistance of peoples such as the Tukano. But mining companies have started work in some indigenous areas, a few colonization schemes have been funded, and the project helped the *garimpeiros'* invasion of Yanomami territories. The army claims the program is necessary for national security. Critics see no significant threat and suggest that it is instead designed to consolidate the army's power in the region.

AVANÇA BRASIL

The Avança Brasil (Advance Brazil) project aims to expand agribusiness, mining, oil and gas in the Amazon. A huge network of roads, dams, railways and waterways are planned, at a cost of US$40 billion. Scientists believe it will significantly increase deforestation, by 14–25 percent every year. Brazil's good environmental laws are rarely applied. Governments have not curbed occupation of the Amazon and concern is mounting over the impact of this scheme. It will encourage land speculation and occupation, logging and mining on a huge and unsustainable scale. The huge funds allocated clash with current rational and sustainable development initiatives by local Amazon populations.

Indigenous initiatives

The territories of indigenous peoples, which cover about 20 percent of the Amazon, contain vast tracts of virtually undisturbed forest. However, as other parts of the Amazon are destroyed, these are suffering from increasing invasion and depletion by loggers, miners and settlers. About 90 percent of territories suffer from some form of invasion. Though populations of most tribes have stabilized, several small, uncontacted groups such as the Kanoê and the Akuntsu in Rondônia number a few dozen or less and are on the verge of extinction, having been relentlessly persecuted by ranchers and subjected to disease. The population of the Yora in southern

Peru has also fallen since 1984, stricken by the diseases of the colonists.

Despite their considerable problems, however, it is the idigenous peoples who are responsible for the most positive events in the region's recent history. Throughout Amazonia, hundreds of organizations representing indigenous communities have taken responsibility for defending their rights and planning their own development: mapping and demarcating their reserves; training their own lawyers, teachers, soil scientists and ecologists; lobbying governments and internationally; reinforcing a sustainable economy. If they succeed, it is likely to do more than

some of their territory. But no administration has yet confronted the problems that continue to fuel deforestation, such as inequalities of land and power, allowing the privileged few to act with impunity and depriving the poor of any options but destructive farming and mining.

Northern culpability

In several respects industrialized nations are to blame for many of Amazonia's problems. The junta responsible for promoting Brazil's disastrous development policies was helped into power by the US and supported from abroad even during its most repressive and destructive

international initiatives to save the forests.

The governments of some Amazon nations have taken imaginative steps towards reducing the destruction there. Brazil's Pilot Plan for the Amazon has resulted in the demarcation of many territories for, and supports projects with, indigenous peoples and rubber tappers. The Colombian government has recognized the land rights of the Nukak Makú, one of the country's most vulnerable tribes. In Peru, hundreds of indigenous communities have managed to secure title to at least

LEFT: construction of the Serra de Mesa dam, Brazil.
ABOVE: evidence of logging is clear all along the Amazon.

years. Though many loans are as much the result of irresponsible lending as of irresponsible borrowing, industrialized nations still insist that Latin American countries repay their debts. The size of such debts makes a significant contribution to poverty and dispossession, promoting destruction in the Amazon. World Bank and IMF loans have forced countries to make severe cuts in environmental and social projects designed to protect rainforests and to recognize reserved territories. Aid made available for Amazonian projects continues to fund misguided development. If the governments of the North are to help save the greatest ecosystem on earth, they must first attend to their own destructive policies. ❏

CONSERVATION STRATEGIES

Experts argue that the best hope for the conservation of the Amazonian forests is to put the funding and control in the hands of indigenous peoples

Many people think of the Amazon as virgin forest, unaltered and uninhabited by human beings. In reality, much of it has been inhabited, shaped and used by people for thousands of years. However, in the last 50 years governments of all Amazon countries have sought to open up vast tracts of forest to exploit its mineral and timber wealth and to solve the pressures of growing and hungry populations by encouraging colonization schemes.

Neither the global environment nor the inhabitants of the Amazon can be helped by attempts to preserve the forests as if there were no one living there. It is now generally recognized that people who have lived in the Amazon are best placed to defend the forest. Indigenous peoples themselves see the recognition of the right to collective ownership of their lands as the best safeguard for the future. Reserves that exclude peoples from ancestral lands or prohibit traditional land use are likely to fail. The celebrated debt-for-nature swap, in which the American foundation Conservation International attempted to save the Chimanes forest in Bolivia, facilitated the invasion of timber cutters, as it took control of the reserve out of the hands of the indigenous people who lived there.

Strategies such as buying portions of Amazonia or exchanging areas for debt relief lead to selective conservation successes, but do not solve the basic problem that much of the rural population continues to be driven by poverty and dispossession into the virgin forest. Nature reserves are often targets for settlement, rather than regions to be avoided, as they contain none of the gunmen that ranchers hire to keep people off their lands. Conservation efforts which depend upon foreigners purchasing or taking control of other countries' sovereign territory tend to convince the inhabitants that conservation is solely a concern for outsiders, and affects themselves only negatively.

If the flow of colonists into Amazonia and the impunity with which timber cutters, mining companies and ranchers can operate are not addressed, then no conservation area in the Amazon, however much money is spent on it, could be considered safe. While stopping big corporations is a question of maintaining pressure on them and on governments in Latin America and the North, the arrival of colonists can be stemmed only by addressing the problems driving them. Without land reform, education, better distribution of wealth and attention to the needs of small farmers, peasants will continue to migrate to Amazonia. Without guarantees that they will not be expelled from their lands and given more sustainable means of farming in the forest, they will continue to move once they get there. While it is important that people in Europe and North America continue to lobby their governments to stop helping Latin American nations sustain the injustices driving people into the forests, conservation initiatives within Amazonia must come not from abroad but from the people living there.

Local conservation

All over the Amazon Basin there are communities of Indians, rubber tappers and peasants taking development initiatives into their own hands. As they frequently make clear, they recognize that the conservation of the forest is essential to their own livelihoods: it is only through lack of choice that many communities have been destroying the forests they live in.

With little money and no training, peasant farming communities have begun to devise the means of living in the forest sustainably. These techniques involve the planting of perennial crops – trees and shrubs – rather than annual species, such as rice and beans. These help to conserve the minerals in the soil and provide a steady source of income, while reducing the amount of labor required. Some settlers are experimenting with bee-keeping or the intensive cultivation of vegetables, exploiting gaps in the Amazonian markets. Long-established

LEFT: a highway, driven through the rainforest, bringing loggers, settlers and miners in its wake; a disaster for the environment and indigenous peoples.

residents, such as the rubber tappers, are trying to identify new forest products, with which they can supplement their low incomes.

The extent to which native forest products, such as nuts, fruits, gums, resins and fibers, can generate income and help protect the forest is hard to predict. Over 48 native Amazonian fruits have been identified with international sales potential. For example, the *açaí* palm produces fruit and palm hearts, and in 1995 almost 106,000 tonnes of wine, worth US$40 million, was produced from the fruit. *Cupuaçu*, a large forest fruit, is used in ice creams and drinks throughout Brazil. Many Indians harvest Brazil nuts and the Saterwé Mawé tribe successfully sells large quantities of guaraná nuts.

Indigenous organizations

The most exciting of all the initiatives come from indigenous peoples' organizations. In Peru, Ecuador and Bolivia, they have combined programs to improve their economic situation with campaigns to strengthen their political position and lobby vigorously for their rights. When governments have failed to recognize their lands, indigenous peoples have taken charge of the process themselves and invited consultants to train them in surveying and mapping techniques.

FOREST ECONOMICS

Some estimates of the economic potential of forest products are optimistic, but preservation of the forest for sustainable use is likely to be economically superior to unsustainable timber harvest. Investigation of potential forest products is important to make available as wide a range of development options as possible. However, indigenous peoples' rights must not be made conditional on their lands' potential benefits to outsiders. National and international markets are notoriously fickle and governments have tried to deny indigenous land rights by claiming the land uses are not "productive" in agro-industrial terms.

In Peru, indigenous organizations such as Aidesep have overseen projects improving the marketing of crafts, building up a transport network to cut out white middlemen. Organizations in Brazil and Peru have taken charge of community health, combining traditional medicinal knowledge with western health training. Many communities in Venezuela and Brazil have schools with teaching in indigenous languages.

Some organizations have begun regeneration projects restoring the damage inflicted by colonists and *garimpeiros* (miners). Amazonian forests regenerate if the land is cleared only once. The trees that return are not as diverse as in virgin rainforest, and the relationships between the plants and the returning animals are much less intricate. However, new trees do act as a "carbon sink". If forest is cut repeatedly, and seeds and shoots from which it would normally regenerate are killed, it is replaced by scrubby grassland. This can be replanted at low cost with closed canopy forest trees. Soil nutrients leached from the surface are still present and regenerating at greater depth. Fortunately, most deep-rooted Amazon trees grow with astonishing speed in old pasture, so regeneration projects should return full forest cover within 25 years.

With spectacular wildlife and waterways, there is potential for ecotourism in the Amazon.

devised by governments that involve local people only as employees. Funding agencies could invest much smaller amounts of money in schemes which clearly benefit the people wanting development, not the developers, and which support their rights. In these cases, the communities controlling the development projects have every interest in their success, and ensure that the resources are preserved for their future use.

By contrast, institutions like the World Bank still finance schemes likely to destroy more than they conserve. Some of these cost hundreds of millions of dollars, but many local initiatives could be funded for less than $10,000.

Some communities have established small-scale tourist ventures such as trekking and canoeing, and selling crafts. However, all too often these projects are exploited or hijacked by middlemen and only really work where the community itself controls all aspects of the project.

Funding issues

One of the best conservation investments for outsiders is to assist projects started by indigenous peoples, rubber tappers and peasants, rather than disbursing millions for schemes

LEFT: sorting Guaraná seeds.
ABOVE: forest produce being harvested.

While there is no shortage of money for big projects, the small ones wither and die, because big funding agencies overlook them.

If these fundamental problems can be solved, then there is a hope of preserving some areas that will never be altered. If any ecosystems are to be saved, it is essential that tree cover is maintained over most of the Amazon, as the trees of the forest generate much of their own rainfall. The water which they put back into the atmosphere through transpiration condenses above the forest to form rain. As the prevailing winds blow from east to west, the the eastern forests are responsible for much of the rain received in the west. Without it they will die through drought or fire. ❑

ENDANGERED ANIMALS

Over the past century, roughly one hundred vertebrate animal species
have become extinct, and today even larger numbers are threatened

There are three main reasons why a species may become extinct. Firstly, it may require a specialized habitat that initially shrinks and then disappears, leaving the animal nowhere to feed, breed, and rest. Secondly, it may be unable to cope with diseases or with other species introduced by humans. Thirdly, animals

Most South American nations have a record of aggressive land conversion: of the Atlantic forests of southern Brazil 90 percent have been destroyed, and only 31,000 sq km (11,970 sq miles) of the remainder is properly protected; meanwhile, the forests of the Western Andean slope in Colombia and Ecuador have disappeared

have been indiscriminately hunted for food, as trophies, or for the pet trade. Fortunately, in the Amazon proper, no known vertebrate species has yet been made extinct for these reasons. But there are warning signs that need to be carefully watched.

Habitat destruction is causing the extinction of countless species, many unknown to science, and many of extremely limited range. For the most part, the larger mammals have escaped, but this may not be the case for much longer. Habitat destruction is likely to be the major reason behind future extinctions; species such as the Maned Sloth and Muriqui have only about 2–3 percent of suitable habitat left.

in the last decades at a rate equaled by few areas on earth. It is here that habitat loss has brought many species to the brink of extinction. Examples in Brazil are the Golden Lion Tamarins, whose numbers are estimated at some 400–600 in the wild; the Muriqui, South America's largest monkey whose population hovers around the 700–1,000 range with only very few in captivity, and which continues to be hunted for meat; and in northern Colombia, the Cotton-top Tamarin.

Researchers havew been addressing the problem of shrinking habitat: long-term experiments look into the minimal size of habitat needed to guarantee survival of a small population of a species. A few hectares may suffice for a small forest bird of the undergrowth, a square kilometer of forest for a group of howler monkeys, while a Harpy Eagle may need a hundred sq kilometers (40 sq miles) to find enough prey.

Fortunately for the Amazon, the sheer size of uninterrupted habitats has generally kept diseases and exotic competitors at bay, at least as far as animals are concerned, although things like mercury poisoning in river systems are threatening this equilibrium in parts, since mercury levels build up, especially in the fats of those at the top of the food chain. Disease and competition with European immigrants have often had a far more devastating effect on indigenous human

populations before the arrival of the Portuguese, but on a sustainable level. The Europeans exported thousands of these animals for meat to Europe in the 17th century. Those surviving came under pressure from the skin trade at the beginning of the 20th century. Today, manatee populations are closely surveyed in some areas, but still killed for food in others.

The skin trade eliminated the Giant Otter from many areas, and it survives in healthy populations only in pockets of the Amazon. Caimans have also been heavily hunted for their skin: large individuals are now a rare sight. Some animals considered food in the Amazon wouldn't be eaten

populations. Excessive exploitation of animals by settlers has had a negative impact on population sizes, but has not caused the extinction of any major species as far as we know – at least not in the Amazon proper: on the southern peripheries, the pet trade has driven several species of blue macaws into or close to extinction.

The impact has been particularly heavy on animals that live in or close to the waterways, like the Amazonian Manatee and the Giant Otter. The manatee was hunted by indigenous

LEFT: a stall in Belém market selling illegal animal skins.
ABOVE: a Black Caiman murdered by poachers.

elsewhere, for example armadillos and monkeys. Thus their populations are depressed, but their survival is not threatened. Other animals that have joined the Giant Otter and Amazonian manatee on Appendix I of the CITES list of endangered animals are the Bush Dog and the Spectacled Bear, of which there are only an estimated 20,000 in its vast range.

Neither the political will nor funds are available to enforce conservation laws. Thousands of Amazonian people live in extreme poverty, and a trapped parrot or the fur of a cat may make a week's income. Nature tourism might help preserve species as a source of steady income rather than eliminating them in a one-time harvest. ❏

THE BIRD TRADE

*Large sums of money change hands in the illegal trade of rare species
and, despite international legislation, the traffic continues*

Within the neotropics, there are 141 species of parrot of which no fewer than 43 are considered by the International Council for Bird Preservation to be globally threatened. Of these, only Golden and Pearly Conures of Brazil are endemic to the Amazon region, but for several other species, notably the

more significant, and indigenous groups often respond to these demands too. The decline of Hyacinth Macaw in the southern fringe of Amazonia might have been due to hunting for plumes, and the Golden Conure is still hunted ruthlessly.

Habitat loss is ultimately the most serious threat for most endangered species. It reduces

large macaws, Amazonia provides one of the last great refuges. The Scarlet Macaw, for example, is now critically threatened in Central America, but is quite common in parts of the Amazon basin.

New World parrots face two threats: habitat loss and trade. Indigenous hunting has taken place over many centuries yet its impact has been limited. Feathers, much valued for religious and ceremonial purposes as well as ornamental value, were traded with Andean and western coastal groups; the preferred species were toucans, macaws, tanagers and eagles. But if an external demand is imposed, for example to supply feathers for tourist trinkets, or when populations are already threatened by other factors, hunting can be

bird populations, making them vulnerable to other pressures such as trapping. Habitat loss is particularly significant for species which have limited ranges (such as the endemic parrots of the Caribbean). Important areas for parrots where habitat loss has been severe include the northern Andes of Colombia and Ecuador and the Atlantic Forests of southeast Brazil, where only 5–10 percent of the original forest cover remains.

With wild populations facing such pressures, trapping for the bird trade imposes an unacceptable burden. Parrots have always been kept as pets by indigenous peoples, but the numbers taken from the wild have been sustainable. This changed once a demand for parrots was established in the developed world. Over 700,000 parrots of 96 neotropical species were imported into the US during the first half of the 1980s. The number of wild birds trapped far exceeded this figure, which is just the tip of the iceberg. The total does not include illegal trade or those birds that died during capture and transport.

To capture birds, trappers use a variety of techniques including limed sticks, decoys and robbing nests. The latter often involves the felling of the nest tree, which not only risks

SEVERE CRUELTY

Once captured, birds spend a long time in confinement, densely packed, vulnerable to the spread of disease and to physical injury. Inadequate food, water or ventilation will cause fatalities.

Little is earned by locally-based trappers and smugglers divert birds out of countries with tight controls and export them from others with less strict legislation: Brazil has banned the export of wildlife, but still loses birds via Guyana and Paraguay. Species with tiny populations are under great pressure. The Hyacinth Macaw persists in just three regions and illegal trapping is the most serious threat it faces. Spix's Macaw, long a prize specimen for trappers, is now extinct.

How can the bird trade be tackled? Much

killing the nestlings but also destroys the nest site. There is increasing evidence that even in pristine forest, parrot populations may be limited by nest sites. Thus trapping inflicts long-term damage on the remaining population. Probably 60 percent of the parrots die before leaving their country of origin. About 20 percent of the survivors are likely to die in transport and quarantine. The bird trade thus involves great cruelty and terrible wastage, which increases pressure on wild populations.

LEFT: Blue-and-Yellow Macaws in captivity.
ABOVE: the Golden Conure is endangered by the illegal pet trade.

effort is needed to encourage people, if they must keep parrots, to purchase captive-bred birds which are healthier and make better pets. At the moment, international trade is forbidden only if it can be shown that trapping is threatening populations. Many conservationists now believe that a system of "reverse listing" should be adopted – those species whose populations are proven to withstand trapping being commercially exploited. Finally, the development of well-planned wildlife tourism can create a source of revenue across local communities, creating conditions in which tourism becomes a viable and attractive economic activity. Such schemes are being tried in Peru and Brazil. ❑

TOURISM IN THE AMAZON

Thanks to mounting pressure from governments and conservation groups,
many initiatives have been taken to encourage sustainable tourism

The dream of many travelers is to take a slow boat down one of the world's greatest rivers, catching glimpses of wildlife and flora, and of the riverine life of indigenous peoples who have little contact with the 21st century.

The reality, however, is a little different, and perhaps the best preparation for a trip to the Amazon is to divest yourself of false expectations, both positive and negative: of easy access to a wilderness teeming with wildlife life, and of life-threatening diseases and massive nature destruction. Instead, determine what you want from the trip and take advantage of the vastly increased amounts of information available to plan a visit to this fascinating region.

In recent years infrastructure has improved and there is a wider choice of flights, accommodation and operators. Inevitably, among the many new "eco-friendly" lodges and companies there are some unscrupulous operators wishing to make a quick buck from the ever-increasing number of travelers and wildlife enthusiasts; therefore, it is worth checking out the many government, conservation and tourism websites for up-to-date guidance on the best places to visit and companies to use.

Wildlife enthusiasts will now find a wealth of worthwhile conservation projects that they can visit and support, while tourists who enjoy their creature comforts will be surprised at the quality of some of the lodges and ships in the region, although some argue that building accommodation in primary rainforest to promote eco-friendly tourism is a contradiction in terms. However, by visiting the Amazon you are directly helping to save the rainforest by showing local communities the global importance of its biodiversity.

Getting there

The larger Amazonian cities are still served by only a few intercontinental flights, and there are not many more international connections. An exception is Manaus, which has a regular service from many South American capitals. It is nor-

mally necessary to fly to a capital city, and to continue from there by domestic flights. Various air passes make flying between countries in Latin America very affordable.

The Brazilian Amazon is best accessed from São Paulo, Rio de Janeiro, or Brasília, which normally means flying 2,000 km (1,200 miles) south of the equator, before heading in the opposite direction. It is important to plan early, since these domestic connections tend to be overbooked.

A glance at a road map of the Amazon will give the impression of a network of highways. Many of these roads exist only in the minds of development-oriented politicians, are impassable for much of the year, or have even never been built at all. Limited transport exists, however, and the daring and budget-conscious traveler may consider embarking on the two- to three-day bus journey from Rio de Janeiro to Belém.

The cities

Having arrived in cities like Iquitos, Manaus, or Belém in the Amazonian heartland, do not expect to find undisturbed forests and large numbers of wildlife close by. These cities are surrounded by a wide corridor of developed land, and without the guidance of a local naturalist, there is little chance of finding access to good sites.

The cities themselves do not fit into the environment sympathetically. Past generations had human development on their mind, without considerations for natural surroundings. So, despite glossy tourist board brochures promoting the hotels and museums of the larger Amazon cities, the reality is often noisy and shabby with no parks and often very few cultural attractions to justify more than an overnight stay.

By contrast, the more remote towns with larger indigenous populations, such as São Gabriel da Cachoeira in Brazil, are generally peaceful and seem to exist more harmoniously with the rainforest. But the real reason to come here is to experience the forest, and there are many ways of getting into the jungle, including by cargo boat, commercial passenger boat, canoe, small turboprop and jet, modern bus and 4x4.

LEFT: the Gaiola riverboat on the Rio Negro.

Language

English is rapidly becoming a world language, but not in the Amazon. Most people in Brazil speak only Portuguese, and those of the surrounding countries, Spanish. Establish a small vocabulary in either language, which are related: while pronunciation is quite different, many words can be understood in either language.

Budget

Most Amazon nations are economically declining, the exploitation of oil and gold, instead of bringing prosperity, has brought disaster to the indigenous peoples and environment, and led to excessive national debts. Most of the population is very poor, and even the better-off have difficulty coping with high inflation and irregular supplies. Nevertheless, it is a mistake to conclude that a tourist will feel rich. Hotels – in particular in Brazil – often have higher prices than in Europe or North America, and transportation prices can be exorbitant.

There are two ways to minimize these difficulties: to visit multiple areas, either arrive by prearranged tour, or try to join such a tour that just happens to have a vacancy. This way, the large expenses for transport are shared among several people. Alone or in a small party, it will

be best to use a local agency for a stay in one of the few hotels or *posadas* that cater to the nature tourist, or to reach one of the few national parks with accommodation and food supplies. In more remote areas of the region, it is always best to check with the local authorities before setting off and, where possible, to hire a qualified guide for invaluable local and wildlife knowledge.

National Parks

The largest part of the Amazon falls within Brazil. Unfortunately, this country has only a few national parks and biological reserves, and none of them has lodges, restaurants and campgrounds like in other countries. Often access is expedition style, by rented boat, with acommodation on the boat or in tents. Food, drinks, and fuel have to be brought along. A large fleet of Manaus-based excursion boats now cater to nature tourism, and many boats offer comfortable, fan-cooled cabins with beds, electricity, good kitchens, and cold drinks. These also make expeditions into the Amazon a possibility for disabled travelers. In Belém and Santarém, however, these developments are still in their infancy.

In other countries, the last decade has seen significant activity in the creation and development of national parks; access has improved although still requires some time and careful planning.

asked foreign travelers to sign forms confirming their awareness of the hardships ahead, before being allowed on the boat. Travelers often end up in population centers rather than close to nature, and during the boat ride, the forest is often nothing more than a dark line on the horizon. For the enterprising traveler, these boats are a good means, however, of getting to small settlements, and finding accommodation with local people within walking distance of undisturbed forest.

Dangers
The most dangerous experience for the majority of tourists will be crossing the road that sepa-

Public transport by boat
The cheapest way to explore the Amazon is by public transport. Boats run regularly between all major and minor centers, and it is possible to tour most of the Amazon and its tributaries in this manner. It takes up to a week to get from Belém to Manaus, and up to two weeks to continue to Iquitos. This will certainly be adventurous, and the conditions, such as sleeping on a crowded deck in a hammock, the noise, and sanitary conditions will certainly appeal only to few travelers. The authorities have occasionally

rates Manaus' shopping malls and the jetties on the banks of the Rio Negro. Of the generally imagined dangers, jaguars are secretive and to spot one remains for most visitors an unfulfilled dream rather than a dreadful nightmare. Vampire bats generally attack domestic animals, and mosquito nets suffice for protection. Piranhas are said to be dangerous in some localities, but in most places the locals take a swim in the same area where they fish for piranhas – follow the example of the locals. And leeches, a major nuisance of Southeast Asian rainforests, are actually quite rare in the Amazon.

Some snakes are poisonous. It is impossible to avoid an occasional encounter, and even less

LEFT: a boat trip allows you to observe the forest.
ABOVE: the Madre de Dios airstrip, Peru.

possible to carry a selection of antisera on a longer trip. The best comfort is that snakes try as hard as they can to stay away from us. Most attacks occur when people run carelessly through undergrowth, dig with bare hands under branches or fallen tree trunks, or approach a snake that makes every effort to stay by itself. The best protection is to walk slowly, to stay on trails or on open ground, and to keep your hands away from places where snakes may hide.

Getting lost is a serious danger in tropical

MOSQUITOS

Places overwhelmed by mosquitos tend to be boreal or temperate, rather than tropical. Black water rivers like the Rio Negro are nearly free from mosquitos – the acidic water prevents larval development.

Diseases

While it is rare for the itchy bites from mosquitos to become a real problem, their potential to transmit contagious diseases should not be underestimated. Yellow fever, a mosquito-borne viral infection, still strikes in some localities. Fortunately, vaccination is cheap, legally required and provides full protection for about 10 years.

Malaria is more problematic than yellow fever. This is a mosquito-infected disease caused by Plasmodium, a single-celled

forests. The tree trunks are very uniform, and, standing dense, quickly close off the view of the horizon, of the river, or of a clearing in the forest. To determine the direction, you may try to look for the sun; this, however, is barely visible through the dense canopy, is for most of the day in a nearly vertical position anyway, or is covered by clouds. The slope of a hill? There are not many in one of the world's largest plains. A call, or the sound of a motorboat? The incessant rattling of the cicadas will drown that. The best protection is to enter the forest along trails, leave trails or the rim of the forest only for the shortest distances, and generally to take time in the exploration of an area.

animal. With no vaccination, most doctors prescribe a combination of Chloroquine and Paludrine (Fansidar is best avoided). Unfortunately, Plasmodium becomes increasingly drug-resistant, and the best prevention is to avoid being bitten. Use a fan or mosquito-net at night, and wear long pants and long-sleeved tops during the day; repellent will provide added protection.

What to expect in the forest

The Amazon is the world's richest ecosystem, but it would be a major mistake to conclude that a stroll through the forest will lead to easy observation of a large number of mammal and bird species, and flocks of colorful butterflies. On the contrary, one's first impression of a tropical rainforest anywhere on earth is that of a quiet, green hall, with little to marvel at except trees. Even in optimal habitats, a whole hour may pass without the sight of a single mammal, and at best the faint call of an unidentified bird.

A good starting point for nature observation are the obvious large trees, with their big buttressed or stilt roots. More than a hundred different tree species may occur in a single hectare. On the forest floor, leaf-cutting ants carry bits of greenery. A sudden flurry of activity in the undergrowth and lower canopy may point to the approach of a mixed foraging flock of birds: there

where wild mammals are regularly fed. In these places, up to a dozen mammal species may congregate, and may become so tame that they feed out of people's hands, although conservationists question such direct intervention by humans.

Various strategies will help you find different birds: for parrots, be at a place with a wide view at sunrise, when they descend from their roosts to their feeding grounds. For hummingbirds, wait at flowering bushes, for tanagers, at fruiting trees. And try to go out at night with a flashlight, when white, green, or pink reflecting points in the darkness may be the eyes of a caiman, a frog, a Boat-billed Heron, or a Spectacled Owl.

may not have been a single bird for two hours, and then 20 species pass by within a few minutes, leaving little time for identification. Scanning the upper canopy for the origin of some bird-like calls will identify a group of Squirrel Monkeys, and this may be the only mammal observation of the day, not counting the bats that come out at dusk. To avoid disappointment, it is important to appreciate these details.

What is the best way of increasing the chances of seeing wildlife? There are lodges and park headquarters throughout the Amazon

LEFT: exploration on foot is also an option.
ABOVE: weary travelers.

Photography
Three problems should be taken into consideration. Frequently, the danger to camera and film in tropical climates is exaggerated. Over a year or two, films will spoil, and even well-coated camera lenses may be overgrown by fungi – closed containers with water absorbing chemicals provide important protection. Normally, however, over two or four weeks no problems occur. For the best results, a sky-light filter is necessary from 9am to 3pm, and for animal and plant photography, a tripod is essential as less than 1 percent of the sunlight reaches the forest floor. Many visitors take a camcorder or mini-disc to capture the sounds of the forest. ❏

AMAZON PLACES

A detailed guide to the entire region, with principal sites
clearly cross-referenced by number to the maps

There are about 4 million sq km (1.5 million sq miles) of rainforest in the Amazon, but getting to them can be a more demanding enterprise than expected, simply because much of the region is still so remote. Many of South America's developed and accessible national parks are outside the Amazon region, at the periphery of Caracas, Bogotá, or Rio de Janeiro, where land had to be legally protected to retain its natural wealth.

Useful stopovers on the way to the Amazon, these cities and their parks are a good introduction to tropical nature. For this reason, some of them are included in this book. Excursions to the Pantanal, for example, or the Iguaçu National Park, are offered by many tour operators in Brazil's cities.

Very few parks in the Amazon fulfil the high expectations for comfortable access, good accommodation, and fine food formed in North America or East Africa. But the situation is changing: Canaima in Venezuela can be reached by jet plane; access to Amacayacú National Park in Colombia and Manu National Park in Peru has improved, and showcase ecotourism projects have sprung up in Bolivia and Ecuador that reflect a general trend to help conserve the Amazon's biodiversity and support the culture of the indigenous peoples.

So far, the Brazilian Amazon has no parks with facilities, but a fleet of excursion boats stationed in Manaus has developed over the past decade and is still expanding. Private hotels are springing up in natural settings, and *ranchos* starting to cater to nature tourism, especially in the Pantanal where the wildlife viewing is world-class.

The following pages describe some of these places and excursions with further details given in the Travel Tips section at the back of the book; although, given the pace of new developments, and the unpredictable political and economic situations in most Latin American countries, it is difficult to give up-to-the-minute advice.

For many, a trip to the Amazon is a once-in-a-lifetime journey, and to avoid disappointment it is worth considering what you want from the trip, the importance you place on comfort, what you hope to see and experience, and what impact your visit will have on the rainforest. Fortunately, there is a increasingly wide choice of tour operators, accommodation and transportation, making it far easier than in the past to enjoy one of the world's great wildernesses. ❑

PRECEDING PAGES: a Venezuelan *tepuis*; a tributary meandering through lowland forest; tropical coastline backed by Atlantic forest, Brazil.
LEFT: a canyon at the southern edge of Amazonia.

WESTERN COLOMBIA

Map on page 152

Although this is potentially the most rewarding region in Latin America for observing wildlife, the continuing conflict makes Western Colombia too dangerous to visit at present

Tlthe Pacific slope of the Western Andes is not geographically part of the Amazon basin, but it is included here because of its wealth of endemic fauna and flora. It lies off the beaten track and is one of Colombia's least developed regions.

This zone, known as the **Chocó**, extends from the Darién Gap to northern Ecuador. The political department of Colombia which also carries this name occupies the northern third of this area. It has a very high, evenly distributed, rainfall and is the wettest place in the Western Hemisphere, with an average of over 13 meters (43 ft) of rain a year in some central localities. The heaviest rain falls in the coastal areas, up to about 250 meters (820 ft) above sea level.

The natural vegetation is tropical wet forest, with a high diversity of trees, epiphytes and palms. At higher altitudes, mist and low cloud hang over the forests, which drip constantly with condensation: this habitat is aptly named "cloud forest". The trees are festooned with layers of mosses, orchids, bromeliads and other epiphytes.

● **Although we include this chapter for the sake of completeness, much of Colombia is very dangerous because of increases in drug-trafficking and the resumption of war between the government and left-wing guerrillas. Intending visitors should check the latest travel advice with the relevant authorities before planning a trip.**

The tanager coast

All the important neotropical bird families are present in the Chocó zone, but the community composition shows some difference compared with that of Amazonia. Most striking is the strong representation of tanagers, indeed the region is popularly described as "the tanager coast". Many of these are of the genus *Tangara*, small and brightly colored birds which travel in mixed-species flocks. The Chocó zone is one of the most important centers of high biodiversity in the neotropics, and the high level of endemism shown here by the birds is also reflected by other groups.

Springboard for the region

The best base to use for an exploration of the Pacific slope is the private reserve of **La Planada ❶**, situated in the Colombian department of **Nariño**. It is managed by FES (Foundation for Higher Education), which is based in Cali. It has also received support from a number of international bodies, including the WWF.

The journey to La Planada starts in the capital of Nariño, **Pasto ❷**, a small city of about 350,000 inhabitants set at the base of the **Galeras volcano**. The journey to La

PRECEDING PAGES: paddling through flooded forest. **LEFT:** a Brazil nut tree in tropical rainforest. **RIGHT:** a rufous-tailed jacamar.

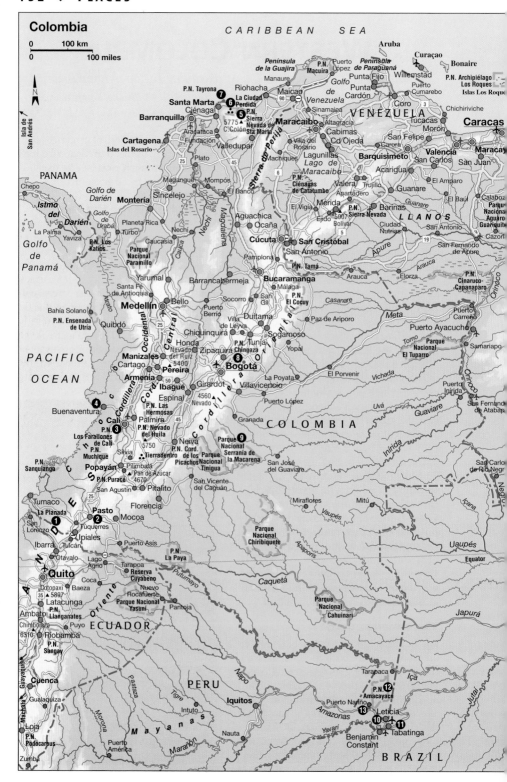

Colombia

0 — 100 km
0 — 100 miles

CARIBBEAN SEA

Aruba
Curaçao
Bonaire

Península de la Guajira
P.N. Macuira
Puerto López
Península de Paraguaná
Punta Fijo
Punta Cardón
Willemstad
P.N. Archipiélago Los Roques
Islas Los Roques

Manaure
Golfo de Venezuela
Puerto Cumarebo
Coro
Chichiriviche

P.N. Tayrona
Riohacha
Maicao
Sinamaica
VENEZUELA
Tucacas
Morón

Santa Marta
La Ciudad Perdida
P.N. Sierra Nevada de Sta Marta
Maracaibo
Altagracia
Cd Ojeda
San Felipe
Carora
Barquisimeto
San Carlos
Valencia
San Juan
Maracay
Caracas

Ciénaga
Barranquilla
5775 C. Colón
Cabimas
Lagunillas
Lago de Maracaibo
Valera
Acarigua
El Amparo

Cartagena
Islas del Rosario
Aracataca
Fundación
Valledupar
Villa del Rosario
Machiques
Trujillo
Guanare

PANAMA
Plato
Magangué
Mompós
El Banco
Ciénagas de Catatumbo
P.N. Sierra Nevada
Barinas
Apartadero
El Baúl
Calabozo
Parque Nacional Aguaro Guariquito

Golfo de Darién
Istmo del Darién
Golfo de Uraba
Monteria
Sincelejo
El Vigía
Mérida
Ejido
Ciudad Nutrias
San Antonio
LLANOS
Cazori

Chepo
Planeta Rica
Aguachica
Ocaña
Bolívar
San Fernando de Apure

La Palma
Yaviza
Turbo
Nechí
Caucasia
Cúcuta
San Antonio
San Cristóbal

Golfo de Panamá
P.N. Los Katíos
Parque Nacional Paramillo
Pamplona
P.N. Tamá
Arauca
P.N. Cinaruco-Capanaparo

Bahía Solano
Yarumal
Barrancabermeja
Bucaramanga
Málaga
Casanare
Elorza
Puerto Carreño

Santa Fé de Antioquia
Socorro
San Gil
P.N. El Cocuy
Paz de Ariporo
Meta

P.N. Ensenada de Utría
Medellín
Bello
Puerto Berrío
Villa de Leyva
Duitama
Orinoco
Puerto Ayacucho

Quibdó
Chiquinquirá
Tunja
Sogamoso
Yopal
Samariapo

Honda
Nevado del Ruiz
Zipaquirá
P.N. Chingaza
El Porvenir
Parque Nacional El Tuparro

PACIFIC OCEAN
Manizales
Cartago
Pereira
Bogotá
La Poyata
Vichada
Puerto Inírida
San Fernando de Atabapo

Armenia
Ibagué
Girardot
Villavicencio

Buenaventura
Cali
Palmira
Espinal
Puerto López
COLOMBIA
Uvá
Guaviare

P.N. Las Hermosas
Granada
Inírida
San Carlo de Río Negr

Los Farallones de Cali
P.N. Muchique
Silvia
Neiva
P.N. Cord. de los Picachos
Parque Nacional Serranía de la Macarena
San José del Guaviare

Popayán
Pilimbala
Tierradentro
Parque Nacional Tinigua

P.N. Sanquianga
P.N. Puracé
Pan de Azúcar
Pitalito
San Vicente del Caguán
Miraflores
Mitú
San Carlo de Río Negr

Tumaco
San Agustín
Florencia
Parque Nacional Chiribiquete
Ipana

La Planada
Pasto
Mocoa
Vaupés
Equator

San Lorenzo
Yaquerres
Ipiales
Puerto Asís
P.N. La Paya

Ibarra
Tulcán
Lago Agrio
Tarapoa
Reserva Cuyabeno
Putumayo
Caquetá
Uaupés

Otavalo
Coca
Nuevo Rocafuerte
Parque Nacional Yasuní
Pantoja

Quito
Cotopaxi
Baeza
Latacunga
P.N. Llanganates
Puyo
Parque Nacional Cahuinari
Japurá

Ambato
Chimborazo
Riobamba
ECUADOR

P.N. Sangay
Tarapaca
Içá

Cuenca
PERU
Iquitos
Puerto Nariño
Amacayacu
P.N. Amacayacu
Leticia
Tabatinga

Gualaquiza
Intuto
Amazonas
Yavarí

Loja
P.N. Podocarpus
Nauta
Benjamin Constant

Zumba
Puerto América
Marañón
BRAZIL

Machala
Guayaquil
Morona
Pastaza
Tigre
Mayanas
Napo
Jutaí

Map on page 152

Planada takes about four hours along the main Pasto to Tumaco road, climbing up to 3,200 meters (10,500 ft) above sea level through spectacular scenery.

The reserve lies 1,300–2,100 meters (4,265–6,900 ft) above sea level, 7 km (3 miles) above the village of **Chucunes**, near **Ricaute**. Temperatures are moderate, averaging 10°–17°C (50°–62°F), with about 4.5 meters (15 ft) of rain annually. This falls mainly in the late afternoon and at night. Most mornings are dry and very clear, with a good view of the **Cumbal volcano** possible for a short while after dawn.

The reserve has an administration building with restaurant and meeting room, and a scientific center with accommodation and a museum. There is also an orchidarium with over 360 species, all collected from La Planada (which has one of the world's richest orchid flora). La Planada has also been the center for a captive-breeding program for the endangered Spectacled Bear – a 2-hectare (5-acre) enclosure has been set up and breeding has already taken place

quite successfully. Research on the Spectacled Bear in the wild is recorded from the reserve and its surroundings.

The reserve contains a range of vegetation types, with small pastures, some secondary growth and extensive primary forest that has hardly been modified by man. At the lower levels, the forest is particularly luxuriant. The flora is a mixture of both low and high elevation elements and shows a high level of endemism.

Bird life at La Planada

The avifauna has been well studied by Colombian and foreign ornithologists and over 240 species have been identified. Of these, no fewer than 24 are endemic species of restricted range (less than 50,000 sq km/ 19,300 sq miles). The main families are the tanagers, with 30 species known (of which 11 are of the genus *Tangara),* the tyrant flycatchers with 32 species, hummingbirds with 26 species, and furnarids with 15 species.

Near the buildings of the reserve are flowering shrubs that are visited by humming-

BELOW: the rare Spectacled Bear survives at La Planada.

birds, especially early in the morning. Large mixed-species flocks of birds are a particular feature of La Planada. These can be spectacularly varied and pass rapidly over the canopy. Fortunately, because of the relief of the land, the path often follows ridges and canopy flocks can be watched from above. The parties contain tanagers and flycatchers, flower-piercers, furnarids and, from late September until April, North American warblers, such as the Black-and-White Warbler and the Black-burnian.

The forest-covered gorges provide haunts for the Andean Cock-of-the-Rock. The resplendent males of this species are bedecked in red, black and gray and gather in courtship leks in the subcanopy. Away from these arenas, they lead quite solitary lives and can prove difficult to see.

The best bird-watching times are from dawn to late morning. Some activity continues in the afternoon, but the onset of rain puts an end to productive watching. However, if the rain dies away before dusk, there is often a late flurry of activity.

Home of the Awa Indians

The area extends as a forest belt into northwest Ecuador and is the home of the Awa people, who live mainly in small settlements below 1,500 meters (5,000 ft), cultivating maize, beans, sugar cane and bananas. Their great knowledge of the forest enables them to utilize a wide range of forest products.

The objective of FES is to develop an international integrated land management project that crosses the frontier into Ecuador, drawing together the interests of both the Awa people and the conservation of this important center of endemism. Overall, such a scheme could protect over 7,500 sq km (2,900 sq miles) of forest habitat and the livelihoods of some 12,000 Awa people. Already, the Awa are closely involved with the management of the area.

At a higher altitude, above the tree line on the **Chiles volcano**, a second research station has been built. This both manages an Andean Condor reintroduction program and encourages scientific research into the

BELOW: tree fern canopy.

Map on page 152

altitudinal variation and movements of the wildlife. A third cabin closer to sea level completes the series.

North of La Planada

Cali ❸ can be used as a base to explore a zone of the Pacific slope to the north of the La Planada region. Attractive and well-organized, this is the third largest Colombian city with a population of around 1.7 million. It is situated in the **Cauca valley** and in the heart of the main sugarcane growing belt. Its climate is pleasant, with an average temperature of 24°C (75°F).

Cali's good system of buses can be used to visit many important wildlife sites of southwestern Colombia. The start of the Pacific slope lies on the outskirts of the city, 18 km (11 miles) along the road to **Buenaventura ❹**. An expanse of forest on the ridge overlooking the city, this easily accessible site provides an excellent morning's birding, with over 50 species to see, including the Multi-colored Tanager and the Bronzy Inca Hummingbird. Other patches of forest worthy of investigation lie nearby, scattered between cattle pastures and fine houses set in large grounds.

Alto Anchicayá

The best-known site on this section of the Pacific slope is **Alto Anchicayá**. It lies about 60 km (40 miles) from Cali on the old Buenaventura-Cali road. Access is from the town of **Danubio**, where there is a security post at the head of the road up to the Anchicayá watershed.

The whole area is under the protection of the *Corporación Autónoma del Cauca* (CVC). A reservoir drives a hydroelectric plant, and, with permission from CVC (obtained in advance in Cali), it is possible to stay nearby. The area is extremely attractive, holding a large tract of intact forest which remains undisturbed. Scenically it is superb, with forest-clad hills, waterfalls, rocky streams and gorges.

The climate is tropical but the high rainfall gives a feeling of freshness. Most mornings are dry and clear, with the deep valleys filled with clouds and ridge tops exposed. As the sun rises, so do the clouds; by the afternoon there is usually rain.

Most bird-watching can be done from the roads. These climb the hillsides, affording the walker good views of the forest canopy below. There are some trails, particularly radiating out from the settlement, which are useful for finding some of the more skulking species of the forest floor and undergrowth. The birds are spectacular and mostly quite easy to see. An experienced observer might well see over 100 species in a day. As at La Planada, a strong component to the avifauna are the mixed-species flocks. These are dominated by tanagers and tyrant flycatchers, with accompanying woodcreepers and woodpeckers.

The tanagers present do not show a very close overlap with the La Planada avifauna, in part because the area is at a lower altitude than La Planada. Many of the Pacific slope specialties have a very narrow altitudinal range and may also show restricted latitudinal distributions along the slope.

Despite being one of the most visited sites of the region, surprises do still turn up. For example, the recent sighting of the

RIGHT: a Mantled Howler monkey.

Banded Ground-Cuckoo, a species that had not been observed before, and which was known only from museum specimens. Alto Anchicayá remains one of the best areas in which to see the Long-wattled Umbrella-bird, similar in appearance to the related Amazonian Umbrellabird, but known only from the Chocó zoogeographical province and considered to be globally threatened.

At a lower elevation (200 meters/650 ft above sea level) on the Pacific slope near the village of **Aguaclara**, is a cabin belonging to the Cali-based *Fondación Herencia Verde* (FHV). It lies amid forest-covered hills, at the confluence of the Río Tatabro (which gives its name to the site) and the Anchicayá river.

Birdlife of the Río Tatabro

Although the forest has been modified and the influence of man has led to the local disappearance of large birds such as the cracids, the area around the **Río Tatabro** is excellent for birds, sharing many species with Alto Anchicayá. Footpaths lead along ridges on the slopes behind the cabin, offering good views of the coastal plain and the Buenaventura bay, as well as providing plentiful encounters with bird parties. These are dominated by the colorful tanagers, such as the Blue-whiskered, the Scarlet-and-White, the Golden-hooded and the Gray-and-Gold, all of which are accompanied by Slate-throated Gnateaters, Lesser Greenlets and Slaty-capped Shrike-Vireos.

Some of the species move around in mainly single-species groups, the common Tawny-crested Tanager and the Dusky-faced Tanagers being good examples. Not all birds will be associating in groups – the Black-breasted Puffbird spends most of its time perched quietly in exposed situations, waiting to collect its large insect prey from nearby foliage or from the ground.

A commonly heard call is that of the Fulvous-bellied Antpitta, an elusive ground-dwelling bird which haunts the thick vegetation around natural tree falls.

The Tatabro region is of great importance for other wildlife besides birds. A survey carried out on butterflies in the area recorded 520 species along a 30-km (18-mile) transection.

Río Tatabro is lower than Alto Anchicayá and this is reflected in its higher than average rainfall. Practically every day there is heavy rain. However, as is the usual pattern elsewhere, this occurs mainly in the afternoon or evening, leaving the visitor free to make the most of the mornings.

Integrating people and nature

The FHV, the conservation body running the base near Aguaclara, is seeking to establish schemes of working with the local communities in order to achieve conservation through integrated land use management, thereby encouraging the sustainable use of forest materials.

Although the forest along the old Buenaventura Road appears undisturbed, it has been logged intensively. As the supply of large trees diminishes, workers must travel further and further inland to find suitable materials. The limit is the distance that can be reached by a loaded mule – beyond that the forest remains essentially undisturbed.

Further disruption comes from gold panning. Along streams, there is much evidence of active prospecting. The net result is that people spend less and less time cultivating their plots and tending animals. The rural economy is changing and much damage is also being done to the environment.

In an attempt to stem the tide, FHV is setting up an initiative to reinforce the family structure and small-scale farming, using forest products in a sustainable way. It is hoped that the economic pressure imposed on the farmers by external commercial interests can thus be marginalized. So far, pilot schemes are under way with four families. FHV is also working with WWF and COAGRITA, a community-based agriculture co-op, to improve banana production and marketing techniques.

The highly committed staff of these organizations work alongside the Department of the Environment (Ministerio del Medio Ambiente, MMA) and the universities on studies within the existing national parks and protected areas network. Their work already receives some international support and recognition, and merits more. For more information, check out www.latinsynergy.org ❑

RIGHT: Lobster-claw heliconia.

Map on page 152

SANTA MARTA

Outstandingly beautiful and with a rich range of birdlife,
the Sierra Nevada de Santa Marta national park can
be visited from the nearby town of Minca

The **Sierra Nevada de Santa Marta ❺** is the highest coastal range of mountains in the world, and also a national park. The tallest peak is **Cristóbal Colón**, 5,775 meters (18,946 ft) above sea level and just 45 km (28 miles) from the coast.

The Kogi people who inhabit the region live a life of self-imposed cultural isolation from the outside world, even though they may come into regular contact with others in the course of their activities. According to them, they represent the Elder Brother, the rest of mankind being the Younger Brother whose disregard of nature has pushed the planet to the edge of ecological disaster. During the 1970s a lost city, **La Ciudad Perdida ❻**, built by the Tayrona people about 700 years ago, was discovered in the Sierra Nevada. The Kogis are probably the direct descendants of this culture.

Exploring the region

The Sierra Nevada de Santa Marta is also a region of great natural beauty and biological interest, holding 12 species of endemic birds including a parakeet and two hummingbirds. To enter the heart of the Sierra Nevada and travel to the Ciudad Perdida involves a trek with guides and takes a week to get there and back. It is an extraordinary experience.

Alternatively, it is sometimes possible to reach the area by helicopter. Fortunately for the visitor with less time or money **BELOW:** a tarantula waits on a leaf.

Map on page 152

to spare, all but one of the endemic bird species can be seen on the northern side of the Sierra Nevada, along the ridge of **San Lorenzo**, which is accessible by jeep from the town of **Minca**.

Minca lies in the coffee-growing belt and the road from Minca to San Lorenzo is about 35 km (22 miles) long, steep and muddy, and without a vehicle it is a tough tiring haul. However, this route offers a tantalizing glimpse into the heart of the Sierra Nevada and its wildlife, without the risk attached to entering the interior.

Unfortunately, parts of the massif can be dangerous to enter because of the activities of drug smugglers and guerrillas. Advice should be sought from the MMA (Department of the Environment) office in Santa Marta, from which a permit must also be obtained.

It is well worth staying overnight (accommodation is available in MMA cabins on the San Lorenzo ridge). The mornings are chilly at this altitude (about 2,400 meters/8,000 ft above sea level), but the views are fantastic. To the north, a panorama reveals the Caribbean coast from the city of Santa Marta westwards. Southwards are a series of ridgetops, extending to the snow fields of the peaks. Once the clouds in the valleys have risen, these views are lost for the rest of the day.

Close to the Sierra Nevada is another national park: **Tayrona** ❼, consisting of dry to humid coastal forest. It contains the ancient town of Pueblito, which is easily accessible at just a three-hour walk from the main park center, though birdwatchers will take longer as there is a good variety of birds to see, such as White-bellied Antbirds, manakins and tyrant flycatchers. The area is also a good place to see the King Vulture.

Much of the coast is covered by bus along the Santa Marta to **Riohacha** route; however, hiring a car is almost essential for a trip to the San Lorenzo ridge, and a vehicle does allow greater freedom to explore the arid country to the east towards Riohacha.   ❑

BELOW: this caterpillar's spikes are poisonous.

Map on page 152

FROM BOGOTA TO THE AMAZON

The Andean moorlands offer excellent walking, with plenty of opportunities for bird-watching, while lower down lies some of the most impressive tropical forest in the Amazon

The northwest limit of the Amazon basin is marked by the eastern slopes of the Colombian Andes. The watershed is high up above the treeline. To take a complete view of the Colombian Amazon, you should first of all pack some warm clothes and head for the Andean moorlands or *páramo*, to find out where it all begins.

The nearest *páramo* to **Bogotá** is in the **Parque Nacional Chingaza** ❽, three hours' drive from the city center. (Check locally to see if the park is open since it may be closed for security reasons.) The Park straddles the eastern *cordillera*, providing on one slope the source of Bogotá's water supply and on the other streams that head eastwards to the great plains of the Llanos. This water eventually leads to the Río Orinoco, not the Amazon, but **Chingaza** provides an excellent example of the *páramo* landscape and is certainly the most accessible of the sites in the eastern Andes.

Páramo is a distinct type of habitat found in the northern Andes from Venezuela south to Ecuador and parts of Peru. It is open moorland, cold and wet, with tall grasses, characterized by species of composites of the genus *Espeletia,* tall hoary-leaved plants with yellow flowers. The *páramo* starts at about 3,400 meters (11,000 ft) above sea level. Above 4,000 meters (13,000 ft), the vegetation is dominated by grasses, although some *Espeletia* persists. At lower altitudes and in sheltered areas are patches of temperate woodland with rather stunted trees, heavily laden with mosses and other epiphytes. Out on the open moorland are boggy areas and small lakes.

It is excellent walking country, although newcomers should not over-exert themselves because of the risk of altitude sickness. The climate is variable and can change dramatically in the course of a few hours. If the sun is shining and there is little wind, a T-shirt and pants will be sufficient clothing. However, with low cloud and a strong wind or rain, it can get very cold, so visitors are advised to carry a waterproof and several layers of clothing. At night, the temperature may drop below freezing.

Within the park is a large reservoir that serves Bogotá, and just outside a cement works processes limestone mined within the park. However, the park is so large that visitors can enjoy their surroundings oblivious to this industrial feature. Several well-marked tracks lead off the principal road to the reservoir, and these provide excellent hiking trails. Small paths also run off the tracks, but walking through the moorland vegetation itself can be very tiring.

Hummingbirds

On the *páramo*, probably the most spectacular birds are the hummingbirds. Chingaza has a good hummingbird community, including the Black-tailed Trainbearer with a tail almost twice as long as its body. The Bearded Helmetcrest is quite common at the higher altitudes and is a regular visitor to *Espeletia* flowers, often feeding from a perched position rather than hovering. This may be a strategy to conserve energy. Unusually for a hummingbird, it can also be seen walking on grass tussocks, making short flights to catch insects.

The Great Sapphirewing also visits *Espeletia* or terrestrial bromeliads. One of the largest hummingbirds (at over 16 cm/ 6 inches long) it hovers with rather slow, bat-like wing-beats. Most extraordinary of all is the Sword-billed Hummingbird, with a long, straight bill (as long as its body). It occurs in some of the woodland patches that border the *páramo* at lower levels.

The most widespread bird is the Great Thrush, which occurs from the *páramo* well into the temperate zone: indeed, it is a

LEFT: a Brazil nut tree emerges from the canopy.

common bird of the Bogotá suburbs. The *páramo* and shrubby areas also support furnariids such as Many-striped Canasteros and White-chinned Thistletails, and seed-eaters such as the Plumbeous Sierra-Finch.

Some of the most colorful birds are the Mountain-Tanagers. You should expect to see the Scarlet-bellied Mountain-Tanager, a striking black and red bird with a sky-blue rump. The woodland edges provide opportunities for seeing small bird parties including Masked Flower-piercers, Golden-fronted Redstarts, White-throated Tyrannulets, tanagers and Rufous-browed Conebills (a Colombian endemic).

While watching the undergrowth and trees for small birds, it is important not to ignore the sky. Black-chested Buzzard-Eagles are impressive broad-winged raptors that take to the wing mid-morning to soar high above the moorland. Look out also for the Andean Condor. This bird was formerly extinct from this part of the Andes, but MMA and San Diego Zoo have initiated a reintroduction program.

Eastern Colombia

Eastwards, the *páramo* gives way to the Andean foothills, thence to the plains of the Llanos to the north and the Amazon forest to the south. Much of the forest of the eastern base of the Andes has been cleared.

The southern portion of eastern Colombia comprises a flat tract of forest, the relief interrupted by the massif of the **Serranía de la Macarena ❾**. This isolated range is over 100 km (60 miles) long and, in places, 25 km (15 miles) wide, reaching an elevation of 2,500 meters (8,200 ft). It is a relic of the mountain ranges that existed in South America before the Andes and is an area of high biological interest, much of it still unexplored. It is estimated that over 450 species of bird inhabit the area, including several species of cracids as well as a rich primate diversity. It has a spectacular landscape of cliffs, waterfalls and forest.

Access to the National Park is difficult, but the biggest obstacle is the guerrilla activity in and around it. At present, foreigners are strongly discouraged from

BELOW: Chingaza National Park.

Map
on page
152

traveling to La Macarena and no one should consider making a journey without taking up-to-date, reliable advice from MMA.

To Leticia

The finest introduction to the Colombian Amazon takes place on board the 1¼-hour flight from Bogotá to **Leticia ❿**, the capital of Amazonas province, on the northern bank of the River Amazon itself. After passing the eastern Andes and the Macarena mountains the plane heads southeast, crossing the equator, and passing over a great sea of forest. An unbroken canopy of trees extends to the horizon. It is then that the true enormity of the Amazon rainforest reveals itself.

The plane lands at Leticia in the early afternoon. The temperature outside will be well into the 30°C (86°F) and the climate very humid. Crowds of locals wait to greet arrivals and everywhere there is the rhythm of Latin and Brazilian music.

Leticia is a small administrative town of just 23,000 inhabitants. It sits in the south-east corner of the part of Colombia known as the "trapezium", a curious piece of territory sandwiched on three sides by Peru and Brazil. It is well worth spending one or two nights in Leticia.

The waterfront is the hub of life, a bustling center of commerce to which the local communities come to sell vegetables, fruit and fish. From one of the terrace bars, you can relax and watch the comings and goings of the traders, traveling in long, narrow canoes with simple outboard motors. There are floating houses, some of them acting as gasoline stations for the river craft, or as mooring places for larger boats. Across the river lies the frontier of Brazil and Peru, whereas just downstream is the Brazilian border town of **Tabatinga ⓫**.

It is only a few minutes' walk into Tabatinga, and there is great interchange between the two towns. Colombian currency is acceptable in Tabatinga and most people understand and speak Spanish. Many Colombians do their shopping there, taking advantage of favorable prices for many

BELOW: a tree frog with its calling pouches inflated.

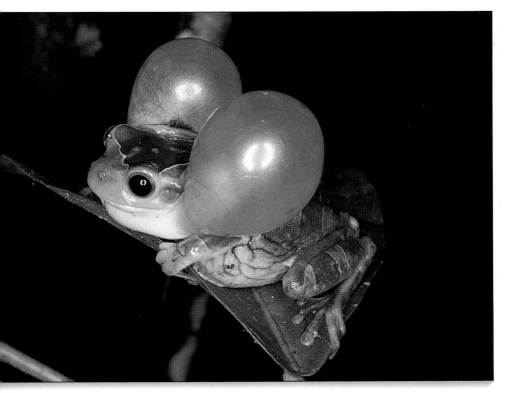

consumer products brought upstream by boat from Manaus, and the Brazilians are certainly happy to accept Colombian currency, which is stronger than their own.

From Tabatinga it is possible to travel by boat to Iquitos by "luxury" cruise once or twice a week, taking three days, or on regular *rápido* services that take only eight hours; and to Manaus (leaving twice or three times a week and taking four day, although schedules can be unreliable). However, Leticia and the Colombian Amazon can provide an excellent and very inexpensive base for exploration.

It is possible to enter forest from Leticia itself, although a 10-km (6-mile) hinterland, mainly given over to ranches, has to be crossed first. This can be done by taxi. A far better and more satisfying way of seeing the forest is to travel upstream and stay at the **Parque Nacional Amacayacú** ⓬. This involves a boat journey of 60 km (40 miles), which can be done by making an arrangement beforehand with MMA. Failing that, boats can be hired from the water-front, or a ticket bought for one of the public boats which ply the Colombian side of the river two or three times a week. This is the cheapest way of making the journey and perhaps the most interesting, although it is also the slowest, taking anything between six and eight hours.

The Amacayacú National Park is a lozenge-shaped area in the center of a trapezium, reaching down to the Amazon between the tributaries of the **Matamata creek** and the Amacayacú river. At the mouth of the Matamata is the Visitors Center, which has facilities to accommodate up to 30 visitors, although very rarely are there more than a handful of people staying. The center has a restaurant, a museum and library, craftshop (run as a cooperative venture between the park and the Tikuna communities living in and around the park) and meeting room. There are also facilities for scientists. including living quarters and a laboratory. The site is the main administrative center for the park with MMA offices and accommodation for park officers.

BELOW: a Blue-backed Manakin.

Map
on page
152

Three other stations elsewhere in the park provide bases for officers. All are in radio contact with one another.

The Visitors Center is situated on the riverbank with an excellent view across the river to **Mocagua Island**, one of a series of large mid-stream islands. Connecting the Center and the Park offices is an elevated walkway, 3–4 meters (10–13 ft) off the ground. This is to allow access during the four or five months of the year when the river level is high enough to flood adjacent low-lying ground. This period usually starts in January and is a response to heavy rains far away in the Peruvian catchment.

During this period a belt of forest along the riverbank, about a kilometer (0.6 miles) wide, is flooded. This seasonally flooded forest is known as *várzea*. In parts of Peru, the area of *várzea* can cover many hundreds of square kilometers. During the inundation period, access into the forest is possible only by dugout canoe, which can be hired at the center.

The elevated walkway provides an excellent vantage point for bird-watching in the environs of the center. This region has secondary vegetation, now allowed to regenerate. The more open aspect of this woodland provides a wonderful opportunity to become easily acquainted with some of the birds and butterflies. Indeed, spend part of the day here. It offers the best views of parrots and macaws, particularly in the early morning and late afternoon, when birds are making flights between their roosting and feeding areas.

Late afternoons, in particular, are very rewarding, when Plum-throated Cotingas, Umbrella-birds and Bare-necked Fruit-Crows may be observed and, as dusk descends, Short-tailed Nighthawks, Bat Falcons and Common Potoos.

The best time of the year to visit this area is from late July to early September. This is the flowering season of the *Erythrina* trees, whose peach-orange blossoms attract large numbers of hummingbirds, troupials, caciques and species of parrots

BELOW:
a Red-legged
Honeycreeper.

such as the Dusky-headed, Tui and Cobalt-winged Parakeets.

Northwards from the park offices is a path which leads through a kilometer of *várzea* forest, before rising somewhat into *terra firme* forest, which never floods and is the classic diverse lowland Amazon rainforest. The *várzea* has a fairly open ground layer and the trees are bare of epiphytes for the first 2–3 meters (6–10 ft), marking the height of the annual flood. There are also frequent gaps caused by natural tree falls, which fill rapidly with shrubby thickets. Although not as faunistically or floristically rich as the *terra firme,* this terrain nevertheless hosts a number of species which are rarely if ever found in *terra firme* forest. These include a number of antbirds such as the Bare-spotted Bare-eye, a "professional" ant-swarm follower whose nasal call is a distinctive sound.

As the trail rises into the *terra firme*, there is a side path leading to a platform overlooking an open, vegetation-choked swamp. This is an excellent place to watch Horned Screamers, Yellow-tufted Woodpeckers, Bat Falcon, and a variety of parrots and macaws. The *terra firme* forest provides the most challenging birdwatching of all and is the area where time spent will pay off in terms of the number of species seen.

Several species of monkey can be seen in the park, including Squirrel Monkeys, tamarins, sakis and titis. Look out for the puncture marks of sap-tapping on some of the tree trunks; these are made by Pygmy Marmosets. Visitors will almost certainly see agouti, various squirrels and other small mammals. Tapirs, peccary, deers, otters, Lesser Anteaters, two species of sloth and cats including jaguar, jaguarundi and ocelot, are all known in the park, although luck is required to see any of these. The butterflies are spectacular, especially the huge Morpho butterflies which patrol sections of the path with slow, heavy wingbeats.

The path runs for about 12 km (7 miles) to the Amacayacú river, and there are numerous side trails. Back at the park offices, there is a short path leading to a huge *Ceiba*

BELOW: a Passion Vine Butterfly.

Map on page 152

tree bearing a ladder. Scaling this, it is possible to climb 45 meters (150 ft) to the crown of the tree where it emerges from the canopy, affording breathtaking views of the canopy and across the Amazon river to Mocagua Island.

You can arrange with a guide from the neighboring village of Mocagua to visit the island. This is made up of a series of ridges or *várzea* forest with swamps and open water between them. There are also two large lakes. On the island it is possible to find the *Victoria Amazonica*, the largest water lily in the world with a leaf diameter of over 1.5 meters (5 ft). The island also supports a population of the extraordinary hoatzins and is also an important roosting area for parrots and a range of distinctive birds not readily found on the mainland. In addition, the swamps contain remnant populations of the threatened Black Caiman, while the lakes are the home of the world's largest freshwater fish *Arapaima gigas* or Pirarucú, which can sometimes be seen coming to the surface to gulp air.

Fifteen kilometers (10 miles) upstream of Amacayacú is the small town of **Puerto Nariño** ⓭. It is accessible only by boat and has about 4,000 inhabitants, a hotel, shop, restaurant and clinic. It marks the entry to the **Loreto Yacu** river, along which the extensive system of the **Taropoto lakes** can be explored. Again, arrangements can be made through MMA for hiring a boat, or through the hotel at Puerto Nariño. Not only are the lakes very beautiful but they also provide the best opportunities to observe both species of freshwater dolphin that occur in the Amazon basin, as well as a number of waterbirds, raptors and macaws.

During a two-year study by the British Ornithologists' Union, more than 490 bird species were recorded, almost all of them around the Visitors Center and along the first few kilometers of the forest path.

The best period to visit is between mid-July and September, when there is less rainfall, the trails are not flooded, and the *Erythrina* trees are in blossom. ❏

BELOW: a Swallowtail Moth.

Map
on page
172

VENEZUELA

*With some of the largest stretches of virgin
rainforest in the world, Venezuela is a prime
destination for wildlife tourists*

More than 60 percent of Venezuela's territory is untouched by the modern world, and pristine forest is plentiful since over 10 percent of the country is protected in national parks.

Venezuela offers the visitor three very different regions: **the Orinoco-Amazon basin**, **the Caribbean Coast**, and **the Andes**. The Caribbean Coast and the Andes are not closely related to the Amazon biogeographic region. But visitors may wish to explore these areas from **Caracas**, where most international flights arrive from Europe and the US, before moving south.

The coast and islands

Venezuela has 2,200 km (1,400 miles) of coast, in addition to its innumerable islands, which shows a diversity of scenery hard to find elsewhere in the Caribbean. Two hundred kilometers (125 miles) west of Caracas, the mangrove forests and coral reefs of **Parque Nacional Morrocoy** ❶ have white sandy beaches, a diverse bird fauna, although in recent years there has been a die-off of the coral. Three hundred kilometers (185 miles) to the east of Caracas lies **Parque Nacional Mochima** ❷, a windprotected bay containing tiny islands with mangrove forests and rocky coasts alternating with small sandy beaches. Mochima has a large variety of sea fauna, with good fishing, and excellent beaches for camping.

The most spectacular Venezuelan islands are undoubtedly the **Islas Los Roques** ❸, a national park and located some 150 km (95 miles) north of Caracas. They have lobsters, Botutos *(Strombis gigas)*, coral reefs, a huge variety of fish, and some sharks in their waters. There is supereb diving here and their sandy beaches alternate with the coral reefs, forming a paradise environment. The Brown Pelican, Masked Booby, Brown Booby, Magnificent Frigatebird, and others inhabit some of the many islands. The

small fishing village at **El Gran Roque**, next to a small airstrip, offers some facilities including lodging and boats.

Paria, **Araya** and **Paraguaná** are three peninsulas on the Venezuelan coast. **Paraguaná** ❹, west of Caracas, is mainly arid, separated from the mainland by a strip of desert with enormous nomadic dunes. The single, central, volcano-like mountain has some endemic forests, but the peninsula is predominantly inhabited by goats and migratory birds. These lands are rich in archeological remains, and, in some areas, dinosaur fossils and those of gigantic turtles and crocodiles are being excavated.

Three hundred kilometers (185 miles) to the east of Caracas, **Araya** ❺, also a desert peninsula, has what was once the greatest salt-mine on the continent. In contrast, Paria ❻, some 100 km (60 miles) east from Araya, guards dense forests and a national park, where rivers flow from the forest to the sea. The forest at Paria has fauna and flora similar to the Orinoco-Amazon basin and that found on Trinidad and Tobago, although some endemic species do exist here.

The coastal mountains

The mountain range of northern Venezuela, facing the Caribbean Sea, is extraordinarily rich in tropical cloud forests. **El Avila**, **Guatopo** and **Henri Pittier** (or **Rancho Grande**) are three of the national parks that guard some of the diversity in flora and fauna of this area. **Parque Nacional El Avila** ❼, limiting the expansion of Caracas to the north, and separating it from the sea, gives the city its distinctly linear form.

The eastern part of the park has been spared from frequent fires and therefore has an exuberant deciduous forest, while at higher altitudes on the mountains is cloud forest. Three important peaks may be climbed at **Parque Nacional El Avila**: **Avila proper**, **La Silla de Caracas** and **Pico Naiguata**. At the summits of La Silla

PRECEDING PAGES: the cliffs of Roraima. **LEFT:** road through the coastal mountains.

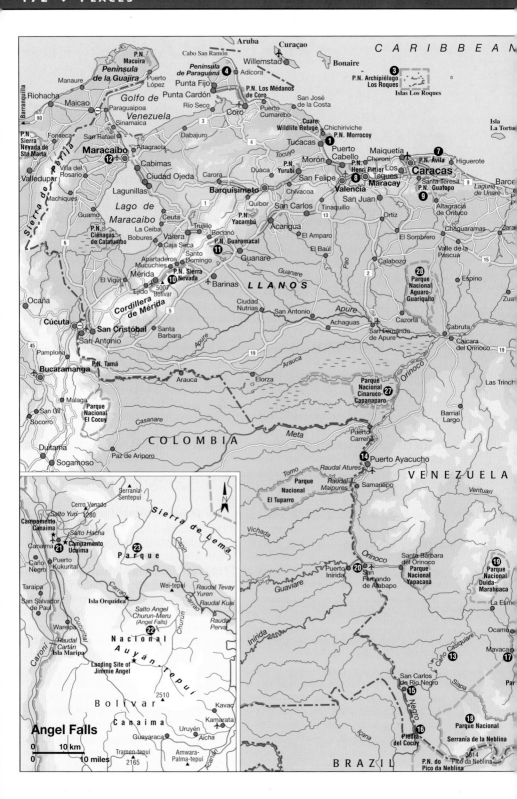

Angel Falls

0   10 km
0   10 miles

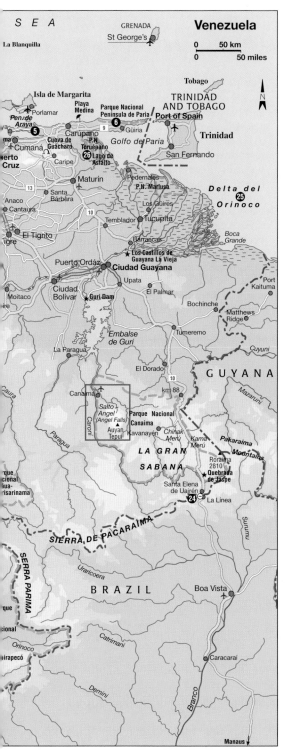

de Caracas, and Pico Niguata, *páramo* vegetation substitutes the cloud forest. Thus in a one-day walk from Caracas, a succession of tropical ecosystems is accessible, including savannas, rivers, forests and *páramos*.

Parque Nacional Henri Pittier ❽, 100 km (60 miles) west of Caracas, offers some facilities for the nature lover and exceptionally rich avifauna. It is the gateway of most migratory birds and insects entering central Venezuela and has one of the most botanically species-rich cloud forests of the world, it is rich also in snakes, insects, bats and, of course, birds.

Thirty kilometers (19 miles) further north, the ocean has excavated many small bays, where coconuts are still commercially grown. These bays are excellent sites to experience the Caribbean Sea. The most beautiful bays are **Turiamo**, **Cata**, **Cuyagua** and **Choroni**, which are easily reached by car from **Maracay**.

Guatopo ❾, 50 km (30 miles) south of Caracas, formerly an area of coffee plantations, was declared a national park in 1958. The forests here have suffered very little human intervention and thus have also very few footpaths. The road to **Altagracia de Orituco** traverses the park from north to south, reaching the Llanos at Altagracia.

The Venezuelan Andes

The Andes mountain chain originated in the tertiary period and is responsible for much of the continent's topography. The mountains rise up to 6,000 meters (19,680 ft) and extend from Alaska to Tierra del Fuego in southern Chile, although the range is known as the Andes only from Panama to Chile.

In Colombia, this mountain range splits, so that a secondary range extends to the east and into Venezuela. The mountains here reach a maximum height of 5,007 meters (16,420 ft) at **Pico Bolívar** in **Estado Mérida**. The Venezuelan Andes, as elsewhere in South America, are densely populated, but some national parks conserve the natural vegetation. The largest park, **Parque Nacional Sierra Nevada ❿**, encircles the highest peaks in the country including glaciers, spectacular *páramo* vegetation and high-altitude cloud forests. In this park, the highest cable car in the world takes visitors from **Mérida city**, at 1,631 meters (5,350 ft) to **Pico Espejo**, at 4,756 meters (15,600 ft) in about 60 minutes, covering a strip of 12.5 km (8 miles).

The dominant flora are *Frailejones* (*Espeletia* sp.), which expose hairy and succulent leaves, protected from frost, to the sun. The plants are very diverse with many endemic species, so that many *páramos* or high-altitude savannas (from 2,500 to 3,000 meters/8,200–9,800 ft) have their own endemic species of *Espeletia*. Other national parks are **Páramo Piedras Blancas** and **Páramo El Aguila,** near Mérida city, and **Páramo Guaramacal** ⓫, near **Boconó**, Estado Trujillo.

The Andes mark the limit of Orinoquia-Amazonia. Thier waters flow south and east into the Amazon or Orinoco rivers, while to the north or west, they flow into the Caribbean or the Pacific Ocean. The fauna of the Andes is not very rich but peculiar. The Spectacled Bear, near extinction, still inhabits some of the Andean forests. The flora and the panoramic views are the most conspicuous features of the Andes; however, small colorful mountain villages also form part of the beautiful Andean scenery.

Perija

Between Venezuela and Colombia, a 300-km (190-mile) long mountain range extends across the center of the **Perija peninsula**. This peninsula, west of **Maracaibo** ⓬, has mangrove forests, deserts with xerophytic vegetation, and mountains with peculiar forests. This area is proposed as a biogeographical refuge for fauna and flora, particularly for the Spectacled Bear.

Explorations in the area are dangerous because of continuous fighting beween drug-traffickers and the Colombian and Venezuelan military. The resultant isolation of the area helps the local peoples maintain many of their traditions. The Añu Wayu, Yukpa and Bari still preserve their language and customs, in spite of contact with *criollo* culture at markets in **Machiques**, **El Rosario, El Mojan** and **Maracaibo city**.

BELOW:
Mérida, tucke
into the Andes

Meeting the Orinoco

Map on page 172

The Amazon river is connected to the third largest river in South America: **the Orinoco**. The Orinoco starts at **Serra Parima**, 1,074 meters (3,520 ft) above sea level, but by the time it reaches **Mavaca**, about 200 km (125 miles) from its source, it has dropped to an elevation of 250 meters (820 ft). At about halfway from its source, at **Puerto Ayacucho**, some 1,000 km (620 miles) downstream, it has dropped only another 150 meters (490 ft).

The **Río Negro**, 180 km (110 miles) from **Ocamo**, runs at the same elevation as the Orinoco, but separated from it by swampy flatlands covered with evergreen tropical rainforest. These conditions explain the existence of an unusual hydrological feature – the **Casiquiare Channel** – which connects the Orinoco with the Amazon. The 354-km (220-mile) long "**Caño Casiquiare**" ⓭ was seen in 1744 by Jesuit missionaries, and was admired and described in detail by Alexander von Humboldt during his 1800 Orinoco expedition.

The Casiquiare flows from the Orinoco to the **Río Guainia**, which becomes the Río Negro after joining with the Casiquiare at **San Carlos de Río Negro**, which in turn flows into the Amazon at Manaus. The Casiquiare occasionally flows from the Río Guainia to the Orinoco. Geologists have still to uncover the channel's history, but it is probable that, at some point, the Upper Orinoco flowed – or will flow – completely to the Amazon, and that the lower Orinoco may start with the **Río Cunucunuma**.

Thanks to the Casiquiare, the Orinoco and Amazon basins are connected, enabling plants and animals to disperse, making the two river basins a single macro-ecosystem: the Orinoco-Amazon basin. It is possible to sail from the **Orinoco Delta** to the Amazon, via the Upper Orinoco, over to the Río Negro by the Casiquiare, and then continuing to the Amazon. One obstacle stops commercial river traffic between the two rivers' mouths: the rapids of **Atures** and **Maipures**, which divide the upper from the lower Orinoco at Puerto Ayacucho.

LOW: snow the *páramo.*

The Orinoco basin, which covers nearly one million sq km (386,000 sq miles), is divided into four distinct geological and ecological areas: the **Venezuelan Amazonia** and the **Upper Orinoco**, including the Casiquiare and Río Negro; the **Guayana Highshield** at the south and east of the Orinoco; the lowlands or Llanos to the north and west; and the Orinoco Delta, 2,150 km (1,340 miles) from the spring.

The Casiquiare and the Upper Orinoco

True tropical lowland rainforest is found in Venezuela only around the Casiquiare and Río Negro. **Puerto Ayacucho** is the most important gateway for travelers, scientists and missionaries to the state of Amazonas. This state has nearly 6.3 million hectares (15.5 million acres) protected by the Ministry of the Environment, but, most of the region is protected by its sheer remoteness. An irregular commercial flight or a more reliable charter connects Puerto Ayacucho to **San Carlos de Río Negro** , a small town at the mouth of the Casiquiare and Guainia rivers, start of the Río Negro. The flight takes over one hour and crosses 400 km (250 miles) of largely undisturbed forest, one of the few remaining areas of the world that has resisted human invasion. Travel in the area is either by air or the river.

Exploration by foot through the jungle is practically impossible, for both humans and large animals, which may explain the absence of large mammals in these forests. Although jaguar, deer and tapir inhabit the region, they are very rare and much more common in the unforested Llanos. Walking is restricted to a few earth paths used by the locals, who use them to tend their *conucos* or small farms; they establish them in the jungle, cultivating them for five to ten years, and then letting the forest reclaim them.

Indigenous peoples and *criollos* cultivate mainly banana and manioc, but maize, tobacco and a series of native tubercles and roots are commonly found in *conucos*. With the exception of the Yanomami, all tribes have mastered boat (dugout canoe) building

BELOW: a well camouflaged stick insect.

Map on page 172

and river navigation. The Yanomami inhabit the southern borders of this lowland forest, up to the mountains south of the Casiquiare and to Sierra Parima, where they maintain footpaths connecting them with other Yanomami in northern Brazil.

Insects and birds dominate this rainforest. Large swarms of noisy macaws and parrots glide over the forest, while clouds of mosquitos protect this sanctuary from most visitors. Among the animals to be seen are toucans, Guianan Cock-of-the-Rock, hummingbirds, parakeets, curassows, and mammals such as tapir, paca, jaguar, ocelot, and oncilla, and more than a 100 kinds of bat. The **Siapa**, a tributary of the Casiquiare, has, until now, avoided complete exploration; the first expeditions were carried out only in 1986 and 1989. The upper Siapa, not accessible by boat, is the last refuge for the Yanomami (or Waicas) escaping the attentions of Afro-European colonizers in Brazil and the Upper Orinoco.

The area can be accessed from the village of San Carlos de Río Negro, from which river trips using a *curiara* (dugout canoe) or *voladora* (aluminum speed boat) can be arranged. The *curiaras* are carved from a single tree trunk, which is then further opened with fire. The few crevices formed during the drying and processing of the trunk are then sealed with natural tars.

At San Carlos, the indigenous inhabitants are mainly Bare, Corripaco and Baniwa, all three with strong Spanish and Brazilian influence. Yanomami are rarely seen in San Carlos, and are more commonly seen on the Siapa river, a day tour from San Carlos. Trips from Puerto Ayacucho (Puerto Samariapo) may take over two weeks to reach San Carlos, and another week to Manaus.

San Carlos is near the frontier of three countries. Just across the Río Negro is Colombia, while Brazil is a few hours away by *curiara*. The **Piedra del Cocuy ⑯**, a day trip from San Carlos, is an impressive granite stone or *laja*, marking the common border of Brazil, Colombia and Venezuela.

At the other end of the Casiquiare is **La Esmeralda**, a village inhabited by Catholic

BELOW: a Two-striped Forest Viper *Bothriopsis lineata).*

missionaries and Yecuana (also called Maquiritare), at the border between the Yecuana and Yanomami. Upstream on the Orinoco are **Ocamo** and **Mavaca** ⓱, where Catholic missions have interfered with the Yanomami for nearly 200 years.

Places south of Mavaca, accessed only by air, are where the notorious Protestant New Tribe Mission replaces the Catholics. After Mavaca, strong rapids make river transport on the Orinoco practically impossible. In fact, the spring of the Orinoco has been rarely visited since a Venezuelan expedition first went there in 1951. Today, access to the area is further hindered by strict regulations, which control visits by foreigners in order to protect the Yanomami from devastating illnesses like common flu, malaria and syphilis. Illegal gold mining by *garimpeiros* penetrating the area from Brazil, is the major cause of environmental destruction. It is difficult to control, and new illegal *garimpeiro* camps with small airstrips are frequently discovered and dismantled in and around Sierra Parima.

South of the Casiquiare is one of the most pristine and isolated parts of the world. The most spectacular feature is the **Parque Nacional Serranía de la Neblina** ⓲, covering 13,600 sq km (5,250 sq miles) on the Venezuelan part of the most southeasterly and, at 3,000 meters (9,840 ft), the highest *tepui* (sandstone table mountain) in the continent. Neblina, a Precambrian *tepui* mountain is accessible by helicopter or a week-long expedition from Brazil.

The area has many other impressive *tepui* such as **Aracamuni**, granite monuments such as **Aratitiyope**, mixed granite and sandstone formations like **Tapirapeco**, and tertiary mountains, such as the **Sierra Parima**. Situated at the source of the Orinoco, the last is a beautiful, non-*tepui* mountain which emerges from the forest to reach a height of 1,500 meters (4,920 ft). Its climate is excellent and it is the home of most of the Yanomami. Access, however, is limited by the military and government.

West of the village of La Esmeralda, **Duida-Marahuaca National Park** ⓳

BELOW: many waterfalls interrupt the forest rivers.

Map on page 172

protects three practically unknown *tepuis*: **Duida**, **Marahuaca** and **Huachamakare**. The Duida was first explored in 1928–29 by the Tyler Duida Expedition, whereas Venezuelan expeditions using helicopters visited the Marahuaca in 1975 and 1983–85. The steep vertical walls of Marahuaka and Huachamacare resisted climbers until 1984 when a Venezuelan team reached both summits for the first time without the aid of helicopters. The Duida is easier to reach as it has many gentle slopes, one near to La Esmeralda on the Orinoco river, which makes it possible to reach the plateau of the mountain in a one-day excursion. These three *tepuis* are sacred places in Yecuana mythology, in which Marahuaca symbolizes the tree of life, the origin of all species on earth.

Any exploration of this area has to start in Puerto Ayacucho. The city has a good airport and is a melting pot of northern Amazonian peoples, Africans and Indo-Europeans. The indigenous people of the area are the Piaroas, who still inhabit the surroundings. Puerto Ayacucho offers good excursions to the **Río Cataniapo** (inhabited by Piaroas), to the "**Tobogán de la Selva**" (a natural swimming pool), or further south to **San Fernando de Atabapo** [20], where four important rivers meet: the **Atabapo** (black water river), the **Inirida** and **Guaviare** (white waters from Colombia) and the **Orinoco**. A river excursion from San Fernando de Atabapo to **Santa Bárbara del Orinoco** takes visitors to the mouth of the **Río Ventuari**, a river running through beautiful forests and savannas, inhabited in its upper part by the Yecuana.

Visitors to the area will note the color of the black and the white water rivers. Black water is a clear and clean, but acid, water (pH < 5). The rivers are reddish if not very deep, and are black with white foam over 1 meter (3 ft) deep. White water is opaque, brown, and filled with sediments.

White waters carry large amounts of clay from the Andes and the western Llanos, which makes up most of the sediment of the Orinoco river. Black waters come from

BELOW: female Marsupial Frog (*Gastrotheca ovifera*) releases froglets from her pouch.

the Guayana highlands and the *tepuis*, owing their color to tannins and other plant chemicals, which produce high acidity. The lack of nutrients in the sandstone, exposed to over two billion years of rains, make it difficult for bacteria and other biodegrading organisms to survive in these waters, and allow the chemicals to accumulate.

The Guayana Highshield

The *tepuis*, made famous by Arthur Conan Doyle's *The Lost World*, are one of the continent's greatest spectacles. These are sandstone table mountains with long vertical walls. Rivers from the plateaux form gigantic waterfalls, the highest in the world. The flat summits are covered with endemic flora and fauna, including four carnivorous plant genera, many birds, amphibians and reptiles, and exceptional arthropods, including water crickets and gigantic non-poisonous red spiders. The *tepuis* are difficult to visit without a helicopter – not easy to get in the area – and visits are restricted by Inparques (the institute responsible for national parks).

Even so, every year many tourists come to the area, leaving litter on the trails which threatens to disturb the fragile ecology of the area. (See www.thelostworld.org for more information.) **Roraima-tepui** and **Kuke-nan-tepui** may be climbed on foot, with a guide, on a trip that can take from five days to two weeks. Tours can be booked in Caracas and guides arranged in Paraitepui, near San Ignacio de Yuruani. You can also admire the *tepuis* from **Santa Elena de Uairen**, southwest of Venezuelan Guayana, near the Brazilian and Guyanan border, and **Canaima**, at the center of the Guayana Highshield, accessible only by air.

Canaima ㉑, gateway to Angel Falls and with regular air connections, now has several places to stay including Campamento Canaima, the intimate Campamento Ucaima and Campamento Parakaupa, and some basic *posadas*. A three-day expedition navigating the **Río Carrao** upstream (June to October), or a 15-minute flight takes the visitor to see the highest waterfall in the world: **Churun-Meru** or **Angel Falls** ㉒.

BELOW: Pierid Butterflies gathering at a river bank.

Map on page 172

Here, the water drops 972 meters (3,188 ft) from the **Auyan-tepui**, where it forms a second smaller cascade, so that altogether the water falls over 1,000 meters (3,280 ft) into a spectacular jungle setting. Made known to the outside world by the North American pilot Jimmie Angel in 1923, the Pemon call the falls Churun-Meru, or the fall of the Churun river.

Canaima, like most places in the Highshield, is surrounded by savannas and forests. The average height of the lowlands of the Guayana Highshield is 800 meters (2,600 ft) above sea level, making the flora and fauna different from that found in the rest of the Orinoco-Amazon basin. All rivers here have crystalline black waters and spectacular beaches and scenery. They are the source of many tropical aquarium fish.

Canaima lies in the heart of **Parque Nacional Canaima** ㉓, a 30,000-sq km (11,600-sq mile) protected area including several of the more spectacular *tepuis*. Auyan-tepui, the biggest of these *tepuis*, carrying the Angel Falls, raises its steep walls near the village. The national park is a reservoir of the water resources which could supply most of Venezuela and neighboring countries for the foreseeable future. Its waters are also used to produce electricity at **Guri dam**, where the huge Embalse de Guri allows for excellent fishing.

The fauna is not abundant, although tapirs, armadillos, Giant and Tamandua Anteaters, deer, jaguars, turtles, snakes (some of them highly poisonous), lizards and frogs may be sighted with some luck. Birds and insects are plentiful and butterflies show exuberant colors and forms. The plantlife stands out, and the varieties of orchids (notably *Cathleya* and *Catasetum*), bromeliads, lianas, fungi and ferns will be the most delightful finds. Over 10,000 plant species are estimated to grow in this area, and most of them are known thanks to the efforts over 40 years by Julian Steyermark (1909–88), although many more species remain to be discovered. Rocky areas, savannas with views over the *tepuis*, and diverse kinds of forests cover the area.

ELOW: the ghshield has any dramatic ags.

Explorations of the Guayana Highshield are fairly recent; some of the first were carried out by Robert Schomburg, starting from British Guyana, in around 1910; and Auyan-tepui was first climbed and explored by a Venezuelan scientific expedition in 1956. At Canaima, the *Fundación Terramar* – a charitable private foundation – promotes research on nature conservation and experiments with diverse management techniques in search of a sustainable conservation plan, which should guarantee the preservation of this unique area.

Santa Elena de Uairén ㉔, founded in 1931 by Capuchin monks, is today the southern door to the Guayana Highshield. It is a center for tourists, miners and merchants from Brazil (15 km/9 miles away), Guayana and Venezuela, who reach Santa Elena by air or road. Two roads connect Santa Elena to the rest of the world, one coming from **Boa Vista** (Roraima–Brazil) 200 km (125 miles) away, and another reaching Santa Elena from **Ciudad Bolívar**, 600 km (370 miles) to the north.

Santa Elena offers trips to Roraima-tepui, to rivers and *tepuis* such as **Quebrada de Jaspe**, **Kavanayén**, **Río Aponguao**, **Kamoiran** and **El Pauji**, and to diamond and gold mining sites at Kilometer 88, **El Dorado**, **Las Claritas** or **Icabaru**.

Roraima-tepui and Kukenan-tepui are undoubtedly the most impressive features. The former was first climbed in 1884 by Everard F. Im Thurn, curator of the Museum of Georgetown, and H.I. Perkins. They started from Georgetown in Guyana, but the route they used for climbing Roraima is still not completely clear.

Roraima, of *Lost World* fame, is one of the most beautiful *tepuis*. It has an average height of 2,500 meters (8,200 ft) and forms rocky labyrinths containing lakes, cliffs and spectacular views. Its vegetation is characterized by the carnivorous plants of the genera *Heliamphora (Sarraceniaceae)*, *Droseras (Droseraceae)*, *Genlisea* and *Utricularia (Lenti-bulariaceae)*, and by *Odocarpus* trees, *Speletia*, orchids, etc. The flora of Roraima, and, to a lesser extent, that

BELOW: Angel Falls

Map on page 172

of Kukenan, are suffering heavily from tourism, and may disappear, like the quartz crystals which once completely covered the ground of several valleys on the *tepui*.

The savannas near and around Santa Elena extend over sandy soils covered with termite hills: the head of the termite soldiers can be eaten by biting them off from the thorax and separating them from the mandibles. They taste very refreshing and aromatic. The termites feed on grasses, concentrating nutrients in their hills that allow seedlings to start or expand the existing forests. Another interesting ecological feature is the existence of ant gardens, ant-plants and carnivorous plants: ants build nests on trees, carrying seeds of epiphytes which will grow only on these "garden-nests". Other plants, such as the *Melastomatacea Tococa*, form specific structures to attract ants, which inhabit them, feeding the plant and protecting it from herbivores. Ant-termite associations are also present, as well as endemic spiders, a variety of *Coleoptera*, and poisonous butterfly larvae which cause

a heavy haemorrhage on contact with the skin and may even cause death.

The area north of Santa Elena, leaving Canaima National Park, has been devastated by gold mining. Some mining towns on the road to **Puerto Ordáz** and **Ciudad Bolívar**, such as **El Callao**, **Tumeremo** and **Guasipati**, have Anglo-Caribbean influences, since laborers from Trinidad and Barbados were imported by an English mining company in the 19th century.

The city **Puerto Ordáz**, or **Ciudad Guayana**, was founded in 1961. Its heavy industry uses cheap electricity from Guri, water from and river transport on the Orinoco, iron deposits at **Cerro Bolívar**, and aluminum deposits at **Los Pijiguaos**. The environs of Ciudad Guayana, however, do have some nature sites, such as **Parque Cachamay**, **Parque Loefling**, **La Llovisna** and **Los Castillos de Guayana**.

Two navigable rivers, the **Río Caura** and **La Paragua**, offer excellent excursions into the south of **Estado Bolívar**, with views of the exuberant flora and some of the fauna.

BELOW: a luminescent caterpillar.

Special river camps for tourists are flourishing with the expansion of eco-tourism. Tourism and gold mining are the area's main private economic activities.

The Orinoco Delta

The **Orinoco Delta** ㉕ is a fan of alluvial deposits bounded on the south by the Río Grande and on the west by the Río Manamo, which branch off the main river at **Barrancas**, the apex of the Delta. From Barrancas it is about 180 km (110 miles) along the Río Grande to the Atlantic, and an equal distance along the Manamo channel to the ocean. Most of the vast tidal swamp of the delta is irrigated by channels, called *caños,* that form a network of navigable waterways across the entire delta.

The climate is hot (average 26°C/79°F) and humid (average 70 percent relative humidity) and, together with the fertile soils that are inundated each year by new sediments from the Orinoco, this gives rise to rich and exuberant forests. Most of these forests consist of red mangroves *(Rhi-zophora mangle)*. At the center of the drier islands, the mangrove dies and permits the invasion of useful palms such as *Moriche (Mauritia flexuosa)*, with its exquisite fruits, *Manaca (Euterpe* sp.) and *Temiche (Manicaria saccifera)*. This area is inhabited by a distinct tribe – the Warao – who possibly arrived from Florida, via the **Antilles**. The Warao live in wooden *palafites*, built over swamps or rivers, allowing them to escape the annual floods caused by the rising of the water level of the Orinoco during the rainy season, from May to October.

The forests are very dense, making the river the only means of communication. The forest is under pressure from rice plantations, water buffalo breeding, and the commercial exploitation of natural fibers. *Palmito*, the heart of certain palms, is commercially harvested, slowly decimating the natural palm populations. Many birds, especially macaws, parrots, and toucans, are illegally exported to French Guiana, Aruba, Curaçao or Trinidad, from where they are shipped to Europe or the United States.

BELOW: the endemic plant life has adapted to the poor rocky soils and damp climate.

Map
on page
172

Mosquitos are plentiful in the deltas, unlike in the Guayana Highshields, but the area has a beauty of its own. **Tucupita**, the main city in the delta, is best accessed by car since there are no longer flights from Caracas. From there, river expeditions go through the numerous *caños*, past small Warao villages, to the Atlantic.

The delta's waters are rich with life; plentiful fish, and various turtles and crocodiles inhabit the rivers and channels. Fishing requires knowledge of the different *caños*, as each *caño* has specific fish species, including river catfish (*Meglonema* sp. and *Cetopsis* sp.), pavon (*Cichla* sp.), and palometa (*Pygopristis* sp. and *Mylossoma* sp.). Insect and bird life is very diverse and little explored. Several mammals are found in the delta: some South American marsupials, monkeys and rodents are common terrestrial animals; freshwater dolphins, Giant River Otters, and manatees, on the brink of extinction, represent the aquatic mammals. At the mouth of the *caños*, the brackish, sediment-rich waters support many species.

The **Guanoco asphalt lake** (Lago de Asfalto) ㉖, north of the delta, is the largest asphalt lake in the world. Its black surface of over 2 sq km (21,000 sq ft), spiked with plant-islands, is surrounded by spectacular forests with rare flora. Access is difficult and takes three days on foot, or by mule, from **Guanoco** village, near **Caripito**.

South of the delta are little explored forests which are slowly being logged. The **Serranía de Imataca** separates the Orinoco basin from that of the Esequibo. This area has a reputation for sheltering numerous jaguars, but gold mining, extensive deforestation and poaching are taking their toll.

The Llanos

The plains north and west of the Orinoco, covering the central parts of Venezuela and Colombia, form an ecosystem rich in vertebrate fauna. The Llanos can be classified into the upper and lower Llanos. The lower Llanos are in the south of Colombia and Venezuela but mostly north of the Orinoco. They are floodplains with very few trees.

BELOW:
an endemic
Drosera from
the Highshield.

The upper Llanos, further north and nearer to the Andes and the Venezuelan coastal mountain range, have more trees and small circular forests called *matas*, which are thought to be initiated by collapsing leaf-cutting ant nests. The Llanos are best visited during the dry season (December to May), as dirt roads will be transitable and the fauna will be concentrated around rivers and lakes.

The most conspicuous feature of the Llanos are the bird colonies formed by thousands of Great Egrets, Scarlet Ibises, Wood Ibises, Jabirus, and small numbers of Great Tinamous, Little Blue Herons, chachalacas, currassows, hoatzins, etc. The rivers of the Llanos contain pavon (*Cichla* sp.), piranha, cachama (*Colossoma* sp.), and river catfish species (*Meglonema* sp., *Rhamdia* sp., *Cetopsis* sp.), as well as caimans *(Caiman crocodilus)*, iguanas *(Iguana iguana)*, anacondas and river turtles *(Podecnemis vogli)*. Caimans, called *babas* locally, lay their eggs in nests made of grasses near lakes and rivers. Sadly, they are commercially exploited under the supervision of the Venezuelan Ministry of Natural Resources. Manatees may still exist in the few nature sanctuaries of the Llanos, the national parks: **Aguaro-Guariquito** and **Santos Luzardo**, also called **Cinaruco-Capanaparo ㉗**, and the "**Esteros de Camaguan**" near **San Fernando de Apure**. Lesser Anteaters, Red Howler Monkeys, *Cachicamos* or Nine-banded Armadillos, common opossums, capybaras, White-tailed and Great Brocket Deer, and Brown-throated Three-toed Sloth are a few of the common mammals here.

The Llanos are sedimentary, formed mainly from sand brought down by rivers from the Guayana Highshield by the Orinoco and from the Andes, over the last few million years. The area has distinct dry and rainy seasons. In the dry season, water exists only in a few rivers and lakes, and the fauna concentrates around it. Savanna fires are common, and the Llanos may be crossed with or without dirt tracks. In the wet season, rivers emerge and large areas

BELOW: a Spectacled Caiman.

Map on page 172

are flooded, isolating many villages for two to five months. The animals disperse, making observation more difficult.

The Llanos harbor huge grasslands which are used for extensive cattle farming. The cattle are collected once or twice a year for veterinary inspection and left free to feed and breed without further human intervention. This kind of management has little effect on the natural ecosystems. *Chaguaramas* (royal palms) and colorful trees, most of them flowering during the dry season, give the region a picturesque appearance. However, intensive cattle breeding has changed some of the natural vegetation.

The most conspicuous mammal of the Llanos is the capybara, the world's largest rodent. Capybaras, locally called *chiguires*, have developed a complex social behavior allowing them to successfully defend themselves against jaguars, pumas and crocodiles. The piranha (also called Caribbe fish), present south of the Orinoco in less aggressive varieties, is present in most rivers and lakes in the Llanos, making bathing unwise.

San Fernando de Apure, **Maturín**, and **El Sombrero** are the best cities from which to access the southern low Llanos, the eastern middle savannas, and the central upper Llanos respectively. To the northwest, the Andean town of Mérida has excellent operators offering tours of the region, while San Juan de los Morros is another convenient gateway to the central upper Llanos. Flights go to most places, and ranchers often have no choice but to fly to their *haciendas* during the rainy season and even part of the dry season. Some of these *haciendas* are good places for nature lovers to stay. The owners of these farms have created small sanctuaries with minimal facilities and variable degrees of comfort, where the fauna of the Llanos can easily be observed.

Aguaro-Guariquito National Park ㉘ is a good place for watching the wildlife of the Llanos. Located between the two rivers, it has been protected from excessive hunting, and retains important populations of monkeys, caimans, deer, anteaters and dolphins. Visitors to the park must camp. ❑

BELOW: savanna of the Parucito river valley.

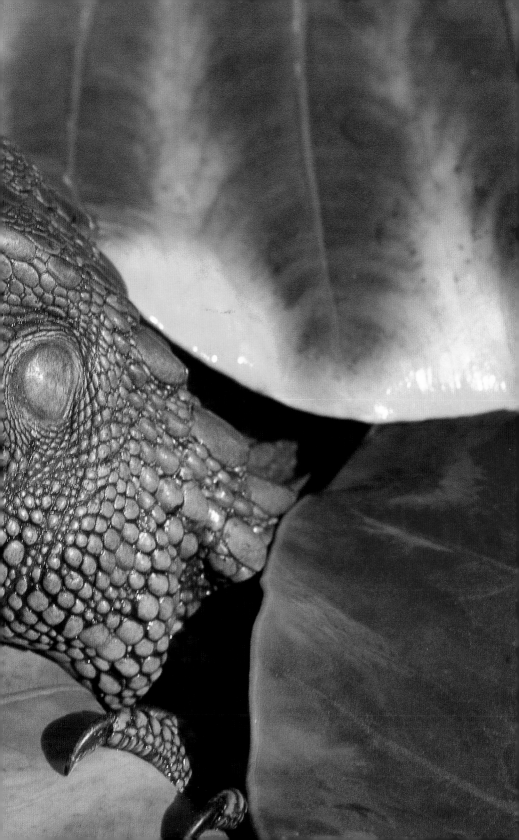

Map on page 204

THE GUIANAS

With most of the population living along the narrow coastal strips of these three adjacent countries, the interior has developed as a wonderful haven for wildlife

The three Guianas – **Guyana** (formerly British Guiana), **Suriname** (formerly Dutch Guiana) and **Guyane** (or French Guiana, a *département* of France) – vary little in their geography. A narrow strip of flat, swampy land runs along the coast, behind which are isolated savannas, marked by expanses of white sand, grassy plains and occasional sparse woodland. Further inland are extensive uplands whose higher points are separated by narrow, swampy valleys.

This central Guianese shelf, 250 meters (820 ft) above sea level on average, is made of ancient crystalline rock. Isolated inselbergs and granite peaks, formed by erosion of surrounding formations, rise abruptly to between 300 and 800 meters (985 and 2,625 ft). The weathering on bare rock near the summits and steep walls bears witness to countless downpours. The peaks are often carpeted with bromelia, and afford magnificent views of virgin forests covering the gently undulating landscape. Rising to the southwest is the **Guianese Plateau**, which reaches a height of 900 meters (2,950 ft) in the **Tumuc-Humac** mountains.

The highest point in the Guianas is **Mount Roraima ❶** (2,810 meters/9,220 ft), a red sandstone table mountain on the borders of Guyana, Brazil and Venezuela.

All the important rivers of the Guianas rise in the mountains and flow north to the Atlantic. The many rivers with their frequent rapids give the region its name: Guyana means "land of waters". Ten major rivers flow into the Atlantic along French Guiana's 345 km (215 mile) coast. In many cases rapids lie near the rivers' mouths, posing a barrier to travel into the interior.

The hill and mountain regions are still sparsely populated, and there are few roads leading into the interior. During the main dry season from August to November, and February to March, when the rainfall is low, the rapids become streams flowing through fissures in the rock.

In Guyana, many travelers visit **Kaieteur Falls ❷**, the world's highest single-drop waterfall, where the Potaro river plunges uninterrupted for 226 metres (740 ft) of its 250-metre (820 ft) fall. It is possible to see silver fox, tapirs, monkeys and ocelots in the surrounding unspoilt forest.

The streams and rivers are a hunting ground for two types of otter. The smaller, light-colored Southern River Otter fishes at night, while the gregarious, dark brown Giant Otter – which, with a tail of around 80 cm (30 inches) can grow up to 2 meters (6 ft) long – searches the river banks for fish, small mammals and birds during the day. **Karanambu Ranch ❸**, in the heart of the Rupununi Savannah in Guyana, is the home of Diane McTurk, famous for her work rehabilitating orphaned Giant River Otters. It is also a superb place for birding.

On the smooth-flowing lower rivers, where tidal influences make themselves felt, the vegetation is dominated by mangroves. Here, among the mangroves, live Four-eyed Fish *(Anableps anableps)*, which grow to about 20 cm (8 inches) in length and swim so close to the surface that their protruding eyes are half out of the water. A division in the eye at the waterline enables them to see above and below the surface at the same time. Because they have no tear glands, they must regularly dip their heads below the water to lubricate the upper part of their eye, which presumably serves to keep watch for airborne predators. It is easy enough to see them at night, since their eyes reflect the beam of a torch. The Four-eyed Fish is viviparous, that is, it bears living young, the eggs being carried in the body of the female until they hatch.

Several beaches in the Guianas are well-known laying grounds for marine turtles. Some of these are officially protected (including Wia-Wia and the Galibi Nature Reserve in Suriname, and Les Hattes and Aouara in French Guiana).

PRECEDING PAGES: a Guyanan Caiman Lizard *(Dracaena guilanensis).* **LEFT:** a Malachite Butterfly *(Siproeta stelenes).*

Shell Beach 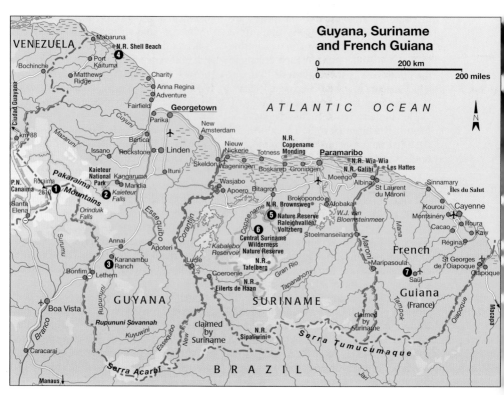, a protected stretch of Atlantic coastline in Guyana, is one of the best places to view turtles. Recently, the local Arawak family that manages the eco-friendly Shell Beach Conservation Camp, was awarded the Neotropic Award by Conservation International for their work in ensuring that turtles are safe to nest here.

Apart from four species of true marine turtles *(Cheloniidae)*, between June and July you might also find the giant leatherback *(Dermochelys coriacea)*, which lays its eggs at night. Up to 2 meters (6 ft) in length with a weight of 600 kg (1,300 lb), leatherbacks are the largest turtles in the world. They spend most of their lives in temperate and tropical seas, feeding mostly on jellyfish. Marine turtles leave the sea only to lay their eggs. Certain sections of the beach above the high-tide mark have been used for centuries. The females dig holes approximately 50 cm (20 inches) deep in which they lay around 100 eggs. They then cover up the hole and smooth the surface before returning to the sea.

Voltzberg Nature Reserve

The Guianan hinterland is home to extensive rainforest and its fauna. One reserve well worth a visit is the 560-sq km (216-sq mile) **Raleighvallen/Voltzberg Nature Reserve ❺**. A direct flight goes from **Paramaribo** to **Foengoe**, an island in the **Río Coppename**, in the reserve. A cheaper way of getting there is the four-hour drive to **Bitagron**, on the Coppename, followed by a four-hour trip up the river to **Foengo Island Lodge**.

As well as high rainforest, the reserve has less common environments such as mountain savanna and liana forest. It is home to the eight species of monkey found in Suriname, the Yellow-handed Marmoset, the Common Squirrel Monkey, the Weeping and Brown Capuchin, the Red-backed and White-faced Saki, the Red Howler Monkey and the Black Spider Monkey.

There are many other mammals in the Voltzberg region. Agoutis are common on the forest floor, as are the similar Acouchis. *Edentata* are represented by the Giant -

Map on page 192

Armadillo, Nine-banded Armadillo, Tamandua, Giant Anteater and two types of sloth. Among the larger animals are Collared Peccaries, tapirs, Jaguars and Ocelots.

The forest canopy is home to the brightly-plumed Red Fan, Orange-winged and Mealy Parrots, Scarlet Macaws, aracaris and toucans with their characteristic fluting call. Larger birds include the Black Curassow, Tinamous, the Marail Guan and the Grey-winged Trumpeter.

Another bird is the plain, gray Screaming Piha, whose whip-like "pi-pi-yo" cry is one of the most striking sounds of the rainforest. Of particular interest to the zoologist is the group courtship ritual of the Cock-of-the-Rock, a fairly common bird in the Voltzberg region. The males, which grow up to 35 cm (14 inches) in height, congregate at traditional grounds and perform theircourtship dance to attract the more dowdy females. Their plumage is bright golden orange, enhanced by a bonnet of feathers around the head. In the dance they show it off by springing into the air, shaking their heads and wings, spreading out their tails, making rattling noises with their beaks, and calling while producing a buzzing sound by vibrating their feathers.

In 1998, the government of Suriname undertook to protect one of the world's largest tropical forests. The new 1.6 million-hectare (6,200-sq mile) **Central Suriname Wilderness Nature Reserve ❻** covers about 10 percent of the country's territory. Much of the area was saved from logging. Conservation International secured private funding for management costs and is involved in developing sustainable agroforestry and non-timber products. One project seeks to use liana cane to make furniture and to extract its resins and oils for personal care products. Research is also made into medicinal plants in the reserve.

In French Guiana, there is no comparable reserve to the Raleighvallen/Voltzberg Reserve. Adventurous tourists with time for a trip to the interior may visit **Saül ❼**, where they will find a network of well marked paths through the rainforest. ❏

BELOW: an Orchid Bee (with pollen on its back) visits a local orchid.

Map
on page
198

TRINIDAD

*Connected by a land bridge to the Amazon
Basin during the last ice age, Trinidad's flora and fauna
has much in common with that of Amazonia*

Trinidad, the southernmost island of the Caribbean, supports a unique wildlife link between the Amazonian and the Caribbean biotic regions. Trinidad lies only 12 km (7miles) from the South American mainland at its closest point. However, 10,000 years ago a great deal of the Earth's water was locked up in ice, and sea levels were much lower than they are now. Trinidad was then connected to the Paria Peninsula of Venezuela. Wildlife of all kinds invaded Trinidad, and many species flourish there to this day. The rugged peaks of the **Boca Islands ❶**, submerged mountains that dot the channel between the two islands, are remnants of that connection.

Trinidad's biogeography is still profoundly influenced by South America. The close proximity of the two lands gives Amazonian species a reasonably easy opportunity to colonize the island. Competition between Amazonian and Caribbean species is intense, with new invaders recorded annually as others decrease and then disappear.

A powerful agent aiding immigration by Amazonian species is the Río Orinoco, whose outflow swirls northward like an immense muddy river past eastern Trinidad, depositing debris of all kinds on the beaches. Flood-borne trees from Venezuela serve as life rafts for wildlife, and they often root themselves once ashore.

As in the *tepui* region of Amazonia, elevation in Trinidad dictates the dominant vegetation. With increased elevation comes greater rainfall. The highest and thus wettest areas are the mountains of the **Northern Range**, which reach a height of 940 meters (3,000 ft) and are believed to be an extension of the **Andean Cordillera**. Some peaks receive more than 400 cm (157 inches) of rainfall annually, in sharp contrast to the low-lying areas in the west and southwest – areas of rainshadow that receive only 150 cm (60 inches) annually. The difference in available moisture results

in lush rainforests carpeting the mountains while thorn-covered scrub struggles to survive in parts of the west and on islands. At low to moderate elevations, there are grasslands, wet savannas, and semi-deciduous woodlands to be found. These habitats prevail where they have not been destroyed for sugarcane or rice cultivation or for the ever-increasing spread of settlements.

The "edge" effect, a phenomenon in which more species are found along interfaces between habitats than within habitats, is partly responsible for the diversity of Trinidad's wildlife. Trinidad lies at the northern edge of Amazonia and at the southern edge of the Caribbean. There are thus two "edges" at work, the first derived from interfaces between diverse habitats in Trinidad, the second from the interface between the Amazonian and Caribbean faunal regions. The combined effect yields a greater diversity of species in Trinidad than would normally be expected.

Agreeable lifestyles

For a small island, the variety and extent of habitats are remarkable, with montane elfin forest and rainforest, lowland rainforest, savannah, freshwater and saltwater swamps, freshwater lakes, ocean beaches, and open ocean, as well as a variety of cultivated areas. Perennially available fruits, flowers, and seeds from both Amazonian and Caribbean plants provide sustenance for resident species as well as for migrant birds from North and South America.

The birdlife is astonishingly diverse. Amazonian families are strongly represented, along with many Caribbean families and a few exotics. About 25 percent of the South American species present on Trinidad are found nowhere else in the Caribbean. The presence of Amazonian as well as Caribbean species results in the number of bird species in Trinidad being nearly twice that of any other Caribbean island.

Insect diversity, tremendous in Amazonia, is reduced on most Caribbean islands. With few niches to fill, competition quickly eliminates potential colonizers that are poorly suited to the environment and which lack adaptivity. Most studies of island biogeography find only about 10 percent of the number of species on islands as on similar mainland areas. Fierce competition among species in the few available habitats renders islands under-populated not only of insects for pollination: the species that survive and eventually colonize islands are often larger and heavier than their counterparts on the mainland, a phenomenon resulting from the negative survival value of being easily windblown out to sea.

In contrast to the more usual reduced diversity of insects on Caribbean islands, Trinidad's insect diversity closely resembles that of Amazonia, being among the richest in the world. For example, some 650 species of butterflies have been catalogued.

Many Trinidadian insects, such as Giant Katydids and Iridescent Orchid Bees, show unmistakable links to their Amazonian counterparts. Yet certain anomalies confound scientists. For example, populations in Trinidad of snail-killing flies which are known elsewhere only in Central America. Moreover, despite recent connections with the mainland, Trinidad has many endemic species. Such confusing distributional data prevent researchers from generalizing about the origin of groups of the organisms.

One striking aspect of Amazonian biology particularly conspicuous on Trinidad is the association between birds and a few species of mammals with ants. Ant-following birds such as antbirds, antshrikes, ant-tanagers, antthrushes, antvireos, and antwrens glean columns of army and leaf-cutter ants from the ground as they scour the jungle. The fact that army-type ants are lacking on other Caribbean islands may be due to a lack of tracts of rainforest large enough to host viable populations, or perhaps better adapted species immigrate in numbers too small to become established.

Amazonia is clearly the origin of the

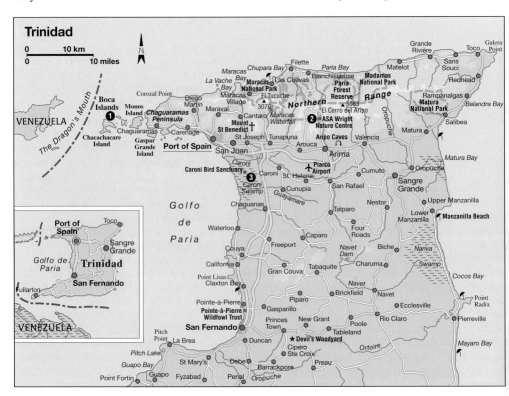

Map on page 198

larger mammals in Trinidad. Large mammals are uncommon anywhere in the Caribbean; in fact, howler monkeys, Agoutis, Prehensile-tailed Porcupines, Silky Anteaters, and Crab-eating Raccoons are found nowhere else in the Caribbean except for Trinidad. Unlike insects, plants, and small rodents, large mammals are unlikely to drift on a log, later to be washed ashore on some distant island. Many rodents, however, especially mice and rats, bear close affinity to Caribbean species.

Reptiles and amphibians

These are difficult to characterize here as either Amazonian or Caribbean. Lizards are abundant throughout the Caribbean, particularly on the drier islands. Similarly, iguanas, which are chiefly arboreal herbivores, are also distributed throughout the neotropics wherever favorable habitat is found, but are far more abundant in Amazonia. Other widespread groups that defy characterization as either Amazonian or Caribbean include anoles, basilisks, and geckos. Most

snakes also are widely distributed, with the exception of the Anaconda, which in the Caribbean region inhabits only Trinidad.

Trinidad has a higher human population density than most other regions of tropical South America, and therefore the concomitant disturbance of natural habitats. This does mean that sites are easily accessible, and have nearby accommodation catering to nature tourists. Travelers should ensure, however, that they have as little impact as possible on this pressured environment.

An excellent hotel in the Northern Range is the **ASA Wright Nature Centre** ❷, 12 km (7 miles) north of the city of **Arima**. Great observation of rainforest wildlife is possible in the estates surrounding the hotel, and in particular along the nearby road from **Arima** to **Blanchisseuse**, which cuts through the forests of the Northern Range. To see wetland wildlife, such as Scarlet Ibis or Boat-billed Heron, the place to go is the **Caroni Bird Sanctuary** ❸ at the Caroni Mangrove Swamp, south of Port of Spain, which has over 130 bird species. ❑

BELOW: a Rufescent Tiger-Heron fishing at the edge of a lake.

Map on page 204

ECUADOR

Although this is one of the smallest countries in South America, the most varied and extreme habitats manage to exist here side by side

With an area of just 283,520 sq km (110,000 sq miles), Ecuador has some of the best wildlife-viewing and ecotourism in the Amazon basin; and an atrocious record of oil exploration, with concessions still being granted to the national oil company within its parks.

The country is bisected, geographically and culturally, by the Andes, the backbone of Ecuador. They form two parallel ridges, often referred to as the Avenue of Volcanoes: over 30 snow-capped cones rise from these ridges, of which at least eight are still active. Between the ridges a fertile trough, divided into several basins, drains both east to the Amazon and west to the Pacific.

Ecuador's climatic zones are also complex. The cold Humboldt current to the west, and the steaming jungle to the east contribute to the clouds that perpetually condense around the Sierra. The eastern slopes are very wet, and long periods of mist and rain occur even in the relatively dry months of November/December and May/June. Average temperatures are determined by altitude, and severe cold at high altitudes greatly restricts what will grow.

Ascending the Andes from the Oriente

In lowland forest, buttress-rooted, columnar trees rise unbranched to a closed canopy some 30 meters (100 ft) above. A dense understory consists of flat-topped miniature trees (*Piper* sp., *Coussarea* sp., *Clidema* sp.) and prop-rooted palms (*Socratea* sp.) grasping the thin soil. Filmy liverworts and mosses are largely confined to the lower parts of smooth-barked trees. Leafy *Polypodium* ferns sprout from their trunks and compete with many species of climbers. Here, in the dry forest, bromeliads and orchids are relatively rare. Numerous monkey species inhabitat this area, including Pygmy Marmosets, Saddle-back Tamarins, Goeldi's Monkey and the Monk Saki.

In the foothills, riverine forests merge gradually with montane species. Subtropical rainforest ranges in altitude from 800 meters (2,625 ft) to 1,800 meters (5,900 ft), with daily temperatures between 18° and 24°C (64–75°F). Typically very cloudy, the rainfall amounts to some 3 meters (10 ft) annually, sustaining a huge variety of plants.

The pale, buttressed trunks of *Higerion*, typical of the lowland, are mixed with the *Podocarpus* trees of the cooler slopes. Fruiting fig trees, *Ficus* sp., and "*ovilla*" (wild grapes) are abundant, providing good places to observe feeding birds. A mid-story layer of palms and tree ferns (*Cyathea* sp.) intercepts the light. At ground level, blood-red begonias and the aptly named "*Lobias de Novea*" (Lovers' lips) flower among fallen trees covered with ferns (*Asplenium* and *Polypodium* sp.), mosses (*Lycopodium* sp.) and lichens. The scarlet, barbed flowers of Heliconia, stretching above the dense ground cover are visited by hummingbirds.

Tracks of ocelot and margay dot the forest trails. Rooting marks of Collared Peccary are common. Fewer primate species are found here, but the Night Monkey, Common Woolly Monkey and Dusky Titi may be encountered. A wide range of birds may be seen, including the Green Jay, Andean Guan, Golden-plumed Conure, and Grass-green and Golden-crowned Tanagers.

Montane forest extends from 1,800 to 3,000 meters (5,900 ft–9,850 ft). These steep slopes have somewhat lower temperatures and are slightly less moist. Due to the instability of the steep, young soils, landslides are a common feature of the terrain, although forest clearance and overgrazing exacerbate the problem. A constant cycle of recolonization, and slower growth rates in the cold, means that montane forest is typically younger and less complex than the mature rainforest below. But the vegetation is no less dense, and as the canopy becomes lower and the foliage smaller with increasing altitude, an undergrowth of herbs and shrubs thrives in the increased light. Bamboo (*Chusquea* sp.) is abundant in the higher zones. Characteristic flowering plants from the families *Asteraceae, Araliaceae* and *Gunneraceae* include, the bright yellow bladder-shaped flowers of *Calceolaria* and the clustered bells of *Bomarea* sp.. Gray-breasted Mountain Toucan and Rainbow-bearded Thornbill may be seen here.

From around 3,000 meters (9,850 ft), the rising humid air condenses into mist. The cloud forest is watered more by direct condensation rather than by rainfall, providing ideal conditions for epiphytic lichens, ferns

and mosses, which clothe every branch and twig of this forest. The tortured, slow-growing *Polylepis* tree provides a framework. Dense yellow asters (*Gnoxis* sp.) and the tubular orange flowers of *Tristerix longebracteatus* adorn the canopy above the tangle of flowering creepers and herbs.

Above 3,000 meters (9,850 ft), the terrain consists of finely dissected ridges and more open valleys at the top of the *cordillera*. The damp valley floors support grassland and marsh, while the steeper gullies shelter rich cloud forest which thins into stunted elfin forest and heaths on the exposed ridge lines. Dwarf tree ferns and a

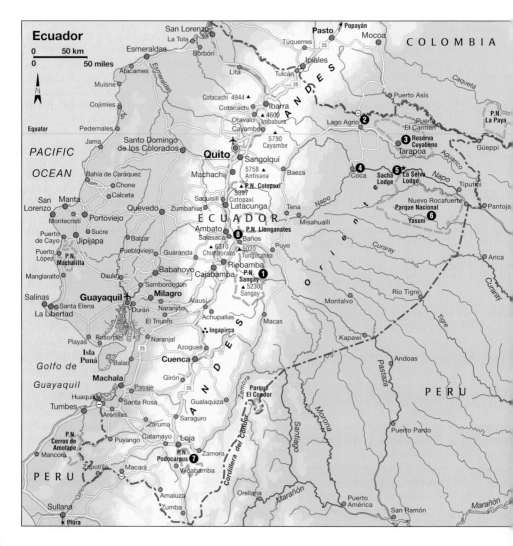

Map on page 204

great variety of bromeliads and heathers are common. The elfin forest is a riot of color and texture. Hummingbirds frequenting this area include the Collared Inca.

At the most exposed and windswept heights, shrubs give way to open *páramo*, dominated by the dense tussocks of *Stipa*, *Calamagrostis* and *Festuca* grasses. Daily temperatures range between 6°C and 12°C (43–54°F), the moisture is not so constant and night frosts are common. From a distance the *páramo* seems uniform, but the tussocks shelter a delicate garden of gentians (*Gentianella* sp.), *Ranunculus*, lupins (*Lupinus* sp.) and the sessile, yellow blooms of *Hypochaeris sessiliflora*. Carunculated and Mountain Caracara skim over this low vegetation for carrion, fruit and insect prey.

On the emergent cones of volcanoes a desolate zone of ash and lava marks the end of the moorland. Here, there is little vegetation apart from lichens. The peaks of the mountains are usually snow-capped, and their glaciers descend to about 4,700 meters (15,420 ft).

Sangay National Park

About 200 km (125 miles) south of the capital, **Quito**, **Parque Nacional Sangay ❶** covers some 2,700 sq km (1,700 sq miles) from the volcanic vents of **Sangay**, **Altar** and **Tungurahua** to the eastern *cordillera*. It has examples of all Andean bio-zones, from subtropical rainforest, through montane and cloud forest, stunted elfin forest and desolate open *páramo* to mountain *tundra*. However, it is threatened by farmers burning areas to provide grazing for cattle.

Riobamba is the most common gateway, and there are various routes into the park. Most visitors head for the volcanic peaks, entering from high mountain villages to the west. Sangay, one of the world's most active volcanoes, may erupt at 20-minute intervals. On a clear night, glowing lava and ash provide a dazzling display. But, as the volcano is almost constantly shrouded in cloud, such vistas are rare. Also, ascent may be prohibited by the weather and activity of the volcano; nevertheless, the trek is spectacular and varied.

ELOW: moss-covered elfin rest in ngay ational Park.

The first leg climbs out of the U-shaped **Alao valley**, across **La Trancha pass** at 4,000 meters (13,100 ft) and down into the broad valley of the **Río Culebrillas**. Local porters can walk the route in five hours; however, the views are worth lingering over. At night, the moaning calls of Spectacled Bear come from the *Polylepis* forest. They occasionally raid maize crops, but are more at home in the cloud forest fringing the *páramo* in the remote areas of the park.

Crossing the pass, you are likely to see the Andean Condor, gliding on updraughts from the valleys on a 3-meter (10-ft) wingspan as it scours the hillsides for carrion.

The *páramo* of the Alao valley and beyond to the edge of the park is grazed by cattle and managed by frequent burning. This strategy simplifies the grassland and destabilizes the steep slopes, but in this region the flora is still quite complex. There is little evidence of mammals here: the occasional paw-print of an Andean Fox, or the skull of its most common prey, the Sacha Cuy, wild ancestor of the domestic guinea pig. White-tailed Deer, once common here, are now hunted out, their lands over run by cattle. Descending from La Trancha, the *Polylepis*-dominated cloud forest starts at about 3,500 meters (11,500 ft). At its upper limit, the hardy bromeliads, *Puya*, and the spiky *Tillandsia* are common, whose remains testify to a good population of bears, who like to eat its succulent heart.

The **Culebrillas valley**, at 3,000 meters (9,850 ft), is relatively flat and drained by the fast-flowing Culebrillas river, surprisingly devoid of fish. Not even introduced Rainbow Trout *(Salmo gardineri)* can survive the acidic water running off the volcanic slopes. The wide floodplain shows the intensity of the river's periodic floods.

The boggy valley floor, although illegally grazed by cattle, still provides a habitat for Andean and Noble Snipe. Cattlemen have erected a couple of grass-thatched huts and a corral, which serves as a convenient camp-site for hikers and alpinists en route to Sangay volcano. Beyond the camp the valley narrows as it climbs towards the moun-

BELOW: a Dusky Titi Monkey.

Map
on page
204

tain. The river is fed by numerous streams that cut steep ravines in the ridges, clothed in a dense blanket of the giant cabbage-like paraguillas (*Gunnera* sp.). Secluded pools and rapids provide foraging for the White-capped Dipper and the rare Torrent Duck.

The mountain tapir

Tracks of the Mountain Tapir are commonly found along the river's fringing sandy beaches, and the prints of the diminutive Pudu are also occasionally seen.

The tapir inhabits the high pluvial Andean mountain forest and wet *páramo* of Colombia, Ecuador and Peru, and possibly western Venezuela. It has a stout body (weight about 200 kg/440 lb and 1 meter/3 ft tall at the shoulder), with a wedge-shaped profile for pushing through dense undergrowth. Its diet consists of selectively browsed plant shoots, frequently ferns, plantains and *Chusquera* bamboo. Like so many animals in this region, the tapirs are under constant threat from habitat modification, illegal cattle grazing and hunting.

Mammals are rarely observed here by day since most seek refuge in the fringing cloud forest, venturing out only at night or under the protection of swirling mists. Andean Fox and marauding puma are the main predators. Birds of the *páramo* include White-collared Swifts, tapaculos (family *Rhinocryptidae*), hummingbirds and fringillids. Flocks of Glossy Black Thrushes are common, as are raptors such as Red-backed Hawk *(Buteo polysoma)* and the Aplomado Falcon.

The route up Sangay eventually emerges on an exposed ridge of elfin forest, stunted by the low temperatures and high winds. At **La Playa**, the forest finally gives way to the unearthly landscape of the volcano.

The Oriente

The three main towns in the Oriente from where jungle excursions can be organized are **Coca**, **Lago Agrio** and **Misahuallí**. Most established tour agencies have offices in Quito, where you can arrange the trip. Short excursions from these towns used to be productive, but disturbance from oil

BELOW: a Guianan Cock-of-the-Rock *(Rupicola rupicola).*

exploration now means that there is little to be seen. Better is an excursion of several days along the **Río Napo**, the **Río Aguarico** and their tributaries.

Lago Agrio ② is the starting point for trips into the **Cuyabeno Reserve ③**, to see Pink River Dolphin, Amazon Manatee, Giant Catfish and four species of caiman. **Coca ④** and Misahuallí are the gateways into the **Napo** area, including the **Huaorani Reserve** which, in theory, protects Huaorani people. A good base – **La Selva jungle lodge** – is situated 100 km (60 miles) down the Río Napo from Coca. Close to the palm-fringed brown **Garzacocha (Heron lake)**, the lodge is an ideal base for anyone who wants to plan their own itinerary or add to the 500-strong bird list.

Forest trails lead to **Mandicocha (Water -hyacinth lake)**. From small canoes it is possible to see hoatzins clambering around the marginal vegetation, and to find roosting Proboscis Bats on fallen trees. Occasionally, rare Harpy Eagles, nocturnal currasow, Long-tailed Potoo *(Nyctibus grandis)* and Zig-zag Heron have been seen here. A canoe trip along the **Mandiyacu stream**, which rises from the lake, may reveal some of the area's 14 species of monkey.

Another trail takes in a ridge area, with views across the canopy. From here, Red Howler Monkey and Spider Monkey can often be spotted, as well as fleeting glimpses of mixed groups of tamarins and marmosets in the flowering and fruiting trees. It may be possible to catch a glimpse of a sloth or Tamandua Anteater, although they are normally well hidden. Look for the tracks of White-lipped Peccaries. Padmarks and a pungent aroma indicate that a jaguar followed their trail. Scarlet Macaws and Mealy Parrots may be seen overhead and toucans in the dense vegetation. High above the canopy are Greater Yellow-headed and King Vultures.

Close to La Selva, two hours from Coca, is **Sacha ⑤**, Ecuador's prime birding and wildlife lodge. Surrounded by black-water lagoons, creeks and forest, it has a 43-meter (140-ft) observation tower close to the Yasuní Parrot Lick. Another eco-friendly place is **Sani Lodge** (www.sanilodge. com), three hours by river from Coca.

Neighboring **Yasuní National Park ⑥** is under serious threat from oil extraction and the construction of roads and settlements. In the late 1990s, UNESCO threatened to review the park's status as a World Biosphere Reserve, forcing the government to change its boundaries and introduce legislation to protect some areas from development, but its future is still uncertain.

In general, the Oriente of Ecuador is still sparsely populated and undeveloped except in the oilfields near the Colombian border and Coca, and the rapidly expanding goldfields at **Zamora**, adjacent to the **Parque Nacional Podocarpus ⑦**, in the far south, where many rare animals such as the Spectacled Bear and Mountain Tapir live. Elsewhere, roads are few and colonization is mostly limited to their immediate vicinity with a handful of ports beyond. Few mountain passes access the area; the most spectacular descends from **Baños** to **Puyo**.

Let the buyer beware

The tiny town of **Baños ⑧**, named after its hot-spring baths, is a popular gateway to central Oriente. Travelers should take care when booking tours, however, since many local agents and rafting guides are unreliable and unqualified.

Below Baños, the **Río Pastaza** enters a breathtaking gorge. Above, lush forests ascend for over 1,000 meters (3,000 ft), dissected by numerous waterfalls.

The road descends to Puyo, then heads north through forest with scattered settlements. Many rainforest birds can be seen, but mammals have largely been hunted out. The journey continues along a dirt track road whose verges provide a wealth of flora. Lowland birds are occasionally seen in the fringing vegetation or soaring above.

Generally, agricultural development has scarred only forest close to the road, and vast expanses of primary forest are accessible with good opportunities for birding. Lowland forest is dominated by oil palm and *Cecropia*. The terrain is hilly and vantage points give views over the canopy. Flowering yellow and flame-red emergent trees are a food source for pollinating birds and the more common species of primates – Squirrel Monkeys and capuchins. ❏

Map on page 204

OPPOSITE: the Passion Flower *(Passiflora coocinea)* is found throughout the Amazon.

Map on page 214

PERU

Containing some of the most spectacular environments in Latin America, Peru's terrain runs from the High Andes to the Amazonian basin

The Amazon, by name, begins 4,000 km (2,500 miles) from its mouth in the northeast corner of Peru, where the waters of the **Río Napo** and **Río Ucayali** meet, joining the eastern catchments of most of Ecuador and Peru. But the geographical source of the river is defined as the point from which water flows furthest to the mouth. At this junction, the Ucayali is longer than the Napo. But exactly where it can be deemed to rise is still a disputed point.

A short distance upstream, past Peru's major Amazon port of Iquitos, is the confluence of the **Ucayali** and **Marañón**. Rising near the peak of **Yerupaja** (6,634 meters/21,760 ft), in central Peru, the Marañón winds northward before descending east to Iquitos. It often carries more water than the Ucayali and was long argued to be the Amazon's source, but by most current reckoning, the course should continue southward with the Ucayali. As the river is traced to the foothills of the Andes, its identity is confused, and it takes a new name at the confluence of each major tributary, several of which may compete for the distinction of "the source." One is the **Urubamba**, which drains the mountains near **Machu Picchu**.

In the 16th century, Spanish explorer Juan Salinas sought the headwaters of the Marañón and Ucayali. The drainage of the Ucayali extends south into the **Cordillera del Chile** in the Department of Arequipa. Some believe its source is in this glaciated mountain range, others that it spills from **Laguna Vilafro** on its 7,000-km (4,350-mile) journey to the Atlantic Ocean.

Iquitos

Iquitos ❶ is the biggest city of the Amazonian lowlands of northeastern Peru, about 1,000 km (620 miles) by air from the capital, Lima, and 1,200 km (745 miles) from Machu Picchu in southern Peru. Iquitos has grown from a village of 80 residents in 1814 to a thriving center of jungle exploitation with a population of approximately 600,000. The main stimulus for the town's development was the collection of latex from rubber trees. The rubber barons grew rich at the expense of the native community but they endowed Iquitos with some excellent colonial architecture, which remains despite the demise of the industry. Modern Iquitos is a major port and the center for oil exploration.

Frontier town

Access used to be difficult: the overland trek from **Lima** ❷ took 10 weeks through dangerous territory. Merchants preferred shipping up the Pacific coast, through the Panama canal and then to Belém and up the Amazon to Iquitos. A road across the Andes from Lima to **Pucallpa** ❸, on the upper Ucayali, was completed in 1943. Pucallpa is a rapidly-developing frontier town of some 400,000 people, reliant on timber and oil exploration. Near Pucallpa is **Yarinacocha**, a lake formed from an ancient meander of the Ucayali. Part of it is a nature reserve.

From Pucallpa it is a three- to six-day boat journey to Iquitos, but the Ucayali is too wide to make bank-side observation of wildlife possible from mid-stream. Because of disturbance from development and oil exploration, little wildlife remains within about a 50-km (30-mile) radius of Iquitos. The large mammal species were hunted out many years ago, although plentiful flocks of birds still commute across this riparian corridor.

But Iquitos is still an island in the wilderness, and fruitful natural history excursions are possible. A three-day cruise down the river reaches **Leticia** (in Colombia) and **Tabatinga** (in Brazil) near **Tres Fronteras** ❹ – the borders of the three countries. The "jungle tours" involve hiking, canoe

PRECEDING PAGES: the Río Urubamba, ne of the eadwaters of the Amazon. **LEFT:** jaguar tracks.

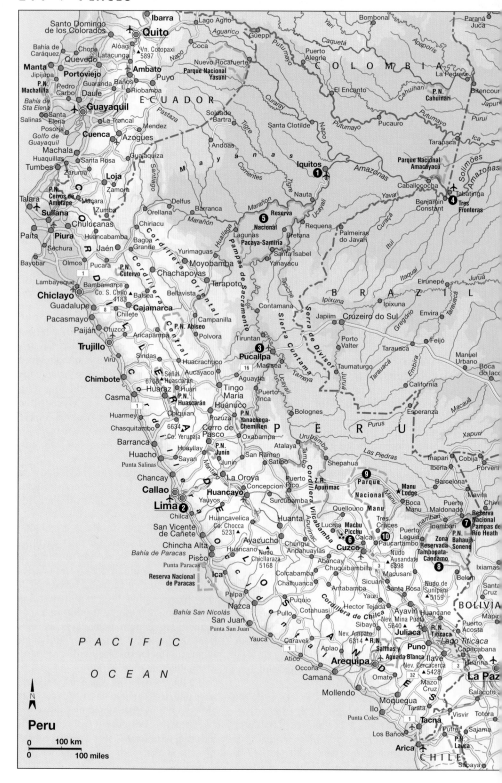

Peru

0   100 km

0   100 miles

Map
on page
214

excursions up tributaries, night-time caiman spotting and fishing for Red Piranha *(Serrasalmus nattereri)*. Good views of birds and occasionally primates, including Black Howler Monkeys, are possible along the densely forested river margins.

Alternatively, boats can be taken from Iquitos to the **Río Marañón** or the **Río Tahuayo**, which is also surrounded by dense, lush rainforest. In the **Yanayacu** area it is possible to see the Boto or Amazon River Dolphin. The rare Amazonian manatee favors quiet tributaries with areas of dense vegetation. Infrequently seen, their presence is revealed by floating droppings – voluminous quantities of fibrous balls resembling those of the horse.

Aiding conservation efforts

Southwest of Iquitos is the vast **Reserva Nacional Pacaya-Samiria ❺**. The Peruvian organization Pro Naturaleza has designed an eight-day program during which visitors can contribute to the work of the reserve. In June, Charapa Turtle conservation work involves collecting turtle eggs along the river beaches and moving them to protected sites. The hatchlings are released back into the rivers in October. Efforts are also being made to protect Giant Otters, manatees and Pink Amazon Dolphins. The area has as many as 700 dolphins in the high-water season.

There are two species of river dolphin in the Amazon area: the Boto and the Tucuxi. The latter is more like a seafaring dolphin, and tends to stay around the main bodies of rivers and lakes. The Boto has an extended, whiskered beak used for scavenging along the river bank, and large flippers for navigating among tree roots. It is more adventurous than the Tucuxi and will explore flooded forest areas. Native people do not hunt these creatures as they are superstitious about their powers.

The river corridors are often the best places to observe wildlife and it is now possible to cruise through the rainforest to Pacaya-Samiria on comfortable, early 19th-century style vessels. Caimans and

BELOW: a herd of Vicuña.

turtles are commonly seen basking on the river banks. Neotropical Cormorants, Roseate Spoonbills and Jabiru Storks fish in the shallow waters.

The forest is at its densest along the river margins. Within, the diffused light reduces the density of the understory. Long, trailing lianas are plentiful. These woody vines bind even the tallest trees together in an embrace so tight that even in death they may be prevented from falling. Included here are the strangling figs (*Ficus* sp. and *Clusia* sp.), which start life as epiphytes when birds deposit seeds in cracks in the bark. Aerial roots develop and quickly grow down to the ground, eventually forming a fusing mesh that envelopes and kills the host, leaving only the hollow "trunk" of the strangler.

Much of the bird and mammal life is in the canopy, allowing only tantalizing glimpses from the ground. These creatures have developed striking adaptations to enable them to exist in this precarious wilderness. The prehensile tail of the Gray Four-eyed Opossum and many of the New World primates provides a convenient extra leg with which to grip slim branches. The peculiar development of the claws of the sloth gives it added security.

Machu Picchu

The **Historical Sanctuary of Machu Picchu ❻** was established in 1981 and includes the **Archeological Park** with the fabulous Inca ruins rediscovered by Hiram Bingham in 1911. Also within the sanctuary is surrounding *puna (see page 222)* and cloud forest.

A spectacular railway brings visitors from **Cuzco**, ascending the mountain by a series of switchbacks to the peak at 3,500 meters (11,480 ft). The train descends to the **Secret Valley of the Incas**, following the Urubamba and Vilcanota rivers to **Puente Ruinas** at 2,000 meters (6,560 ft). From here there's a bus to the Inca city. The best way to see nature is on the **Inca Trail**, a spectacular three- to five-day hike from near **Ollantaytambo**. The trail proceeds

BELOW: the Alpaca belongs to the llama family.

Map on page 214

over a pass at 4,200 meters (13,780 ft) and through extensive areas of cloud forest.

The area has been affected by agriculture and hunting since Inca times. Problems still beset the sanctuary, including destruction of the forest by burning and by cattle, leading to substantial erosion.

Access by 160,000 tourists annually has produced litter and general disturbance. Probably as a result, few signs of mammals are apparent, although occasionally the bear-dissected remains of bromeliads *(Puya)* and the tracks of puma and White-tailed Deer may be seen near the Inca Trail. Numerous species of primate have been reported, including the Night Monkey and the Woolly Monkey. In 2001 new legislation was introduced to protect the trail. Measures include restricting the number of visitors to 500 a day on the trail and allowing only licensed operators to sell Inca Trail packages, using authorized guides and approved camp-sites.

Even in disturbed areas, secondary growth can provide a rich habitat for a wide variety of birds. Of the 1,678 species recorded for Peru, some 22 percent (372 species of 49 families) have been recorded in the sanctuary and its immediate vicinity, including the Andean Condor. The Roadside Hawk, Variable Antshrike, White-winged Black Tyrant and the Green Jay are commonly encountered.

Habitat types

The principal habitat types are *puna* or *pajonale* at altitudes above 3,500 meters (11,480 ft), typified by *Stipaichu* grass over which the Puna Hawk and Mountain Caracara hunt. The Puna Snipe may also be seen. Below this, there is a zone of elfin forest comprising mostly *Clusia*, *Gynoxis* and *Polylepis*. This grades into cloud forest, particularly in the vicinity of the ancient monument.

Tree species such as *Cyathea*, *Alsophila* and *Podocarpus*, *Weinmania* and *Nectandra* are common at the top of this zone. Much of the understory comprises tangled thickets of *Chusquea* bamboo.

BELOW: cloud forest *(left)* and *Puya* plants *(right)* near Machu Picchu.

Where the Inca Trail passes through the cloud forest there are characteristic small trees, birch forest *(Aliso; Alnus acuminata)*, terrestrial bromeliads *(Puya ferruginea)* and cacti *(Cereus* and *Opuntia)*. The *Polylepis* forest contains rich bird life, including hummingbirds, such as the Tyrian Metaltail and Great Sapphirewing, the Andean Hillstar, Collared Inca, Bearded Mountaineer, the spectacular Giant Hummingbird and Sword-billed Hummingbird.

Various species of Spinetail are also common. The Andean Guan and Andean Parakeet, Golden-headed Quetzal, Highland Motmot and Gray-breasted Mountain Toucan are occasionally spotted.

There are a few glacial lakes in the sanctuary, on which it may be possible to observe Puna Teal and, in the streams which they feed, foraging Torrent Duck. The **Cuenca de Santa Teresa**, and the Santa María and Urubamba rivers, where the South American River Otter has been seen, are among nearby areas of interest.

The Lowlands of Eastern Peru

East from Cuzco and Machu Picchu lie river systems quite separate from the Ucayali. The **Madre de Dios**, one of the least accessible headwaters of the Amazon, contains vast areas of undisturbed forests. The watershed of the **Río Heath**, although more accessible, is also of considerable ecological significance, containing the **Reserva Nacional Pampas de Río Heath**, part of the **Parque Nacional Bahuaja-Sonene ❼**.

The Department of Madre de Dios has a total population of 50,000, most of which is are based in or close to **Puerto Maldonado**, a town on the Río Madre de Dios close to the border with Bolivia. The town itself has had a chequered environmental history. The area's initial exploration is attributed to Fitzcarraldo, who discovered a route from the Ucayali to the **Río Cashpajali** – a tributary of the Manu giving access to the Madre de Dios. This enabled the exploitation of jungle products, particularly rubber, and led to aggression against the indigenous tribes. With the demise of the Peruvian and

BELOW: an Andean Condo in flight.

Map on page 214

Brazilian rubber industry, the attention of the thriving town turned elsewhere. In the 1920s, hunters entered the area killing cats, Giant Otters and caimans for their skins. More recently, the exploitation of timber, including mahogany, led to the construction of an airstrip at **Boca Manu**, now a convenient gateway for visitors. Today, gold prospecting and cattle ranching attract thousands of new colonists each year.

The Madre de Dios region is recognized as a World Centre of Plant Diversity by the WWF/IUCN. It is believed that the rainforests, which once covered a much larger area of the earth, retreated to a few isolated pockets during the the last ice age. Madre de Dios was one such refuge, from which plants and animals have subsequently recolonized the Amazon basin. It is, therefore, one of the oldest, most stable and most complex examples of neotropical forest. Two major protected areas have been created in acknowledgement of its value: **Parque Nacional Manu** and **Zona Reservada Tambopata-Candamo**.

BELOW: A Collared Gecko from the lowland forest.

Tambopata

The Tambopata Reserve, a small area of only 55 sq km (21 sq miles), located some 40 km (25 miles) south of **Puerto Maldonado**, was created in 1977. But with the demarcation in 1990 of a huge area of surrounding forest, it became the hub of the 1.5 million-hectare (5,800-sq mile) **Zona Reservada Tambopata-Candamo** ❽. This "park" protects the vulnerable basins of the Tambopata and Heath rivers. In addition, the neighboring **Parque Nacional Bahuaja-Sonene** was created in 1996; together with Madidi, across the border in Bolivia, it has more diverse habitats and wildlife species than Manu and Tambopata put together, but is in urgent need of financial support and worldwide recognition of its biodiversity.

Tambopata is a prime location for ornithologists, holding the highest 24-hour bird list in the world: 331 species. The area's total list extends to 570 species, and over 90 species of mammals, including the Giant Anteater, Giant Otter and Giant

Armadillo. Two of South America's rarest canids are also recorded: the Bush Dog and Small-eared Dog. The tourist facility supports an extensive research program conducted by scientist guides.

The reserved zone is designed to allow for various levels of utilization in designated areas. Adjacent to Tambopata, a major ethno-botanical project, AMETRA 2001, combines modern and traditional approaches to medicine. In part this involves recording traditional knowledge of forest practices and disseminating the information through the local community. Other experiments in sustainable production involve fish, butterfly and capybara farming, and Brazil-nut collection.

Manu Biosphere Reserve

Amounting to some 18,812 sq km (7,300 sq miles) of mainly humid evergreen tropical forest, laced with countless miles of meandering rivers and glinting lakes, the **Parque Nacional Manu 9** is the largest biosphere reserve zone and the fourth largest national park on the South American continent. The park was established in 1973, but it was not until 1977 that UNESCO accepted a proposal from the Peruvian government that it be declared a biosphere reserve. It is one of 200 such areas worldwide which represent an attempt to preserve pristine examples of the world's major ecosystems. Manu is the only biosphere reserve that protects an entire unhunted and unlogged watershed. The reserve contains many rare species of animals (including over 1,000 species of birds) and plants that have been eliminated from other areas.

The park is divided into three zones: a 15,328-sq km (5,900-sq mile) core zone, strictly preserved in a natural state, home to a number of indigenous tribes; an experimental or buffer zone of some 2,570 sq km (990 sq miles) set aside for controlled research and tourism; and the 914-sq km (350-sq mile) agricultural zone for controlled traditional use by the indigenous Amarakeri and Yine. Only six ranger stations

BELOW: a clump of lupins at 4,000 m (13,120 ft), with Mount Chopicalqui (6,354 meters/ 20,845 ft) in the background.

Map on page 214

patrol the park; the most effective protection comes from its inaccessible location.

A few companies run tours into the park in motorized dugouts, to camp for a few nights at designated sites on the sandy beaches at river bends. The only permanent tourist facility in the lowland forest of the zone is **Manu Lodge** (elevation 400 meters/1,300 ft), comfortable and sympathetically constructed from mahogany salvaged from **Río Manu**. It is well positioned on the shore of **Cocha Juarez**, some four hours by boat from the mouth of the river, which is reached in 45 minutes by light aircraft from Cuzco. However, the overland and river route, taking at least two days, is a fascinating journey through all the layers of Andean habitat. The precipitous road is traveled in each direction on alternate days, as it is too narrow to allow vehicles to pass.

Andes to lowland Oriente

The forested eastern half of the Peruvian Andes amounts to 60 percent of the country's land area but holds only about 5 percent of the population. The rivers are used to transport produce like timber and rubber, but also as a corridor for colonization.

Few roads penetrate the area; the most important one runs from Lima to **Tingo María** and thence Pucallpa, giving access to the Río Ucayali and the mighty Amazon. Another major route runs from Cuzco over the Andes to the watershed of the Madre de Dios river, accessed through **Atalaya** or the missionary settlement at Shintuya. It is a six- to seven-hour journey from Cuzco (3,310 meters/ 10,860 ft) to reach a suitable camping point in the cloud forest. The paved road gives way to a dirt track leading steeply to the Quechua village of **Huacarani** (3,800 meters/12,500 ft) and descends vertiginously into the **Paucartambo ❿** valley, famous for its many varieties of potatoes.

These habitats are called *ceja de selva* (eyebrow of the jungle) and include the high grasslands and cloud forests. Andean

BELOW: a Tree Boa *(Corallus hortulanus)* from Tambopata.

Fox and White-tailed Deer are occasionally seen in the open grasslands. Herds of llamas and Alpacas look up shyly. Overhead, Mountain Caracara and Andean Lapwing are often seen, and lucky travelers may spot flocks of Mitred Conure.

Gradually agriculture thins out towards the **Acjanaco Pass** (4,100 meters/13,450 ft). At **Tres Cruces**, there is an exhilarating view of the Amazon basin and Manu Park. Craggy peaks soar over the *puna*, a habitat characterized by tussock grass and pockets of stunted alpine flowers, giving way to relict elfin forest and bamboo thicket. The dissected remains of a *Puya* plant may denote the nocturnal foraging of the rare Spectacled Bear, a frequent visitor to these lands. Here Pumas *(Felis concolor)* roam, preying on the White-tailed Deer and tiny Pudu.

The elfin forest grades into misty cloud forest at 3,300 meters (10,825 ft). These often impenetrable forests cover the eastern slopes of the Andes. The high humidity, due to almost perpetual cloud, supports a verdant kingdom of dripping epiphytic mosses, lichens, ferns and orchids which grow in profusion here despite plummeting night temperatures. Water seeps into a myriad of icy, crystal-clear streams and waterfalls. In secluded glades, flame-red Andean Cock-of-the-Rock give a spectacular display and Woolly Monkeys are occasionally sighted, as are mixed flocks of colorful tanagers, Golden-headed Quetzals and Amazonian Umbrellabirds.

At about 1,500 meters (4,900 ft), there is a gradual transition to the vast lowland forests of the Amazon basin, less jungle-like, but warmer and more equable than the cloud forests above. The dense canopy of leaves inhibits growth at ground level. A great diversity of habitats ranges from 10-meter (33-ft) high stands of *Cana brava*, which bind together the river banks, through lush stands of *Heliconia*, to giants like the mahogany (*Cedrela odorata*) and Kapok tree (*Ceiba pentandra*). It is a habitat hung with strangling vines and phylodendrons, among which troupes of

BELOW: a Common Woolly Monkey

Map on page 214

vociferous Squirrel Monkeys, Brown-and White-fronted Capuchins forage. In the high canopy, Black Spider Monkeys perform acrobatics, while, lower down, Saddle-back Tamarins and Emperor Tamarins forage for blossom, fruit and insects.

Annual daily temperatures vary little, with a high of 23–32°C (73–90°F) falling to 20–26°C (68–79°F) overnight. The annual rainfall (2 meters/6 ft) occurs mostly from November to April, when access to the park may be restricted. The rest of the year, at least in lowland areas, is sufficiently dry to inhibit the growth of non-vascular epiphytes and orchids which are so characteristic of the highlands. For a week or two in the rainy season the Manu floods the forest in its 6-km (4-mile) wide floodplain. Canoeing and wading are the only ways of getting around.

The road ends at the **Alto Madre de Dios**. Dugout canoe is still the method of transport, but is now usually powered by a pair of 50 hp outboards. After traveling for four hours up the main river, you enter

the mouth of the Río Manu. The canoe winds between partially submerged trees left stranded after floods.

The meandering course of the river provides opportunities to see herds of russet-brown Capybara, snuffling Collared Peccary and wary Red Brocket Deer. The dipping flight of a toucan demands attention, and the far-off screeching of macaws increases to a deafening cacophony as they pass over the river corridor.

The sinuous meanderings of the river transform the landscape and provide a living story of forest succession. At each bend of the 150-meter (500-ft) wide river, the forest is undermined by currents during the seasonal floods at the rate of 5–20 meters (16–65 ft) a year, leaving a sheer mud and clay bank, while on the opposite bend new land is laid down in the form of broad beaches of sand and silt. Gradually, these beaches become invaded by plants taking advantage of the alluvial soils, which are far richer than the weathered upland areas. A succession of vegetation is

BELOW: a river flowing through Manu National Park.

observed, from the fast-growing, willow-like *Tessaria*, which stabilizes the sand, to tall stands of *Cana brava (Gynerium sagittum)*. This plant grows at a prodigious rate – some 10 meters (33 ft) a year. Finally, within these almost impenetrable stands, the seeds of tall rainforest trees germinate and thrust their way towards the light. The fastest growing is a species of Cecropia which forms a canopy 15–18 meters (50–60 ft) over the *cana*. A little further up this short-lived species gives way to others, such as *Erythrinia, Guarea* and *Sapium*. Two species, *Ficus insipida* and *Cedrela odorata* (mahogany), are capable of outgrowing the others, forming a closed canopy at 40 meters (130 ft) with a lush understory of shade-tolerant, long-leaved mono-cotyledons such as *Heliconia* and ginger. Eventually, even the long-lived trees die off, to be replaced by others, creating a forest of great diversity.

Consequently, the forest bordering the river is of no great antiquity. It is beyond the floodplain, in the zone of *terre firme* forest and on the ridges, that the largest and oldest trees are found. The oxbow lakes, or *cochas,* which remain after the river has changed course are of great ecological interest as they provide abundant fauna, easily observed around the lake margins.

Wildlife at Manu Lodge

Much of the lowland wildlife can be seen in the vicinity of Manu Lodge, including nine species of monkey and over 450 of the 500 species of bird recorded in lowland Manu. In surrounding trees are the pendulous nests of the noisy Yellow-rumped Caciques, and Blue-and-Yellow Macaws visit the swaying palms. Black Caiman *(Melanosuchus niger)* prowl the lake on the banks of which the lodge stands. Dawn is heralded by the gently rising and falling roar of Red Howler Monkeys. Flocks of Pale-winged Trumpeters forage in the undergrowth, while guans and curassows haunt the canopy. Particularly noteworthy are the Black-faced Cotinga, Crested Eagle and the spectacu-

BELOW: a Three-toed Sloth *(Bradypus infuscatus)*, swimming.

Map
on page
214

lar Harpy Eagle, perhaps the world's most impressive raptor, easily capable of taking an adult monkey from the canopy.

From the 27 km (17 miles) of trails, many animals can be seen at close quarters. Knowledgeable guides with high-powered telescopes and who can identify the sounds of numerous birds, ensure a rewarding experience. Having never been persecuted here, animals are bold and unafraid. Tracks of ocelot and jaguar are commonly found, frequently following groups of tourists; about one in seven visitors sees a jaguar.

One of the trails terminates at a *collpa* (mineral lick; another is reached by river). Here, in the early morning, good views may be had of Collared Peccaries, Red Brocket Deer and Lowland Tapir. Occasionally, their predators may also be encountered, as may rare game birds such as Razor-billed Curassows and Piping Guans. The birds need a daily dose of the mineral-rich clay to neutralize toxins in their diet. At dawn there may be 600 jostling birds of up to six species.

The best way to see the forest wildlife is to get into or above the canopy. In Manu, a sharp ridge provides a number of miradors from which, on a clear day, excellent views are obtained across the lowland floodplain to the foothills of the Andes. Flocks of parrots and macaws commute between various trees. Mixed-species flocks are commonly seen in fruiting trees, containing between 25 and 100 birds of perhaps more than 30 species. As there are few individuals of each species in the flock, competition is minimized. Individuals can take advantage of the security of the flock and may benefit by finding prey disturbed by other species.

An even better view is gained from one of the platforms erected in certain trees. The most notable is sited in a giant Kapok *(Ceiba pentandra)* which stretches above the surrounding canopy to 40 meters (130 ft). Access is provided by *jumars* (rope ascenders). From this vantage point it is possible to see a wide range of primates and birds, attracted to the delicate epiphytes

BELOW:
a Giant Otter
swimming
on its back.

and fruiting vines. Plans are in hand to construct a canopy walkway.

One of the best ways of seeing wildlife is to drift round the 2-km (1-mile) *cocha* in a canoe. The silent approach permits you to photograph Fasciated Tiger-Heron, Sunbittern and raucous Hoatzins. Fleeting glimpses can be had of diminutive Emperor and Saddle-back Tamarins foraging on the overhanging fruit trees.

Nocturnal excursions along the forest trails with a powerful torch provide a different perspective. The booming call of the Bamboo Rat reverberates around the forest, competing for attention with the haunting moan of the Kinkajou. The torch beam may reflect the eye-shine of a Night Monkey disturbed from its foraging activity. On the ground a timid Paca rustles in the leaf litter for oil-palm nuts. The warm, humid air is a living soup of insects.

The giant otter

Giant Otters are a particular attraction. Manu is one of the few places where visitors are almost certain to see them at close quarters. Excursions can be made to **Cochas Salvador** and **Otorogo**, some three hours up the Manu. Here, the otters have become partially habituated during years of scientific study, and may be approached, with care. Their population in Manu National Park is estimated at over 100, perhaps one-tenth of that in all South America. It is the world's largest species of otter; males reach a length of 1.5–1.8 meters (5–6 ft) and weigh up to 30 kg (66 lb); females are only slightly smaller. The Giant Otter is gregarious, living and hunting in permanent groups of four to six individuals. Exceptionally, as many as 20 have been seen together.

The body is superbly designed for motion in water but their gait on land appears cumbersome. The dense, glossy coat is chocolate-brown apart from the chin, throat and chest which are streaked with cream or buff in unique patterns, enabling identification of individuals. The tail is dorso-ventrally flattened and gives rise to the Latin name – *Pteronura*. Formerly distributed throughout South America (except Chile), the species is probably extinct in Uruguay and Argentina, and shows a much-reduced range elsewhere.

Observation of the feeding behavior of Giant Otters suggests that they prefer to hunt in shallow water. Principal fish prey are piranha and catfish, but they will also take caiman (up to 2 meters/6 ft long) and anaconda. The otters cooperate in attacking a caiman, darting at it from different directions. When surrounding lowlands are inundated, the fish move into the forest to forage and the otters follow.

Normally, the otters hunt in a group, diving simultaneously into a shoal of fish. This does not necessarily imply cooperation; fish may simply be captured more easily in the confusion of the churning water. This may be the reason for the otter's gregarious behaviour or it may be a defense against predators; young otters are vulnerable to caiman and anacondas.

The Giant Otter is fiercely territorial, frequently scent-marking its bank-side resting places with latrines, where they have trampled their faeces and urine into the ground as a pungent "keep-out" signal. Only one pair of otters in a group appears to breed, producing four to six cubs. The previous year's young stay with the group but do not participate in rearing the new litter. By the age of six weeks they may accompany the group, playing boisterously while the adults fish, and wailing when neglected. They are catching prey themselves by nine months. Adult Giant Otters have a comprehensive range of vocalizations which they use to great effect to threaten, express alarm, and maintain group contact. They may scream in frustration, snort in alarm or give a wavering scream towards an intruder.

Large, gregarious, noisy and inquisitive, Giant Otters were easily located and hunted for their luxurious pelt. It was the demand of the fur trade which has led to their present endangered status. In the period 1946–73, some 24,000 skins were exported from the Peruvian Amazon. Skin trading has been banned since the 1970s, but it is likely that small numbers of otters are still taken. The entire South American population has been estimated at between 1,000 and 3,000 individuals. ❑

Map on page 214

RIGHT: a forest spider spinning its web.

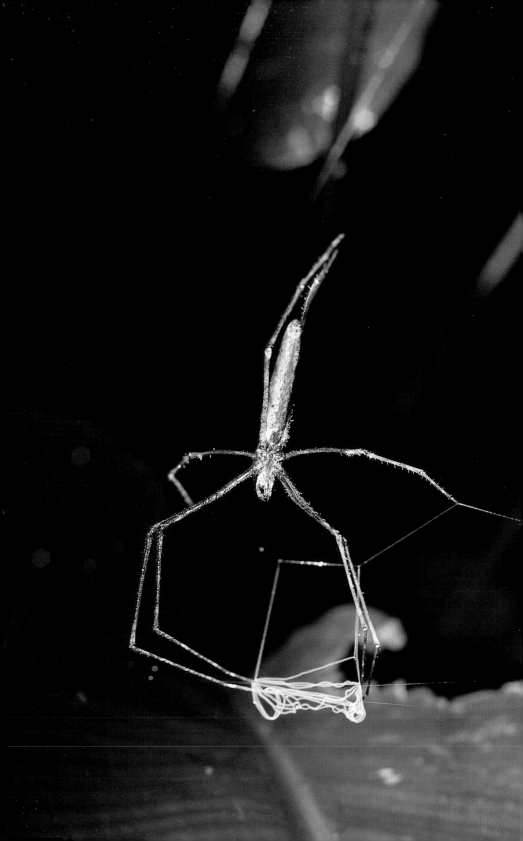

Map
on page
232

BOLIVIAN AMAZONIA

*The world leader in the certified management of
natural tropical forests, Bolivia is working with the Forestry
Stewardship Council to certify more than 1 million acres of forest*

Bolivia is generally associated with the sparse and imposing highland landscapes of the Andes, with the snow covered peaks of the **Altiplano** (highland plateau) providing its signature. However, over half of Bolivia's 1,098,000 sq km (424,000 sq miles) is in the Amazon Basin. Furthermore, the Amazonian lowlands have played an important, albeit less visible, role in the establishment of Bolivia as a nation and still hold great promise for conservation and sustainable development.

The Amazon River Basin tends to be ecologically oversimplified. The traditional view of Amazonia is of a flat, tree-covered tropical forest, where many species thrive in a uniform landscape. In fact, the habitat diversity of this region is largely responsible for its biological diversity. Bolivia's Amazon Basin is unique because it not only contains the lowland tropical forests, but also has habitat types as diverse as montane cloud forests, dry inter-montane valleys in the eastern Andes, tropical savannas, and dry scrub forests associated with the central plateau of the Brazilian Shield.

Three protected areas in Bolivian Amazonia include this wide range of habitats and exemplify the efforts being made in Bolivia to preserve the diversity of life on the planet. These areas are the **Parque Nacional Amboró**, the **Parque Nacional Noel Kempff**, and the **Ríos Blanco and Negro Wildlife Reserve**. In addition to the biological riches they hold, they represent important experiments in the management of protected areas, where successful private-public partnerships have been forged.

These three protected areas are being established, at great cost, to play a significant role in the protection of Bolivia's biological diversity and, at the same time, in the development of the ecotourism industry. Since 1990, 11 national parks and sanctuaries have been created, with another ten being considered. As neighboring countries are plagued by crime and political upheaval, Bolivia remains a peaceful nation that welcomes those who come to learn the beauty of the culture, the people and its nature.

Amboró national park

The Amboró national park is in the center of Bolivia in the Department of Santa Cruz, where the Eastern Andes change direction from northwest-southwest to straight north-south. It currently covers about 6,300 sq km (2,430 sq miles) of habitats ranging from humid montane forest in the south to lowland forests in the north. The park was created in 1973 as the **German Busch Wildlife Reserve**; in 1984 the name was changed to the **Parque Nacional Amboró** ❶ and its territory extended, and in 1990, the park was increased to its current size.

Before its expansion, over 540 species of birds and 120 species of mammals had been recorded in the park, among them the endemic Red-fronted Macaw and Southern Horned Curassow, as well as the rare Andean Cock-of-the-Rock, Harpy Eagle and Crested Eagle, Spectacled Bear, Giant Otter, jaguar, and many others. Many more species are now present following the increase in size, since habitat diversity in Amboró has also increased, especially at higher altitudes. The two best-represented zones are the southern cloud forests and the northern lowland rainforests, described as lower montane wet subtropical forest, and wet subtropical forest, respectively.

Amboró is easily accessible from **Santa Cruz** along two major roads connecting Santa Cruz to **Cochabamba**. The older road goes straight west up the mountains, and borders the south side of the park. The southern headquarters is located in the city of **Samaipata**, about 120 km (75 miles) from Santa Cruz, and it is operated by the Fundación Amigos de la Naturaleza (FAN) in partnership with the Centro de Desarrollo Forestal, CDF (Forestry Development

PRECEDING PAGES: a Giant Millipede.
LEFT: the Kinkajou *(Potos flavus)* belongs to the racoon family.

Center). The southern two-thirds of the park is mountainous and has been significantly less affected by human activity than the northern lowlands. Although pressure is increasing in the south, there are still frequent sightings of the threatened Spectacled Bear and other critical species.

Along the 180 km (110 miles) of road that borders the south side of the park there are over 40 logging and park access roads, built before it was enlarged. Gradually, and with great difficulty, FAN and CDF are beginning to establish a presence in the park to work with the local community to stop invasion and poaching. One of the critical priorities is to develop controlled access to the park so that it can be visited and monitored at the same time. Samaipata has excellent lodging, but adequate visitor access to the park is still limited. Improving the park's accessibility is important, yet the authorities must also prevent poaching, illegal logging, and squatting – the major problems afflicting Amboró.

The northern limit of the park follows closely the Santa Cruz-Cochabamba road, built in the late 1980s. This section of the park is being managed by the Chimoré-Yapacaní Project of the Secretariat General of the Environment, headquartered in

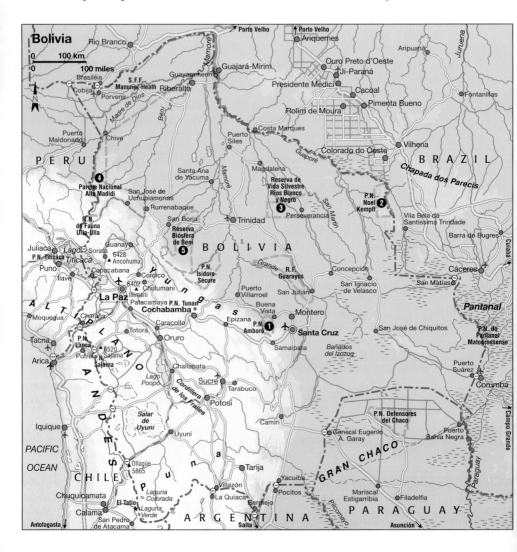

Map on page 232

the city of **Buena Vista**, 120 km (75 miles) from Santa Cruz. The situation on the north side is quite different from that of the south. Being mostly lowlands, and exposed to new road construction, this portion of the park is severely affected by almost 900 families of colonizers who use slash and burn. Great efforts are being made to prevent more colonists from becoming established and to control the damage already done, but the very limited resources of the Bolivian government are not sufficient.

Noel Kempff national park

The **Parque Nacional Noel Kempff ❷** is in the northeastern corner of the Department of Santa Cruz, on the Brazilian border, across the **Río Iténez** from the Brazilian states of Rondônia and Mato Grosso. It covers 9,140 sq km (3,530 sq miles) and a range of habitats. Its center is dominated by an extensive plateau that rises sharply from the lowlands to over 500 meters (1,600 ft). The high diversity of this area – over 550 recorded bird species – is largely due to the climatic and diverse habitats created by the abrupt topography.

On the central plateau, known as the **Huanchaca** or **Caparuch Range**, is perhaps the largest tract of virgin *cerrado* left in the world. *Cerrado* is a sparsely forested scrub savanna found only in Brazil and Bolivia. It has become the preferred habitat for soyabean plantation in Brazil, which has caused it to disappear at an ever-increasing rate. With the unique vegetation also appear rare and endangered species restricted to the *cerrado*, such as the Maned Wolf, Giant Anteater, Giant Armadillo, Marsh Deer and Golden-naped Macaw. To the west, humid lowland forests, drained by the large **Río Bajo Paraguá** and its tributaries, teem with wildlife, including healthy populations of the very rare Giant Otter and Black Caiman.

The **Noel Kempff national park** was created in 1979 as the **Huanchaca National Park** with 5,410 sq km (2,100 sq miles), thanks to the efforts of Noel Kempff, a prominent Bolivian naturalist. Kempff was

BELOW: a sandstone rockface at Amboró.

UNSOLVED MURDERS

Noel Kempff was killed in September 1986 while visiting the Huanchaca National Park. He and his party landed on the plateau to explore the surrounding vegetation. Unbeknown to them, the airstrip was also used by a large cocaine laboratory. What followed resulted in the deaths of all but one of Kempff's party.

The surviving member of his expedition made it out of the plateau alive after hiding until a rescue plane spotted the burnt remains of their aircraft – it had been set on fire to make it look like an accident. It took three days for the police to arrive and start an investigation, by which time the factory had been dismantled and the drug-runners had disappeared.

To this day, the case remains unsolved. Two months after Kempff's death, the park name was changed to the Parque Nacional Noel Kempff and was increased to 7,060 sq km (2,725 sq miles), with an additional 2,080 sq km (803 sq miles) of buffer zone.

also the driving force behind the creation of the Amboró National Park. Dedicating himself to the study and preservation of the flora and fauna of Bolivia, he designed and established the **Botanical and Zoological Gardens** in Santa Cruz. These are full of native *Tajibos*, with brilliantly colored flowers, and *Toborochis*, with their bulging trunks; also look out for gentle Night Monkeys and raucous macaws.

The Amboró park used to be managed by a group of institutions under the leadership of CORDECRUZ, the Santa Cruz Regional Development Corporation. Recently, however, participation of non-government organizations in the management of the park has increased, notably in the acquisition of Flor de Oro, a 100 sq-km (39 sq-mile) holding, by FAN with the aid of the Nature Conservancy. Flor de Oro has been developed by FAN as a research and ecotourism station capable of accommodating visitors and researchers. Los Fierros is a second jungle lodge operated by FAN, which, in conjunction

with Nature Conservancy's Parks in Peril Programme and foreign investment, has made Noel Kempff National Park the best protected park in the country.

Ríos Blanco and Negro wildlife reserve

The **Reserva de Vida Silvestre Ríos Blanco y Negro ❸** covers an area of 14,000 sq km (5,400 sq miles) and is located in the northwestern corner of the Department of Santa Cruz, bordering the Department of Beni. This reserve was created in 1990 specifically to be developed for ecotourism. Management responsibility was granted to the CDF and FAN, who are working with the business community to open the reserve up to visitors.

The Ríos Blanco and Negro wildlife reserve is located in the transition zone between the lowland forest and grasslands to the north and dry deciduous and scrub forests of the **Chaco** to the south. Its inaccessibility has kept squatters and poachers out, shown by the high densities of tame

BELOW: Arcoiris falls in Noel Kempff National Park.

Map
on page
232

large mammals in and around **Perseverancia**, one of two places in the reserve with adequate visitor facilities, accessible only by plane. A few days in Perseverancia are sure to include sightings of tapirs, Brocket Deers, Giant Otters, screamers, Hoatzins, and several species of macaws. The remote **Lake Taborga Biological Station** is located to the west of the reserve on a lake where Guarayo Indians are the local guides.

The success of these three protected areas will, to a large extent, determine the success of many other areas that are only now beginning to become established. The Amboró and Noel Kempff national parks and the Ríos Blanco and Negro Wildlife Reserve are taking conservation action to the field at the same time as they consolidate partnerships between the public and private sectors, thus sharing the responsibility to protect Bolivia's biological diversity with all Bolivians.

A fourth important park, the **Parque Nacional Madidi ❹**, near the border with Peru, stretches over nearly 1.8 million hec-

tares (6,950 sq miles) in the tropical Andes, and has been identified by Conservation International (CI) as one of the planet's 24 biodiversity hotspots. In recent years Madidi and the surrounding area have faced increasing threats from encroaching populations as well as from logging, oil extraction and mining interests.

In response, a new lodge, **Chalalan Ecolodge**, has been built in the heart of the park, supported by CI and the Inter-American Development Bank. This ecotourism initiative was undertaken by the Quechua-Tacana villagers of San José Uchupiamonas, a community inside the park, to generate job opportunities and fend off pressure from logging and oil interests.

Nearby, the **Reserva Biósfera de Beni ❺** is also receiving more attention as a tourist center. Four-day adventure trips from the jungle town of **Rurrenabaque** are on offer, and getting there is an adventure in itself – an 18-hour bus ride from La Paz via Coroico, through beautiful country with towering cliffs and lush cloud forest. ❏

BELOW:
a group of
Blue-and-
Yellow Macaws
(Ara ararauna).

BRAZIL

The largest area of the Amazon rainforest lies within the boundaries of Brazil, a huge expanse where travel is still best undertaken by local boats

The good news is that only a small part of Brazilian Amazonia has been "developed" for human use; the bad news is that an even smaller portion has any protected status. Some areas, like the Anavilhanas Archipelago or part of the Rio Trombetas, have the status of a biological reserve, and other areas, such as Xingu, protect the environment of indigenous peoples. The Brazilian government has also announced an agreement with the World Bank and WWF to establish 25 million hectares (96,000 sq miles) of new protected, zones thus tripling its protected areas.

It is difficult to predict the future of the region. The government no longer pursues a policy of radical development, for example, of cattle ranching. Overwhelming scientific evidence shows that most of Amazonia is unsuitable for intense agriculture. But it is unlikely that the conservationist's hope, that all Amazonia could become a biological reserve, will come to fruition. Research is being carried out to identify the areas of greatest biological richness, which might form the cores of future protected areas. Other research tries to determine the minimal critical habitat that can maintain an isolated population of a particular species. It is likely that many reserves will eventually be gazetted according to the outcome of this research.

Outside the reserves, the region is likely to become a patchwork of areas that have been logged or opened by squatters through slash-and-burn. However, after years of lobbying by environmentalists, in 2001 the Brazilian government bravely announced the cancellation of all but two mahogany logging operations in the Amazon. Paulo Adario, of the Greenpeace Amazon Campaign, welcomed this, saying "the illegal mahogany industry has for years been driving the destruction of the Amazon." Other territories will survive in a nearly natural state and be used sustainably, for example

by the gathering of minor forest products. Locally, indigenous agricultural techniques based on the planting of mixed cultures of useful herbs, shrubs, and trees may expand.

Tourism is still relatively undeveloped in the Brazilian Amazon and there is not the same choice of accommodations or local operators as elsewhere. Travel is still best undertaken by local boats, taking organized boat tours or staying in one of the few accommodations in natural surroundings. Consequently, these chapters make suggestions without giving precise directions. Since most visitors will arrive via cities in southern Brazil, a selection of national parks is included which are either close to **Brasília** or **Rio de Janeiro**, or are easily reached on tours. ❏

Map on page 257

PRECEDING PAGES: on the river at dusk. **LEFT:** endangered hyacinth macaws. **RIGHT:** a blue Morpho butterfly.

Maps
on pages
57 & 244

MANAUS

*This bustling port, founded on the wealth of
the rubber trade, has a remarkable opera house and was
once the richest city in South America*

Manaus ❶ developed from a Portuguese fort, which was built in the second half of the 17th century on the **Rio Negro**, 18 km (11 miles) upstream from the point where it flows into the Amazon. The town reached its peak early in the 20th century, when latex, – a natural substance from the rubber tree *(Hevea brasiliensis)* – brought fame and fortune to this jungle outpost.

The *Hevea* tree grows up to 30 meters (100 ft) high; latex, which consists of about 30 percent rubber, serves to protect it from herbivorous enemies and quickly closes gashes. Untreated rubber is hard when cold, but becomes soft and sticky when warm because temperature changes cause the long, unjoined molecules to change position slightly relative to one another. This substance was of little importance for Europeans until 1839, when the American Charles Goodyear developed the vulcanization process – heating the raw material and adding sulfur, which transforms it into finished rubber with lasting elastic properties. This opened up a host of uses.

The Amazon basin was for a long time the only area producing rubber, and Manaus was able to dictate prices. Increasing demand triggered a rubber boom of immense proportions. Attempts to cultivate the *Hevea* in the Amazon were frustrated by outbreaks of South American leaf blight and, in 1876, the English botanist, Henry Wickham, took 70,000 rubber trees to Southeast Asia, forming the basis of the huge plantation industry. Yet it was not until 1915 that Southeast Asia was able to produce rubber in significant quantities and at much lower prices than Manaus. With falling prices on the world market, this signaled the end of the rubber boom only five years after Manaus had been at its absolute pinnacle, shipping 38,000 tons of rubber in 1910.

The grandeur of the cast iron which adorns buildings is a lasting reminder of the great age of the rubber barons. Schools, hospitals and offices now occupy the great houses originally built by the gentry, preserving at least the façade of the city's architectural heritage. The most imposing symbol of wealth is the 700-seat **Teatro Amazonas** ❹. The design was inspired by the Opéra-Garnier in Paris; virtually all the materials and fittings were shipped out from Europe, and it was European craftsmen who fashioned the stucco and painted the magnificent ceilings.

The citizens of the wealthiest city in South America sought to compensate for isolation by making life as comfortable as possible: Manaus had electric street lighting before London, and an electric tram system before Brussels or Boston.

The port

The floating quays that extend far into the Rio Negro are masterpieces of English engineering. Huge iron air tanks allow ships to dock all year round, even though water levels vary by 15 meters (50 ft) between the dry and rainy seasons. All the port facilities, including the elegant **Alfândega** (Customs House) ❸, were shipped out from England at the beginning of the 20th century. Some distance below the now-closed Customs House are the jetties of the Amazon steamers, where these river craft take on goods and passengers to transport them all over the Amazon basin.

Large enough to have a superstructure, these ships are usually owned by their captains, and ply regular routes which are announced on large boards on the upper deck. Cabins with bunks are few, so a hammock is a vital piece of luggage. It is hung up crosswise, providing a place to sleep for its owner, and a makeshift stowage space for luggage, which is placed either in it or beneath it.

PRECEDING
PAGES:
the Rio Negro
in the rains.
LEFT: Teatro
Amazonas,
Manaus'
Opera House.

The harbor is also where small river boats dock. They bring fruit, fish, cassava flour and jute to the city. The jungle metropolis is linked to the trans-Amazonian network of roads – of which many are now sealed – and has a modern international airport, yet traffic on the river continues to grow. Numerous powerful motor boats, seen around the beaches of the **Ponta Negra** every weekend, testify to the existence of a modern leisure industry.

Free trade zone

Since 1967, when Manaus was declared a free trade zone, foreign firms have been exempt from import duty, and so goods can be sold here at far better prices than in other parts of Brazil. The once resplendent business quarter near the harbor is degenerating into a bazaar for electrical goods. Manaus is decaying and developing explosively at the same time.

The economic decline which set in after the rubber boom, and the population growth and industrial development which started in the 1960s, have left this city and its 1 million inhabitants no time to stop and think. Despite an intensive housing program, a significant proportion of those who have migrated here live in a jumble of wooden huts, with no streets, no drinking water, and no drainage or sewage. The rotting garbage heaps of the slums, markets and municipal tips attract Black Vultures (known locally as *urubu*).

Tourism

The city's sights take little time: the Teatro Amazonas, the harbor quarter, the **Tropical Hotel** (which has excellent swimming facilities, but is 15 km/9 miles from the city center), and a small **zoo** of indigenous animals in its own park. For most visitors, however, the real beauty of Manaus begins as they leave the city, for the most fascinating areas are to be seen during trips on the **Rio Negro** and the **Solimões** and beyond.

At weekends, the bathing places along the Rio Negro and its tributaries are

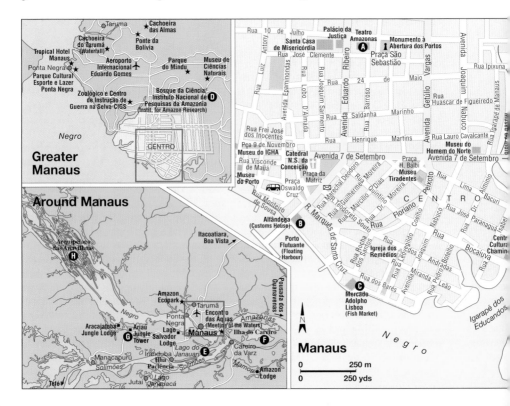

Map on page 244

favorite destinations for local people. The warm, clear, "black" water (actually the color of tea), and the absence of aquatic plants even close to the shore, make for delightful swimming. During *verão* (summer), which lasts from May to October, bathers seek out well-known spots such as the beaches at Ponta Negra or the **Cachoeira do Tarumã-Grande**, a waterfall on the **Tarumã**. Here, there are numerous food stalls selling roast fish and other specialties, soft drinks and beer.

Eighteen km (11 miles) below Manaus, the black waters of the Rio Negro and the beige-colored Solimões join to form the Amazon. Because of their different temperatures and rates of flow, the two at first remain separate rivers, flowing side by side in the same channel. A number of tour agencies offer half-day trips to see this natural spectacle. In good weather, this *"encontra das aguas"* and the extensive river landscape around Manaus can be seen from flights coming in to land at the international airport.

Decades of over-fishing in the waters close to the city have drastically reduced the population of larger fish species. Nevertheless, there is still an easy way to get an idea of the wealth of fish to be found in the Amazon – just visit one of the fish markets (the **cast-iron fish market C** next to the harbor is a particularly good one).

No visitor to the city should miss the chance to try the *peixada* fish soup (be sure to ask for it *sem espinhas*, i.e. without bones). This is prepared with two favorite Amazonian fish, the Tambaqui and the Tucumare. *Tambaqui na brasa*, fillets of Tambaqui cooked on a charcoal grill, is the high point of many barbecues.

Institute for Amazon Research

Manaus has, from its earliest days, been a haven for naturalists recuperating from their exhausting trips into the interior. The establishment of the **Institute for Amazon Research D** (Instituto Nacional de Pesquisas da Amazonia – INPA) in 1952 gave a decisive boost to efforts in Manaus

BELOW: a Black Vulture *(Coragyps atratus)* on the Manaus waterfront.

The Fish Market

The **Manaus fish market** ⊙ stands between the waterfront and the shopping centers of the city's free trade zone, a commotion of stevedores, street traders, shoppers and traffic. Men leave the boats moored to the jetties at a run, carrying bunches of bananas or *pupunha* fruit, sacks of manioc meal or crates of fish. Others hawk on the pavements. In the foul black water around the moorings, vultures fight over offal, and cooks at the foodstalls wash their plates.

The market, a monument in iron and stained glass, is the best place to see some of the extraordinary creatures that contribute to the Amazon's fish fauna. Throughout the year, despite the closed seasons which should restrict the capture of some species, fish such as Tambaqui, Arowana, Tucunare and a great diversity of catfish, can be bought, especially in the early morning, as well as huge specimens of Pirarucú, said to be the biggest freshwater fish in the world.

But while the species on display excite wonder, they also bear testimony to the destructive changes taking place in the Amazon's fishing industry.

Although many of the traditional *caboclo* fishermen (descended from the floodplains Indians) respect the closed season for certain species, the industrial fishing boats now monopolizing the markets openly flout it. From early December to late February no fresh Pirarucú, Tambaqui or Pau – which looks like a large silver coin – should be traded. But the high profits involved, and disputes between government agencies monitoring the industry, mean that most of the forbidden species can be found in the market.

Means of trapping the fish are also changing rapidly – most *caboclos* still use their skill and local knowledge to trap fish individually. Tambaqui fishermen tie their canoes to fruiting trees and attract their prey by hitting the surface of the water with a weight hanging from a pole, to reproduce the splash of a falling fruit. A traditional hunter of the air-breathing Pirarucú, having discovered the fish's lie, waits in his canoe to harpoon his quarry when it surfaces. But the large commercial boats, most of which supply the export market, lay down long monofilament nets in the traditional fishing grounds. These are responsible for an unsustainable level of harvesting, and could lead to the collapse of some of the Basin's riverine ecosystems. This over-exploitation is exacerbated by a tendency to dump the first catch, dead, into the water, if a second shoal of a more lucrative species is found. On the larger rivers of Mato Grosso, the great piranha shoals are being caught, either to be dried and sold as souvenirs (Japan imports dried piranhas by the thousand) or to be converted into fishmeal, for animal feed or fertilizer.

The effects of such over-exploitation, coupled with destruction in areas of the flooded forest on which the fish rely for their food, are already visible – the fish on sale are getting smaller. Certain species bear the brunt of the exploitation, and in time are likely to become extinct. For the people of Manaus, a dearth of fish would be disastrous. It is the cheapest form of available animal protein in the Basin, and the only means by which many poor families survive.   ❑

LEFT: fishing a the meeting c the waters near Manaus.

Map
on page
244

to conduct modern scientific research into the Amazon region. Successful teams specializing in biology and ecology, tropical medicine, agriculture and forestry swiftly developed. In 1962, a link was formed with a tropical ecology team from the Institute of Limnology of the Max-Planck-Gesellschaft in Germany.

At the beginning of the 1970s, INPA moved to its current home, a site of about 25 hectares (62 acres) at **Estrada do Aleixo**, 4 km (2½ miles) from the town center. Here, there are spacious laboratories and offices, a large herbarium with well over 100,000 specimens, accommodation for animals, and a good library. Since 1971 the Institute has been publishing its own scientific journal, the *Acta Amazonica*. Numerous INPA monographs devoted to special topics related to the Amazon region are also for sale.

The forestry division administers two nature reserves (the **Reserva Florestal Adolpho Ducke** and **Walter Egler**) on the Manaus–Itacoatiara highway, and one

(**Reserva Florestal Campina**) on the road from Manaus to Caracarai, but these are not open to the public.

The Biological Dynamics of Forest Fragments project has received considerable international acclaim, which INPA has run since 1976 in conjunction with the World Wildlife Fund and, since 1989, the Smithsonian Institution. Some 70 km (43 miles) north of Manaus, islands of jungle surrounded by grazing land, varying in size from 1 to 1,000 hectares (2,470 acres), serve as natural laboratories. Here, inventories are made of plants and animals and their population sizes, in order to understand how the structure of the jungle and the combination of species are changing in forest fragments of different sizes. The precise reasons for the decline in the number of species and the structural changes of the forest is being investigated as part of this project.

The website www.inpa.gov.br is a useful source of information relating to the reasons for the numerical decline in species and the structural changes of the forests. ❏

BELOW:
looking across the Rio Negro.

FLOATING MEADOWS

*Of great importance to the ecology of the
Amazon, these spectacular expanses of vegetation
support a wide variety of wildlife*

A widespread phenomenon characterizing flood areas of white water rivers is the "floating meadow." As the water level rises in still or slow-flowing reaches, floating plants assemble to form patches extending over several square kilometers. Anyone who makes a boat trip during the high water season will see dense growths of these aquatic plants lining the banks in the stiller river reaches.

The quickest and easiest way to reach this floating vegetation from Manaus is to travel along the channel which leads to the **Lago do Janauari** ❺, a small ecological park equipped to receive many visitors and thereby take the pressure off other areas of the river. Even more extensive floating carpets of aquatic plants, some of which last all year, can be found in the quiet waters around **Careiro** ❻, a large island in the Amazon. Here, visitors frequently come across expanses of the Amazon water lily *(Victoria amazonica)*, best seen from April to September. The black waters of the **Rio Negro** (which are crossed to reach these areas from Manaus) are too poor in nutrients to support aquatic plants, but during the high water season the nutritional substances of the white water of the Amazon reach the channel to the Lago do Janauari.

These meadows are the most important element in the production of primary organic material in the low-lying areas of the Amazon basin. Constantly exposed to direct sunlight, leading to water tempera- **BELOW:** water lilies coming into flower.

Map on page 244

tures of up to 35°C (95°F), they provide an environment for a massive aquatic ecosystem. The term "floating meadow" derives from the dominant plants: the *Paspalum* and *Echinochloa* grasses (growing to 2 meters/6 ft), the extensive carpets of water hyacinths, and a fine layer of aquatic ferns on the surface. Plants like the water lettuce *(Pistia stratiotes)* or the water fern (genus *Salvinia)*, cover the surface in a rich carpet.

Eichhornia crassipes, with its cluster of hyacinth-like violet flowers, is probably the best known of the surface plants. This, and the water lettuce, have been distributed throughout the tropics to cultivators of ornamental pools, from where they spread and become serious weeds in lakes, rivers and paddy fields.

The floating meadows at night are transformed into a hive of vivid, bustling activity. The beam of a torch pierces the floating carpet of vegetation to reveal a glistening patchwork of dewdrops and pick out the tiny, green crystalline eyes of numerous spiders. Spectacled Caimans can be seen as they glide silently through the water and mosquitos rise in swarms. This nocturnal microcosm is dominated by the tumult of colorful tree frogs which seek out this watery biotope for mating and breeding. Grasshoppers cling to the edges of leaves to feed on the abundant vegetation. Along the water's edge flicker the greenish-yellow lights of the glow-worm larvae of the genus *Aspidosoma*, attracting the young operculate snails which hatch at the height of the flood season.

The water beneath the mass of vegetation provides a safe refuge and hunting ground for young fish. The loose roots of the floating meadows teem with insect larvae, planktonic crabs, and freshwater shrimps. The vegetation is a source of food for the Amazonian Manatee, the only herbivorous mammal that lives exclusively in fresh water. All these creatures, unfortunately, are now threatened with extinction as a result of over-hunting and increasing river traffic. ❏

BELOW: the highly endangered Amazonian Manatee *Trichechus unguis).*

FOREST LODGES AROUND MANAUS

*Although Manaus itself is not a prime natural site,
within easy reach are some excellent nature
lodges that give great opportunities for observing wildlife*

The natural attractions of Manaus are extremely limited, so if you have several days to spare, the best choice is to stay in one of the forest lodges. There are now more than 10 of them within a 100-km (60-mile) radius around Manaus, and their number is increasing. These offer an easy, if more expensive, introduction to the rainforest than what's on offer upstream near Tabatinga and Leticia.

Many local agents represent only one particular place, so it is advisable to shop around. Illustrated brochures help you make a decision. None of them is cheap, with prices between US$50 and $150 per night per person, and a similar fee for transport by car or boat. In general, the further you travel from Manaus the greater the chance of seeing wildlife and having a more authentic Amazon experience. Check out www.brazil.org.uk which has excellent links to hotels and state tourist boards, for more information on accommodations.

At the top of the price list is the **Pousada dos Guanavenas**, five hours from Manaus, which has air-conditioned rooms and two swimming pools. Cheaper and closer are the **Amazon** and the **Janauaca Lodge**. On the rustic side, there is a growing number of small "eco-friendly" operators, such as Amazon Indian Turismo, Orquídea Amazônica and Amazon Nut

BELOW: a Red Howler Monkey.

Map on page 244

Safari. They work in conjunction with local people and offer simple, authentic accommodations and camping trips.

For those looking for creature comforts or traveling with children, an attractive option is the jungle tower, **Ariaú G**, with 271 rooms. Promoted by Rio Amazonas Turismo, it is situated on the banks of a small tributary, 35 km (22 miles) west of town on the opposite side of the Rio Negro, and connected with Manaus by a daily boat service or by helicopter.

Ariaú is built into the flooded forest, with minimal disturbance to the environment, standing on wooden pillars in an area that is submerged in water for several months each year. The lodge consists of three wooden towers fitted between the treetops, connected by a fantasy-inspiring network of stairs and wooden paths on various levels above the ground. The balconies of 16 comfortable rooms open directly into the canopy, and there is even the option of a Tarzan house on top of a 28-meter (90-ft) treetop. Two 41-meter

(135-ft) high observation towers protrude from the treetops and permit observation and photography of the canopy wildlife. The restaurants and bar are screened off as protection from mosquitos and from monkeys that demand a share of the dinner.

Be prepared for encounters with animals that have become tame after regular feeding: Coatimundis may gnaw at unattended belongings, and Woolly Monkeys may sling their tails around your neck. Squirrel Monkeys, Red Howlers, Brown Capuchins and Black Spider Monkeys also abound.

Bird-watching here is excellent. Forest birds include Long-billed Woodcreepers, Black Nunbirds, and Cocoa Thrushes. Lesser Yellow-headed Vultures, Yellow-headed Caracaras and Black-colored Hawks are frequently observed raptors.

As you stroll along the river, you may flush out Wattled Jacanas, White-necked Herons or Great Egrets, or study some of the kingfisher species, such as the crow-sized Ringed or the sparrow-sized Pigmy Kingfisher. ❑

BELOW: a Lesser Anteater.

RIVERBOAT EXPEDITIONS: THE ANAVILHANAS ARCHIPELAGO

One of the most pleasurable and fruitful ways of
seeing the rainforest is from the water on
a riverboat tour; these can be booked easily from Manaus

Without question, the most versatile and comfortable way to explore the Amazon is by a privately rented boat. It is home and observation platform at the same time, it permits the transport of heavy equipment and a large reserve of food and drinks, and the itinerary is established according to the passengers' preferences; be they bird-watching, collecting plants, fishing, or simply swimming. Stops are made at idyllic anchorages rather than close to a town, and the accompanying canoe permits the exploration of shallow tributaries or floating meadows away from the larger vessels owned by logging companies and hunters.

There is a fleet of several dozen boats in Manaus. It is worth adopting a flexible approach and making extensive enquiries before choosing one. Apart from large, established operators such as Selvatur and Amazon Explorers, there is a good choice among smaller enterprises such as Amazon Expeditions, Amazon Nut Safaris and Swallows and Amazons. The owners, like Moacir Fortes of Amazon Expeditions, are often captain and nature guide at the same time, and may excel in both skills.

Traditionally, river expeditions were made with 6–8-meter (20–26-ft) long canoes with outboard motor, and the night was spent in a hammock or tent on the

BELOW: the Solimões (left) and the Rio Negro (right) meet to form the Amazon River.

Map on page 244

beach. Canoe trips are still very much alive, but the trend is for larger riverboats, 15–25 meters (50–82 ft) long and built from wood in the traditional style. They have cabins for five to 16 passengers, and additional space for hammocks. The cabins are small, with berths and showers, and are kept reasonably clean. A fan is sufficient to allow sleep during the surprisingly cool nights. In addition to the captain and his helpers, the tour is accompanied by a cook, who may turn the catch of the day, Black Piranha, Rainbow Bass, or Peacock Bass, into an excellent dinner. Soft drinks and beer come refrigerated. The boat may even have a small library with scientific books on the Amazon.

Boat rentals, which include transport, accommodations and food, range from US$500–2,000 a day, depending on the size of the group. This sounds more expensive than it really is: if the group size is right, US$1,200–1,500 per person may be the fare for a two-week trip covering a 1,500-km (930-mile) distance.

These trips can be taken at any time of the year, but boats are booked long in advance for the high season, from March to August, when low precipitation and good weather coincide with high water, and when even small waterways become navigable. The forest is flooded all the way up to the canopy, and monkeys, parrots and toucans can be observed at eye level from the upper deck.

Imagination is the only limit to the objectives of a tour, if there is consensus among the passengers. You may wish to go up the **Solimões** to find the rare Red Uakari on river islands around **Tefé**, and stop en route on the **Lago Janauaca**, an area rich in Giant Amazon Water lilies, to search for birds like Hoatzins or Horned Screamers. Alternatively, you can travel several hundred kilometers up the **Rio Negro** and turn into its tributaries.

A small waterway from the south, the **Rio Caures** is a good area in which to search for the Brown Uakari. The **Rio Branco**, a large tributary from the north,

BELOW: Victoria Water lilies *Victoria amazonica)* on the Amazon.

is also navigable, and a 250-km (155-mile) trip, first by boat, then by canoe, leads to the **Rio Catrimani**, a particularly unspoiled part of Amazonia, with a chance to observe large flocks of at least five species of macaws, or of the rare Harpy Eagle.

The preferred destination for a tour of four to eight days is the **Anavilhanas Archipelago** (Arquipélago das Anavilhanas) ❶, on the Rio Negro west of Manaus. The first islands are reached after 60 km (37 miles), and the archipelago extends another 80 km (50 miles) further upstream. The Rio Negro at its broadest has a width of about 30 km (18 miles) – such a vast expanse that the forest on the southwestern bank is nearly invisible.

Completely different is the mosaic of small landscapes of the Anavilhanas. The river forks into legions of channels, often only 50 meters (165 ft) wide, separated from each other by narrow, spindle-shaped islands, several kilometers long. The channels are often 10 meters (33 ft) deep, and the islands are bordered by steep banks of astonishing height differences, considering the extremely slow movement of the river. During the low water season, it takes a little climb to enter the *igapó* forest on the small plateaus of these islands.

The area is designated a biological reserve, which has at least keeps squatters away. Small populations of large mammals like Jaguar, tapir, Capybara, and manatee persist, but poaching, particularly of the latter species, continues.

Anavilhana is a bird-watcher's paradise. From a canoe that slowly follows the river banks, one will come across a great number of large water birds; Anhingas, Green Ibises, Great Egrets, Great Blue and Boat-billed Herons, to name just a few. There are good chances to find Sunbitterns and Sungrebes, or a flock of Hoatzins resting in a treetop. Interspersed in the network of narrow channels and spindle-shaped islands are some large, shallow lakes. Astonishingly, big, black Muscovy Ducks are the only waterfowl which is abundant in these areas.

BELOW: a Red Squirrel Monkey.

Map
on page
244

To observe parrots, the only thing to do is to get up with the sunrise, when large flocks of Tui Parakeets, Orange-winged and Festive Parrots move from their nightly roosts to their feeding grounds. There are also good numbers of Blue-and-Yellow Macaws, and with luck you may spot one of the other macaw species. The early morning is also the best time to spot toucans or an Amazonian Umbrellabirds, which may hurtle across the channels. Large raptors, such as Yellow-headed and Red-throated Caracara, follow the channels and forage in the open land at the interface between forest and river.

In the heat of the day, forest birdlife does not become as quiet as in the open. A dive through the brush at the margins of the interior leads into more open vegetation and pleasantly cooler temperatures. It is easy to be overwhelmed by the diversity of small bird species. It is not difficult to identify some colorful species, such as the Yellow-tufted and Cream-colored Woodpecker, or the Green Honey-creeper.

As darkness falls, bats and birds of the night, like Blackish Nightjars and Band-tailed Nighthawks, flutter around the boat. There's a startling sight in the perfectly dark night of the Amazon: Manaus, though nearly 100 km (60 miles) away, can be located as a golden glimmer on the eastern horizon.

A trip to Anavilhanas often includes a detour into the **Rio Cuieras**. For a stretch of nearly 50 km (30 miles) along its course, the forest consists mainly of dead tree-trunks. A few years ago, high water levels persisted several months beyond the normal end of the season, exceeding the adaptation of these species and killing most trees. Another detour leads to **Novo Airão**, a small village on the southwestern bank of the Rio Negro, where Amazonian river boats (probably including the one in which you travel) are built from the hardwood trees of the surrounding forest. Further downstream, you can swim at the **Praia Grande** beach, whose white sand would be the pride of any seaside resort. ❏

BELOW:
a sandy river beach on the Rio Negro near Manaus.

PICO DA NEBLINA NATIONAL PARK

Although this is one of the largest tracts of protected forest in Amazonia, its ecological integrity is under threat from ill-conceived "development" projects

Parque Nacional do Pico da Neblina ❷ contains Brazil's highest mountain and is its second largest national park. Together with a park across the Venezuelan border, it forms one of the largest tracts of protected land on earth. It is one of the least accessible areas of the Amazon; its central part being accessible only by helicopter.

The park stretches from the Venezuelan border south to the Rio Negro, which forms its southern boundary, roughly between Uaupés and Tapurucuara (about 700 km/435 miles upstream from Manaus). In Brazil, 22,000 sq km (8,500 sq miles) are gazetted as park, and the ad-joining **Parque Nacional Serranía la Neblina** on the Venezuelan side, has an area of 13,600 sq km (5,250 sq miles).

Most of the Brazilian park is in the lowlands, but the northern sector rises to the **Pico da Neblina** and the **Pico 31 de Marco**, at 3,014 meters (9,888 ft) and 2,992 meters (9,816 ft) respectively, right on the Brazilian/Venezuelan border. They consist of plutonic rocks such as ortho-quartzite and gneiss, which are typical of the Guayana shield; including the Roraima tepuis, which is about 700 km (1,860 miles) north-east of Pico da Neblina. These areas also receive the highest annual precipitation in Brazil (3,000 mm/118 ins).

BELOW: larva of an Owl Butterfly *(Caligo).*

Map
on page
257

Due to the remoteness of the lowland part of the park, good populations of the rare Brazilian Tapir, the Jaguar and even the Giant Otter survive here. An attraction for birdwatchers is the Guianan Cock-of-the-Rock, which is restricted to districts close to the northern border of Brazil.

The real biological interest stems from the large number of endemisms in the montane vegetation, the highest number in the Brazilian Amazon. Botanical expeditions to the upper ranges over the last decades estimate that over 50 percent of the plants are species new to science.

Access to the park is possible by canoe on the **Rio Cauaburi**, a widely meandering river that is the principal drainage of the mountain range to the south into the Rio Negro. A 200-km (125-mile) trip along this river comes within 20 km (12 miles) of the mountains, but a closer approach requires a full-fledged expedition.

This remote park can be reached from São Gabriel da Cachoeira and has better access since the construction of the Perimetral Norte, Brazil's controversial highway BR210. The Perimetral, the northern counterpart to the Transamazônica, provides access to undeveloped parts of the Amazon from the Atlantic coast and the west of Brazil. It was constructed as part of a scheme called Avança Brasil (Advance Brazil), initiated by the country's military governments for spurious "defence" reasons. Scientists have pointed out that it threatens the very existence of the Amazon rainforest.

Scientific and conservation organizations, such as Forest.org, argue that new corridors between densely populated areas will inevitably lead to a drastic increase in deforestation and exploitation of the forest and that the government should be encouraged to pursue environmentally sustainable development. Although the creation of new roads in the region will undoubtedly bring short-term growth and prosperity to the area, in the long run it will threaten Brazilian and global ecological sustainability. ❑

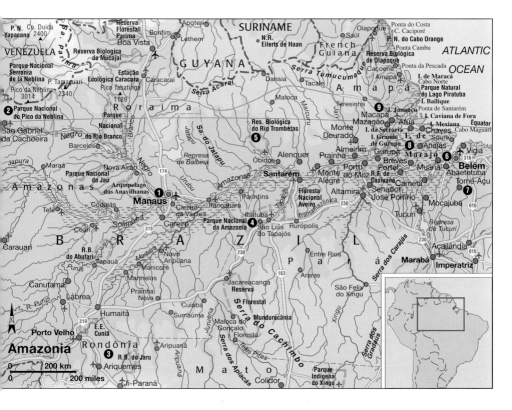

The Tragedy of Rondônia

The development of the state of Rondônia ❸, in the west of the Brazilian Amazon, has been described as the world's greatest environmental tragedy. Until the late 1960s, the state was effectively pristine, and contained some of the most diverse ecosystems on the planet. Today, well over a quarter of the land there has been cleared, and some projections suggest that all of Rondônia's primary forests could be gone by the end of the first decade of the new millennium.

The destruction began when impoverished peasants left central and southern Brazil and followed a new road, the BR364, constructed by the government in 1967 to provide access to the western Amazon. In the early 1970s the government founded settlements to accommodate them. When news of this distribution of land and housing, together with rumors of fertile soils, spread among communities of the

south, colonists began to flood towards the new frontier: in just one decade the population of Rondônia rose from 110,000 to 500,000.

The government colonization agency established more settlements, but these could absorb only a small number of the new arrivals. Others took land for themselves, clearing the forest to secure their ownership rights. Ranchers from elsewhere in Brazil arrived and began to establish properties, pushing the small farmers out. The government gave land to them too, often – since the ranchers applied for adjoining plots with false names – tens of thousands of hectares. The ranchers cleared their land to raise its speculative value, earn government incentives and prevent others from taking it.

Amid the rush for land, and the government's failure to take account of the conditions of the settlers, the infertility of Rondônia's soils was ignored. In contrast to rumors prevailing elsewhere in Brazil, only 9 percent of the land was suitable for agriculture. Since much of this fertile territory fell into the hands of the large landowners – many of whom simply cleared it and left it unfarmed – the majority of the peasants arriving in Rondônia settled on lands which could not sustain their crops.

Most of the land in Rondônia was owned or used by indigenous peoples and peasants collecting rubber and other forest products. The government settlement schemes ignored indigenous peoples, and settlers seeking to establish properties for themselves drove out, and in many cases murdered, the original owners. The colonists introduced new diseases, which caused epidemics. Between 1971 and 1974, for example, the population of the Surui is believed to have declined by half. The Uru Eu Wau Wau, some of whom have had no peaceful contact with the outside world, may be suffering a tragedy of a similar scale, as colonists flood into their territory to mine tin.

Settlers arriving in Rondônia during the 1980s had to buy land from other colonists, or find work as waged laborers. The land was so far from roads that they had no access to markets, schools and hospitals. For them, as for the established settlers, life was, and remains, hard. The infertility of the soil and the abundant pests and weeds of the Amazon mean that crop yields are low and the land is quickly exhausted. Middlemen monopolizing the markets force the settlers to sell their pro-

LEFT: ruined forest in Rondônia.

duce at low prices, and government trading agencies may take so long to supply the money for the crops they buy that inflation renders it worthless. Agricultural laborers can make more money in Rondônia than small landowners.

The conditions of settlers and the original inhabitants of the state deteriorated throughout the 1980s, as the government, with the World Bank, accelerated the development of Rondônia. The government pledged to protect the peoples and environment from the effects of its new program. All these plans ended in disaster.

By 1991, 1½ million people had arrived in Rondônia. Road networks absorbed over half the program's budget, while only 2.5 percent was for protection and aid for the indigenous peoples, and conservation received only 1.4 percent. Reserves were invaded, a new road, the BR429, was built through the fragile ecosystems between the main highway and the Bolivian border, and fertile land continued to concentrate in the hands of big ranchers.

Most colonists were unable to invest in the perennial crops – like rubber and cacao – which they were supposed to farm. These required capital they did not have, and a guarantee that the land would remain in their possession. As peasant settlers feared expulsion from their lands, they farmed as if they would have to leave, planting crops which matured quickly but exhausted the soil, instead of trees whose cultivation might have been sustainable, but which would benefit the land invaders, rather than the planters. Since the colonists failed financially, most agricultural land fell into the hands of large ranchers: by 1985, 66 percent of the land was owned by 1.9 percent of the farmers.

The World Bank is now considering the funding of a new government project, to rectify the mistakes of the old ones. Already the latest proposal has attracted criticism, as it appears to be imposing a development model conceived by planners in Washington and Brasília, rather than the people who are to be affected.

Despite the bad planning, corruption and insensitivity which have accompanied development in Rondônia, some settlers have benefited from migrating. For some, the acquisition of land, however infertile, was an improvement on the landless and laborless circumstances they suffered in the south. Some have succeeded, with the help of the unions they have formed, in defending their properties from

ranchers, and in planting economically viable crops which do not destroy the land. In the oldest of the government-sponsored settlements, at Ouro Prêto do Oeste, the peasants are experimenting with the production of honey, vegetables and treecrops, with some success.

Perhaps the greatest current threat associated with the Rondônian frontier is that of timber cutting, for it extends exploitation into areas which would otherwise have remained untouched. The sawmills now lining the BR364 and many of its feeder roads, having exhausted the high-value timbers elsewhere in the state, are now invading the reserves.

As peasants are forced by poverty or ranchers to leave the Rondônian frontier, they move northwards in search of new lands. The BR364 has been extended deep into the state of Acre, and many of Rondônia's failed colonists, as well as its expanding landlords, have followed it, coming into conflict with indigenous people and rubber tappers as they take possession of their lands. Rondônia has become the gateway to the western Amazon, and the tragedies taking place will be repeated elsewhere. ❏

Map on page 257

PARQUE NACIONAL DA AMAZONIA

This fully protected park is home to many endangered species

Parque Nacional da Amazonia is one of the few areas in Brazilian Amazonia with formal legal protection. No facilities exist and there is no regular access, but the beauty of the park and its proximity to **Santarém** mean that it will probably become a tourist attraction and will help safeguard that city's economy. To enter the park, permission is required from IBAMA at Av. Tapajós 2267, Santarém, or Postudo Fomento Estrada, 53 Bis, km 02, Itaituba (73 km (45 miles) away.

Santarém, with about 300,000 inhabitants, is the biggest settlement between Manaus and Belém, to both of which it is connected by daily flights. The city is on the southern bank of the Amazon, east of the mouth of **Rio Tapajós**. There are several good hotels, and the town is a pleasant place to stay, with colonial squares, simple waterfront cafés and easy access to the forest. Several local travel agents offer Amazon and Tapajós boat tours, lasting from a few hours to four days.

The park, also called **Tapajós National Park ❹**, is about 300 km (185 miles) from Santarém, reached either on the road that detours from the Transamazonian Highway to Santarém, and then continues towards **Itaituba**, or by boat, which reaches the park a few kilometers beyond the first waterfall, **Cachoeira Maranhao Grande**.

The park's area is about 10,000 sq km (3,860 sq miles); most of it extends from the river to the northwest. Tapajós has a full complement of large vertebrate fauna, including endangered species like the Giant Otter, and provides excellent opportunities for observing mammals and birds.

The mouth of the Rio Tapajós, about 700 km (435 miles) from the Atlantic, is one of the furthest points on the Amazon that is reached by the oceanic tides; these cannot run beyond the narrows close to Obidos or up the Rio Tapajós because of its cataracts. The lower part of the Tapajós is broad and slow-running. This expanse of water, up to 30 km (18 miles) wide, is believed to be an inundated valley, created during the Ice Age.

Most Amazon basin rivers are classified as either black water or white water. The former drain massive lowlands, extracting the brown humic acids from the forest soil; the latter originate from the Andes run-off and owe their color and name to suspended soil particles. The Tapajós is in a third category: clear rivers. These drain the extensive highlands of the Brazilian Shield, whose ancient rocks resist quick erosion. Lack of suspended particles or humic acids makes the waters of the Tapajós beautifully transparent. ❏

LEFT: decaying wood forms a habitat for this forest fungi.
RIGHT: a Yellow-rumped Cacique perched above a colony of nests.

THE THREATENED TURTLES OF RIO TROMBETAS

*Now seriously endangered, the Giant River Turtle
used to be common all along the Amazon; today they are
found mostly within one reserve*

During the past three centuries, the Giant River Turtle *(Podocnemis expansa)*, whose shell can be 90 cm (35 ins) long, has been intensely persecuted for its meat and eggs. As a result, its populations have suffered a catastrophic decline. Historical records show that in the 19th century, in the Amazon basin alone, millions of turtles nested annually, and as many as 48 million eggs were harvested by humans each year.

In the 1970s, a survey of 14 Amazonian rivers in Brazil conducted by the Brazilian Institute of Forest Development (IBDF) showed that there remained fewer than 15,000 females nesting annually in Brazil.

The most important nesting area identified by the IBDF survey was at sand banks (or *tabuleiros*) located in the **Reserva Biológica do Rio Trombetas ❺**. There, during the 1970s, an average of 5,500 females nested each year. These sand banks are inundated most of the year, but during the dry season (September to January) they are exposed. When the dry season commences, the reproductive adults leave their feeding areas in the lakes and flooded forests, where they feed largely on fruits, and travel to the *tabuleiros*. They remain offshore for about a month prior to nesting, which takes place during September, October or November, and lasts

BELOW: turtles basking on the river bank.

Map on page 257

about two weeks. The turtles usually nest at night in groups of tens or hundreds, emerging onto the beach at approximately the same point and laying their eggs (about 90 per clutch) within a small area, often where the elevation is highest, to protect the eggs from flooding should the river rise before the end of the incubation period of 44 to 55 days.

Nesting turtles are extremely shy and easily disturbed by light, noise, human activity, and water pollution. Visitors are not allowed on the beach at nesting time, or in the vicinity during the breeding season. Boat traffic on the river can discourage the turtles from laying: during 1984, little or no nesting occurred along the Trombetas, possibly because of unusually heavy boat traffic during the rapid growth of a town 50 km (30 miles) upstream.

Although the species is protected by law, the meat fetches a high price on the black market. The turtles are illegally captured all year round but are most vulnerable when they are nesting. Incidents during the 1985 season dramatically illustrate how far people will go to harvest them. Shortly after nesting began, a group of 30–50 armed men from the impoverished local *caboclo* population drove IBDF personnel away, took control of the *tabuleiro*, and captured between 100 and 300 turtless. The *tabuleiro* was finally liberated by an armed contingent of Federal Police, but the egg clutches left behind were clandestinely excavated by poachers.

The level of protection afforded the turtles improved greatly in subsequent years, but the nesting population has continued to decline. Although the poaching problem on the nesting beach has been brought under control, the remaining turtles are seriously threatened by habitat disruption. Construction of a hydroelectric dam has been proposed upriver of the nesting site. Should the dam be built, the turtle population would be threatened by an increase in boat traffic, possible erosion of the *tabuleiros* induced by unnatural oscillations in water levels, and by water pollution.❑

BELOW: a female turtle burying her eggs.

The Jari Project

T he Jari Development Project, one of the largest in Amazonia, is located on the Rio Jari. The project, second only to the Carajás mining project, lies 500 km (310 miles) west northwest of Belém and extends from the main Amazon River 120 km (75 miles) up the Rio Jari. The project area covers a plateau extending 160 km (100 miles) east to west.

The project was the dream of multi-millionaire Daniel K. Ludwig, who in the early 1960s forecast a shortage of paper. Acting on this prediction, he bought a tract of land of 15,000 sq km (5,800 sq miles) for three million dollars, to cultivate fast-growing timber on the upland and rice in the floodplains.

Approximately 30,000 employees were required to run a project the size of Jari. Most of them were housed in **Monte Dourado**, a new city on the banks of the Rio Jari some 22 km (14 miles) from Munguba, the site of the industrial complex, and provided with a power plant, schools, hospital clinics and roads.

Industry

The Jari Project was based initially on gmelina (*Gmelina arborea*), a fast-growing Asian timber that produces up to 160 cubic meters (5,650 cubic ft) of wood pulp per hectare on a seven-year rotation. Eventually, the Caribbean pine (*Pinus caribea*) was introduced and, more recently, *Eucalyptus deglupta*. Genetically modified eucalyptus root cuttings and pine seeds have since produced results. Experiments with other species are underway. The project also planned to produce rice in the floodplains.

Jari is best known for its pulp mill. Designed in Finland and constructed in Japan at a cost of $270 million, it was built on barges and towed to Amazonia over the Indian and Atlantic Oceans and placed on pilings using the high waters of the flood season and additional dams to float it into place. The power plant is fueled by wood and black liquor, a by-product of pulp production.

Produce

At present, 1,100 sq km (425 sq miles) of forest are planted, mainly with gmelina. The plan-

BELOW: flooded forest.

tations are being increased at about 50 sq km (20 sq miles) per year. This is well within Brazil's regulation stipulating that development projects leave half of the forest intact.

Jari is also a major producer of livestock and is now concentrating on the uses of the water buffalo in the lowland floodplain area, and plans call for building up the present herd of 6,000 head to 35,000. Although food is currently flown in from Belém, plans for producing food at Jari are underway; this includes rice, which is produced in the floodplain area of the main Amazon River. Rice fields were made by clearing areas of the flooded forest and savanna, followed by leveling and diking.

After the project was well under way, an enormous deposit of kaolin was discovered just across the Rio Jari from Munguba. Kaolin is a fine white clay used for paper surfaces and in porcelain, paint and other products. As it is processed, kaolin is mixed in an open pit and pumped as a slurry through a pipeline that goes under the river to the Munguba processing plant. At present, 200,000 tons of kaolin are shipped to Europe each year.

A colossal mistake

The trees at Jari grew much more slowly than was predicted and the project ran into many political difficulties in Brazil. The Brazilian authorities refused to support the infrastructure of the Jari towns, and in December 1981, Ludwig decided to sell the wood and pulp project. This caused him considerable financial loss, and cast doubt on the viability of such large silvicultural plantations in the Amazon region.

As part of a cost-cutting program, 4,000 Jari employees were sacked, and, at the time of its sale in early 1982, Jari was losing $100 million per year. The project was taken over by a consortium of 27 Brazilian companies. Ludwig invested over $1 billion in Jari. It was sold for $280 million ($180 million in debt assumed by the consortium and $100 million payable over three years).

The debt on the $200-million loan for the pulp mill was saddled on the already deeply indebted Brazilian government through the semi-official Banco do Brasil. The government also agreed to provide transportation and communication facilities to serve the project. ❑

BELOW:
Brazil nut trees at dusk.

BELÉM

As an alternative point of entry to the Amazon,
this port provides an adventurous and fascinating
introduction to the region's peoples and environment

Belém is the biggest port of the Amazon delta, and second only to Manaus as the most important gateway to Brazilian Amazonia. It has regular connections by plane with all the major cities within the region as well as with Rio de Janeiro and Brasília, and it is well served by regular bus services from the south of the country.

Belém serves as a gateway to **Marajó Island**, which is the largest river island in the world, but there are no tours from here into the rainforest. Belém is not yet like Manaus, which has numerous reputable small boat companies and is a regular point of departure for passenger boat expeditions. For travelers on a budget, however, there are cheap public boat services to Santarém and Manaus for access to the rainforest in those areas.

Belém ❻ was founded by the Portuguese in 1616 to initiate the colonization of an Amazonian empire. Due to its importance as the leading port of the region, it has grown to today's population of more than one million inhabitants. The city is built on a promontory, 120 km (75 miles) from the Atlantic Ocean, between the **Baía do Guajará** to the northwest and the **Rio Guamá** to the south, which flows around Belém into the bay. Separating Belém from Marajó Island, the bay is 40 km (25 miles) wide and is part of the gigantic Amazon delta formed by the confluence of the **Rio Tocantins** and the **Rio Pará**, a southern arm of the Amazon.

BELOW: a Belém street market.

Maps on pages 257 & 267

Like Manaus, Belém could not be described as a particularly clean or attractive city, but a combination of several city parks, the zoo-botanical garden of the Museu Emílio Goeldi, and sidewalks and markets along the bay, invites a stay for two or three days.

Most of the hotels are downtown, more or less within walking distance of the bay, port and fish market. The **Mercado Ver-o-Pêso** 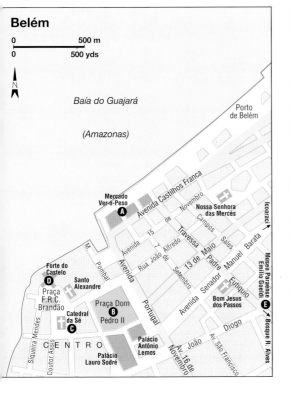 is an interesting starting-point for a walk through the town. The market hall, which dates back to 1688 (when it served as a customs checkpoint), and the surrounding stalls serve as an excellent introduction to the fish fauna of the Amazon delta. After the cleaning of the fish, the innards are thrown on the river's mudflats, a haven for Black Vultures.

It may be a bit strange to find beauty in this scene, but it is somehow reassuring that so many large birds can survive close to human habitation. Belém is actually only their feeding territory. Vultures breed on the ground of closed forests, not in cities. Their population density as a breeding bird is quite low and, consequently, many of the thousands of vultures that feed around Belém are likely to breed several kilometers away. Only the Black Vultures come into town to forage; the Lesser and Greater Yellow-headed Vultures and King Vultures are seen only occasionally, circling high in the sky.

A few species of birds that naturally occur on forest margins have adjusted to life in the city. Palm Tanagers sing from the roof of the Ver-o-Pêso market, and on a stroll through nearby parks and squares, such as the **Praça D. Pedro II** ⓑ, you often see Pale-breasted Thrushes, Blue-grey and Silver-beaked Tanagers. The latter may compete with various Kiskadee Flycatchers for the distinction of being the birds that have gained most from Man's transformation of Amazonian landscapes, since both species are seen virtually everywhere in the region, in gardens, parks, scrubland, and secondary forests.

Strolling through town

Continuing to the southwest, a stroll through town leads to the **Catedral da Sé** ⓒ and the nearby **Forte do Castelo** ⓓ, the first building to be erected after the founding of Belém. The park between the cathedral and fortress, **Praça F.R.C. Brandão**, is a particularly quiet part of town – a place to relax. Many of the impressive trees in this and other parks are mangoes, a species that actually originated in tropical Asia. (Beware of falling objects! The children of Belém throw sticks and stones into the trees to harvest the ripe fruits.)

A safer place to relax is a restaurant in the fortress: enjoy an excellent lunch or dinner of fried fish, while watching the river dolphins surface in front of the restaurant's veranda. The view from here over the huge expanse of water to a remote mangrove should not be mistaken for a view of Marajó Island across the Amazon. Marajó is to the north, beyond the horizon, and this is only the relatively small mouth of the Rio Guamá into Guajara Bay. Adventurous travelers who want to explore this region by public transport will

Belém

0 ————— 500 m
0 ————— 500 yds

N

Baía do Guajará

(Amazonas)

Porto de Belém

Mercado Ver-o-Peso ⓐ

Avenida Castilhos França

Novembro

Nossa Senhora das Mercés

Campos Sales

Travessa

15 de

Rua João Alfredo

13 de Maio

Padre

Senador Manuel Barata

Icoaraci

Euriquio

Museu Paraense Emílio Goeldi ⓔ

Bosque R. Alves

Avenida

Forte do Castelo ⓓ

Santo Alexandre

Praça F.R.C. Brandão

Catedral da Sé ⓒ

M. Pombal

Avenida

Portugal

Praça Dom Pedro II ⓑ

Bom Jesus dos Passos

Diogo

Siqueira Mendes

Doutor Assis

C E N T R O

Palácio Antônio Lemos

Palácio Lauro Sodré

Av. 16 de Novembro

João

Av. São Francisco

find the river boat jetties along the **Siqueira Mendes** road, which runs directly behind the fortress.

Nature in the city

Belém is famous, in naturalist circles, for the **Museu Paraense Emílio Goeldi** , often referred to simply as the Goeldi Museum. This is a center of ecological research in the Amazon, founded in 1866 under the leadership of the naturalist, Ferreira Penna, as the Pará State Museum of Natural History and Ethnology. It owes its present name to the Swiss zoologist, Emil August Goeldi, who was particularly influential around the turn of the 20th century in promoting the institution's public programs.

The museum has research departments that include anthropology, archeology, botany, zoology, geology and geography. Today, most of the research facilities are located far from the traditional site of the museum, on the outskirts of Belém, in the university complex.

Most visitors to Belém will not be interested in these research facilities but rather in the museum's traditional site, the zoo-botanical garden, which is close to the center of town, about 3 km (2 miles) east of the Ver-o-Pêso market. This park, a 5-hectare (12-acre) block of the city, holds about 1,000 plant species, from gigantic emergent forest trees to tiny orchids and water lilies (*Victoria amazonica* sp.).

Sloths, Agoutis and other mammals roam freely in the park, which is also home to forest birds, such as the Canary-winged Parrot. The biggest attraction are the captive animals, a unique opportunity for close observation of well-kept animals of the Amazon that are rarely seen in the wild. An enclosure to the right of the entrance gives a chance to pet a Capybara or a Brazilian Tapir. A pool further ahead to the left is home to an Amazonian Manatee. Other mammals on exhibit are Jaguars, Ocelots, several monkeys, and anteaters.

Among the birds are Harpy Eagles, Golden Parrots, macaws, guans and curassows. The compound for the Black Caimans holds a particularly large individual, and a nearby pond is home to a number of Giant Amazon Turtles. Two hundred and fifty species of fish are kept in the aquarium, nearly 10 percent of those found in the Amazon river system.

An even larger place to explore nature in the middle of Belém is the **Bosque Rodrigues Alves**, 16 hectares (40 acres) of protected forest with about 2,500 Amazonian trees. It is on the **Avenida Almirante Barroso**, 4 km (2½ miles) further east from the Goeldi Museum.

Belém may become a bit tiring after two or three days and the best escape is to one of the beaches. You can choose from **Outeiro**, only 35 km (22 miles) away, or **Mosqueiro**, which is 86 km (53 miles) from the city. The latter, an island accessible by bridge, has numerous hotels, and is quiet during the week but crowded at weekends. Mosqueiro and other beach resorts have a hinterland with scrub and secondary forest that is richer in wildlife than it may appear at first glance. ❑

Map on page 267

LEFT: waiting for the boat to cross to Belém

Tomé-Acú

O f all the settlers in the rainforests of Amazonia, it is the Japanese colonists living five hours to the south of Belém who have developed some of the most exciting new farming techniques.

In 1929, in order to relieve domestic unemployment and population pressure, Japanese farmers were encouraged by their government to move here. In the new colony of **Tomé-Açú** ❼, they tried at first to cultivate rice, much as they had at home. As with crops cultivated by many other newcomers to Amazonia, the rice failed, smitten by poor soils and pests. They then turned to cocoa production, but were again foiled, this time by fungus. Finally, the colony, racked by economic failure and disease, began to depopulate.

But in 1942, when Brazil declared allegiance to the Allies during World War II, the Japanese settlers were rounded up and Tomé-Açú became an internment center. Immediately after the war, the replenished community began to plant pepper, which until then had never been grown in the Amazon. It was immediately successful, and by 1960 these colonists had not only replaced Brazil's pepper imports, but were responsible for 5 percent of world trade.

In 1961, they were hit simultaneously by plant disease and low prices. Once again, many farmers were made bankrupt and forced to leave their land. But those who remained recognized the need to diversify, to avoid the dangers of relying on just one crop.

They began to experiment with trees and shrubs both native to the Amazon and from elsewhere, and soon found that they could create markets for products seldom grown commercially in Amazonia. By planting trees of different heights on the same patch – mimicking the structure of the rainforest – they discovered that they could reduce the area they needed to clear and conserve nutrients and rainwater. This meant that they were not forced to abandon their land because the soil had become exhausted after a few years, as happened to most Amazonian colonists.

The result of this diversification is that the Japanese farmers are now selling at least 55 crop products. Their economy is more robust than that of other settlers, since they are not so affected by the failure of one crop. They process and market their products together, through a cooperative, which ensures that they can maintain prices. The strength of the cooperative is such that the farmers cannot be dislodged from their lands by ranchers. While other colonists are forced by the threat of expulsion to grow crops which mature quickly but exhaust the soil, the Japanese can invest in trees which take 30 years to yield.

The Japanese farmers are anxious to preserve their resources for the benefit of their heirs. But while some of the principles they have developed may benefit others, the fertilizers, pesticides and intensive labor they employ are likely to render their system too expensive for others to copy faithfully.

Employment and ownership in Tomé-Açú and the surrounding villages bear testimony to the success of the farmers. Japanese is the first language spoken here, and in the settlement of Quatro Bocas, in the heart of Amazonia, are two of the reputedly best Japanese restaurants outside Japan. ❑

RIGHT: a climbing *Peperomia* in the rain forest.

MARAJÓ ISLAND

*Within easy reach of Belém, Marajó is the largest
river island in the world. Although used mainly for cattle rearing,
it still has a fascinating selection of wildlife*

The mouth of the Amazon River extends from Cabo Norte over 300 km (186 miles) – about the same distance that separates London and Paris – to **Curuca Island**. The river discharges through two large arms, the main channel to the north of **Ilha de Marajó** ❽ and **Rio Pará** to the south. The estuary embraces several large islands and is in a constant state of transformation, making navigation charts outdated before they are even printed. One feature of this great river is that, unlike the Mississippi or the Nile, it has no delta of accumulated mud which extends into the sea but the amount of sediment in its waters stains the ocean for hundreds of kilometers and turns sandy beaches into mudflats in French Guiana.

The main city on the north arm of the Amazon mouth is **Macapá** ❾. The historic fort of Macapá, on the promontory in front of the city, attests to the defensive role of this settlement in securing the Amazon for Brazil. Today, Macapá is a modern city, in spite of its lack of road connections. A three-day boat trip or 45-minute flight separates Macapá from **Belém**, the principal city on the southern arm of the Amazon.

The Amazon estuary is the site of a tidal bore, locally called the *pororoca*, which has received more than its fair share of publicity. This wave, which runs upriver at a speed of 10–15 km (6–9 miles) an hour, results from the strong spring tides overcoming the river's current in shallow

BELOW:
rounding up
water buffalo
at Marajó.

Map on page 257

waterways no more than 4 meters (13 ft) in depth. When the bore occurs, which is always at the lowest point of the tides, a roaring sound is heard up to 5 km (3 miles) away, followed by the appearance of a wave 1–2 meters (3–6 ft) in height. The phenomenon is most common in January to June on the coast near **Maracá Island** and on the **Rio Araguari**.

Marajó Island, in the Amazon delta, is the largest river island in the world, with an area of over 48,000 sq km (18,500 sq miles; about as large as Switzerland). It is really a complex archipelago. Soils under Marajó are river sediments going down to a depth of over 2,000 meters (6,500 ft); those on the western part of the island are recent deposits. The island's eastern half consists of low-lying natural grasslands that can be underwater for up to four months each year. Throughout the island, high soil fertility persists. The western portion is forested with some of the densest and most handsome vegetation in the whole Amazon. Cattle and the Asian Water Buffalo are found on the large ranches that dominate the region near **Soure**, **Salvaterra**, and **Cachoeira do Arari**. The economic mainstay of the forested part of the island is timber, since rubber extraction can no longer provide an adequate income for tappers and their families.

Ever since colonial times, Marajó, and the associated islands of **Caviana** and **Mexicana**, have been known for Marajorara ceramics, distinctively designed pottery associated with Indian cultures that disappeared before European contact. The finest collections of these ceramics are in the Goeldi Museum in Belém, and the Marajó Museum in the picturesque town of Cachoeira de Arari. Replica vases, pots, plates, and burial urns with painted figures and geometric designs are still produced on Marajó and in Icoaraci, near Belém, but the mystery of Marajó's pre-Columbian inhabitants remains unsolved.

Water forms the background for life in Marajó. River transport is all important, and the people are almost born in the

BELOW:
the Southern
Lapwing is one
of the few
shore birds of
the Amazon.

canoes and boats that are their cars and buses. A river trip to **Breves**, or one of the interior towns of Marajó, follows narrow tidal canals, often lined with houses. The **Strait of Breves**, a short-cut from the southern mouth of the Amazon to the main Amazon channel, is so narrow that ships brush the trees on the banks. Navigation is difficult here and some blind corners require 120° turns in a few hundred meters.

Passengers are always greeted by the inhabitants of the Breves channels with calls of "*Cunardo*." This harks back to the early years of the 20th century, when English ships of the Cunard Line plied these waters. Young children take to canoes when a ship passes, riding the ship's wake like surfers on an ocean beach, to catch small gifts thrown by passengers.

Regular passenger boats run between Belém and Soure and other towns, such as **Ponte de Pedras**, **São Sabastiao da Boa Vista**, **Breves**, and **Gurupá**. Destinations on the northern coast or in the interior of Marajó are more difficult to reach.

Arari Lake, in the interior of Marajó, is only 4–7 meters (13–23 ft) deep in the rainy season and dries to pond-like depths in the dry season. This is one of the principal fishing grounds for the Belém market, but over-fishing has caused a serious decline in the stocks of the Peacock Bass *(Cichla ocellaris)* and the Pirarucú *(Arapaima gigas)*. A brisk trade in ornamental fish for the export market, almost on the scale of Manaus, begins in the small streams near Soure and Ponte de Pedras.

How long this uncontrolled fishing can continue is unknown. The large, well-equipped commercial fishing fleet in **Marajó Bay** (actually the southern mouth of the Amazon) concentrates on river catfish, such as the Piramutaba *(Brachyplatystoma vaillantii)*. On the northern coast of Marajó, fishermen still use fish traps, locally called "corals," to catch fish and Amazon River Manatee stranded at low tide. Cattle ranchers complain that the fishermen are adept at hooking a cow on their line when no one is watching.

BELOW: a Brazilian Tapir swimming.

Map on page 257

Birdlife abounds on Marajó, and several lodges have installed special hides for close-up viewing of water fowl and towers for tree-top birds. One of the finest lodges is the **Fazenda Bom Jardim**, near Soure, where much of the original faunal research for the island has been conducted. The southern margin of Marajó has mangrove forests and extensive mudflats covered with the 6-meter (20-ft) aroid *Montrichardia arborescens*, with giant heart-shaped leaves. This plant cover makes a home for the hoatzin, a relative of the cuckoo, whose young have claws on their wings which enable them to scale vegetation.

Mangrove forest is an experience in itself. The Amazon estuary has mangrove around **Caviana**, **Mexicana**, and **Janauca Islands**, and on the western margins of Marajó. In the southern arm of the Amazon estuary, mangroves penetrate as far as the saline waters. Isolated mangrove trees, such as *Avicennia* and *Rhizophora*, can be found as far upriver as Breves. These mangroves provide the crabs that are eaten so

avidly in Belém. Ocean shrimp, thousands of tons of which are caught by trawlers off the Amapá coast, are believed to pass their larval stages in mangroves too.

Forest is the source of income for most Marajó residents now that the rubber industry is almost extinct. Lumber extraction on Marajó is centered near Breves where dozens of sawmills are located. Only a few tree species are actually sawn, resulting in short-sighted over-exploitation, without reforestation, of the Kapok tree *(ceiba pentandra)*, used in plywood, and *Virola*, exported for fine millwork.

Marajó dwellers consume large quantities of the purple juice of the Acaí Palm fruit *(Euterpe oleracea)*. This palm grows in stands on the banks of streams and sparsely in upland forest. The trunk is almost too slender for its height, but local people will demonstrate how easy it is to shinny up the palm and pick the berry-like fruits. To aid them in this daily practice, they use a vine strap called a *peconha* to hold their feet to the tree.	❏

BELOW: capybara, pig-sized rodents, at home in the water.

Map on page 278

PANTANAL

*This vast, flat plain where the annual rains inundate
the land, leaving only small areas above water,
is the spectacular home of many species*

A UNESCO World Heritage Site since 2000, the Pantanal is one of the world's greatest places to view wildlife. The **Pantanal** ❶ results from a unique blend of species and communities from Amazonia, the dry savannas of central Brazil (known as *cerrado*), and the *chaco* scrublands of Bolivia and Paraguay. It is described as a region undergoing transformation. It is relatively new and unstable from a geological point of view and is dominated by a complex mixture of plants and communities. The combination of floral species includes: Bolivian and Paraguayan xerophytes, savanna species from central Brazil, species from eastern Brazil and Amazonian forests, and hydrophytes which have a wide distribution in the neotropics.

But what gives the area its name (*pantano* means swamp) is that during the rainy season, from December to March, over 150,000 sq km (58,000 sq miles) of it – and area three times the size of Costa Rica – become a flooded lowland with scattered islands, the largest wetlands in the world. At other times the Pantanal is dry, or in various stages of flooding and draining.

Wetland life

The rains start in late November but do not fall uniformly. Near **Cuiabá**, in the north, average annual precipitation is 1,388 mm (55 ins), compared with 1,246 mm (49 ins) in **Corumbá**, and even less further south, where rainfall does not exceed that of the dry central plateau of Brazil.

Thus, the floods can be attributed primarily to the overflow of the shallow river channels as the water rises. At times, the waters rise so rapidly that herons, egrets, and jabirus move in front of the rising waters picking up snakes, small mammals, and other animals as they try to escape the flood. This is reminiscent of what antbirds do as the swarm of army ants moves through the Amazonian lowlands forests flushing small animals and insects from the leaf litter.

At the peak of the flood, the only thing that stays above water are the elongated *cordilheiras*. Smaller *cordilheiras*, called *capões*, look like islands of vegetation. Periodically, however, all but the highest *cordilheiras* get flooded. The topography of the Pantanal is so flat that a variation of only a few centimeters in water level determines whether thousands of hectares are flooded. The Rio Paraguay drops only 30 cm (12 ins) in its 1,300-km (807-mile) journey across the Pantanal.

There are a number of bodies of water that play an important role in the dynamics of the Pantanal. *Bahias* are lakes of varying sizes, from a few hundred meters to several kilometers in diameter, which form between the *cordilheiras*. They are connected to the river or to other *bahias* by seasonal streams known as *corixos*. When the *bahias* become isolated from the cyclical flood patterns, they become saline due to water evaporation during the dry season and subsequent filling with rainwater in the wet season. These isolated *bahias* are known as *salinas* when they have water, and *barreiros* when they dry up and become an important source of salt for wildlife and cattle.

This patchwork of ponds, lakes, streams, rivers, gallery forests, and forested islands forms one of the most exciting ecosystems in the world. The Pantanal teems with life as the *cerrado* and *chaco* blend to create a hybrid ecosystem, where 600 species of fish, over 650 species of birds, approximately 200 species of mammals, thousands of species of plants, as well as an untold number of terrestrial and aquatic invertebrate species, find their home. Each and every one of these species is inextricably linked to a delicate and all-determining water cycle.

The *piracema* (spawning season) takes place between February and April, as the waters begin to recede, when fish migrate from the deeper river channels to the shallower *bahias*. During this period many aquatic plants and animals breed. Apple snails, for instance, lay their eggs in brittle clusters attached to vertical twigs and stems. As the water rises, the larvae mature and hatch into the water where they complete their development.

As water recedes and aquatic life becomes concentrated, the terrestrial animals, faced with an abundance of concentrated resources, start breeding. The wading birds are particularly good examples of this: more than 10 species of herons and egrets, three species of storks, and six species of ibises and spoonbills, start breeding as soon as water levels drop sufficiently to make fish and other aquatic prey easier to catch. Apple snails fall easy prey to Snail (Everglades) Kites. And Wood Storks methodically crisscross the shallow *bahias* with their half-open bills in the water, snapping them shut and catching unsuspecting fish as they swim by.

Mammals and reptiles

Mammals are more visible in the Pantanal than in any neotropical rain forest. Capybaras are a common sight along the water's edge. These large rodents spend their days in family groups feeding on aquatic vegetation. Marsh Deer abound and are generally left alone by the local inhabitants (*pantaneiros*), who prefer beef and the meat of the feral pigs to the leaner deer.

Feral pigs, known as *porco monteiro*, have been roaming the Pantanal long enough to exhibit signs of natural selection. They have longer legs and can be told apart easily from peccaries by their diversity of colors, a trait they have retained from their domestic ancestors. *Pantaneiros* actively manage the feral pig populations and have turned them into an important source of meat. The men hunt down and rope the young adult *porco* males, which are castrated, one of their

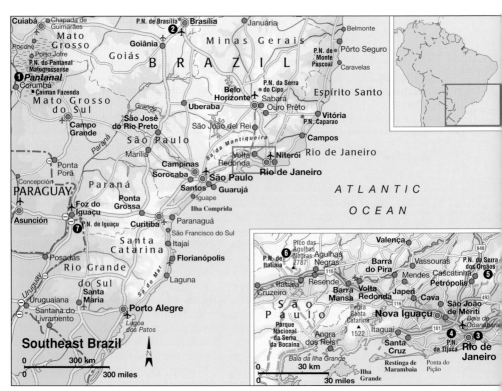

Map on page 278

ears notched, then released back into the wild. These marked individuals are eventually recaptured and slaughtered after they have put on sufficient weight. This is one of the best examples of low intensity resource management that characterizes traditional life in the Pantanal.

Among the reptiles of the Pantanal, Black Caimans are rarely found nowadays because of the unabated poaching that is so typical of the region. They have been hunted so intensively that abandoned poachers' camps are known as *cemitérios* – cemeteries – because of the large quantities of animal remains left behind. Fortunately, other types of caiman are still quite common. A walk at night to the water's edge can bring the hair-raising sight of dozens of pairs of little eyes reflecting the flashlight's beam. But there are few sights that match that of a *sucurí* – anaconda – submerged in the clear water, with up to 7 meters (23 ft) of powerful coils able to move effortlessly in the *bahias* and *corixos* in search of prey.

The Pantanal has been occupied by humans for over two centuries. *Pantaneiros* believe they have struck a deal with Nature, whereby their low-intensity cattle ranching and management of feral pigs allows the region to remain rich and diverse. The threat, they say, comes from the large soybean farms that clear-cut large areas and from the agro-industrial complexes that pollute the waters. Although the expansion of the agricultural frontier and the heavy mercury pollution resulting from alluvial gold mining are major threats, the *pantaneiros'* claim that they coexist with nature in a sustainable way is being contested. Cattle enclosures have been observed to differ significantly from the grazed surrounding, suggesting that grazing does have an adverse effect. What effect feral pigs have on the diversity of plant and animal life is at present not clear.

A more important threat, however, may be the gradual shift in ranching practices, from the production of *boi magro* – thin cattle – to more intensive ranching, with

BELOW: a Swamp Deer of the Pantanal.

more productive and demanding breeds. The *boi magro* practice involves extensive ranching of Pantanal-adapted breeds left to feed on the native vegetation. When the cattle reach a certain age, they are herded or trucked to ranches near major cities where they are fattened before being slaughtered or shipped on for processing.

The new practices involve extensive clearing of *cordilheiras* to plant introduced grasses to increase yield. The loss of the *cordilheiras* is having a major impact on the habitat and species diversity in the region because they harbor a unique ecological complement that includes species with very restricted distribution and narrow ecological requirements.

It is difficult to rate the environmental threats facing the Pantanal with regard to their impact on the biological diversity or the quality of life of the *pantaneiros*. Each and every one of them affects the integrity of this ecosystem in different ways and we are still learning how all these threats affect specific ecological processes.

Mining

Gold mining has become a particularly serious problem. It has gone far beyond affecting the Pantanal and is now reaching human populations inside and outside the region. In addition to the extensive habitat destruction caused by the mining process, there is a tremendous amount of mercury that finds its way to the bottom of the river channels. Heavier than water and not water soluble, mercury concentrations are much higher in mud and consequently are being picked up at higher concentrations by the bottom-feeding catfish that are a major source of protein for families in the Pantanal and in the nearby cities.

Recent studies show that mercury has also been detected in cattle at higher concentrations than expected. The more intensive gold mining operations are restricted to the vicinity of Poconé, in the State of Mato Grosso, leaving most of the Pantanal free of the direct effect of this environmental catastrophe, but not free from the threat of mercury.

BELOW: a pair of Crested Caracaras.

Map
on page
278

Harmful activities

The abundance of large animals makes the Pantanal a hunters' paradise. Avid hunters from the city are not very discriminating – some will shoot everything in sight. Poachers are a very different matter. They go after a few valuable species, such as jaguars, otters, and Black Caimans, for their furs and skins. The number of caimans taken by poachers ranges from 1 million to 1½ million a year. In some parts of the Pantanal the poachers have been associated with the drug cartels and are considered very dangerous by the forces responsible for controlling their illegal activities.

Another threat to the ecology of the wetlands has been over-fishing by international sports fishermen, which has taken its toll on fishing stocks. Permits and closed seasons have had to be introduced but are difficult to impose.

In spite of the difficulties resulting from the extensive areas that require patrolling, the lack of equipment, logistical support, and proper training, there may be hope in sight. Actions taken in the Pantanal by private landowners and government agencies, coupled with vocal and intensive public awareness campaigns, may help significantly in reducing the magnitude of the problem from the supply side, as well as curbing the almost insatiable demand from industrialized nations for the products of poaching. Under pressure from foreign lobbying groups, the Brazilian government has begun to introduce eco-friendly legislation to protect the region and even an environmental "Polícia Forestal", but there are still plans afoot to open up the region to sea-going vessels by deepening the Paraguay River, which would have devastating effects on the ecology of the wetlands.

There is another seemingly harmless activity – tourism – which is having a major effect in some parts of the Pantanal. Tourism has been heralded as the alternative to the non-extractive use of tropical riches. Tourism, it has been said, can provide sufficient income to support local

BELOW: a
Red-billed
Scythebill.

communities, while at the same time providing incentives to maintain the biological systems in their original state. Unfortunately, this is not always so. There are still too many people who are all too eager to see an entire rookery take wing and observe the majestic herons, egrets, and storks fill the sky. There are also too many tour operators who are all too eager to please their misguided clients and scare off breeding colonies of these birds by firing guns into the air. Little thought is given to the hundreds of young that fall to their deaths, or to the severe disruption to the breeding of the "resource" that makes the Pantanal so spectacular. Actions like this kill the goose that lays the golden egg.

Nevertheless, tourism is, in fact, a key solution to the loss of biological diversity and the increasing poverty of rural areas. Although tourists are ultimately responsible for shaping the behavior of the over-eager-to-please tour operators they hire, nature tourism is a viable alternative to more destructive economic activities. There are now some excellent lodges and cattle ranches *(fazendas)* which can be accessed from the gateways of Cuiabá, Campo Grande and Corumbá.

Conservation measures

Currently, there are a number of efforts underway to protect the Pantanal and all the biological diversity it holds. First and foremost, the Brazilian Constitution lists the Pantanal, along with the Atlantic forest, the Amazon, and the coastal ecosystems, as conservation priorities.

The commitment to protect these ecosystems has been followed by action. The Brazilian government has obtained a US$117-million World Bank loan, to be matched by US$50 million in Brazilian funds, to implement the first three years of the Brazilian National Environmental Program. One of the key ecosystems featured in this program is the Pantanal. Bolivia and Paraguay have also made a start by joining Brazil in forming an international working group to address the conservation needs of the whole of the Pantanal.

Map on page 278

Organizations like the Nature Conservancy (TNC) and the World Wildlife Fund (WWF) have been active in the Pantanal for many years and have made significant progress in strengthening the local conservation capacity. The Ecotropica Foundation, a Brazilian conservation organization, is working in partnership with TNC in protecting the Pantanal. Together they have purchased various tracts in the region and designated them as protected areas.

One of these areas is the **Acurizal Ranch** (24,000 hectares/66,690 acres), on the western border of **Parque Nacional do Pantanal Matogrossense**. The ranch contains a large jaguar population, the only remaining semi-deciduous forests and dry vegetation areas in the Pantanal, and the small Serra do Amolar mountain range. A second area is the 32-985-hectare (81,510-acre Doroche Ranch; these two acquisitions together have expanded the amount of protected area in the Pantanal by more than 43 percent.

Conservation International (CI) is also working in the Pantanal area, persuading local landowners to create private reserves. It believes that, since 98 percent of the region is privately owned, creating such reserves is critical to the long-term protection of the Pantanal and its diverse wildlife. CI is working with Brazilian communities and private-sector organizations, with federal, state and local government, and with bilateral and multilateral funding agencies, to develop an ecotourism program for the Pantanal.

In Bolivia, the Pantanal is now also beginning to become a conservation priority. In the private sector the Fundación Amigos de la Naturaleza (FAN) has been evaluating its capacity to work on projects aimed at preserving the biological diversity of the Bolivian Pantanal. They have the support of the General Secretariat of the Environment (Secretaría General del Medio Ambiente – SEGMA) and the National Fund for the Environment (Fondo Nacional para el Medio Ambiente – FONAMA).

In Paraguay, the Fundación Moises Bertoni para la Conservación de la Naturaleze in also pursuing similar objectives, with the support of their government. ❏

RIGHT: Guira Cuckoos.

BRASÍLIA NATIONAL PARK

*The climate and vegetation types change
on the fringes of Amazonia, and the* cerrado *savanna
near Brasília supports a different range of flora and fauna*

Evergreen tropical rainforest can grow only in regions with high precipitation, normally more than 2,000 mm (80 ins), which is fairly equally distributed throughout the year. If there are several consecutive months without rain, trees can only cope with evaporation by losing their leaves. This type of climate is found at the southern rim of Amazonia, approximately 1,000 km (620 miles) south of the Amazon river. With the increasing length of drought, the rainforest gradually blends into a different plant community – tropical deciduous forests. South and southeast of this region the climate becomes even drier, and does not permit the growth of any type of forest. Annual precipitation as low as 1,500 mm

(60 ins), occurring across wide areas of interior Brazil, gives rise to a distinctive type of dense grass cover with scattered gnarled trees, a type of savanna that is called *cerrado*. Some areas in the northeast of Brazil are even drier. Here grows a vegetation called *caatinga*, arid scrub and low woodland with little grass cover.

With the changes in vegetation most of the wildlife changes as well: some mammals and birds, such as the Maned Wolf and the ostrich-like Rhea, are adjusted and restricted to this kind of open country. Others, like the Giant Anteater, which is now very rare in the rainforest, become more common and easier to observe in this open grassland. However, most species of the arboreal monkeys are absent.

PRECEDING PAGES: a Giant Anteater and baby. **BELOW:** dry season in the *caatinga*.

Map on page 278

The *cerrado* is often valuable for agriculture and cattle farming. Consequently, much of the original vegetation is extensively modified by human activity.

The area is easily accessible, so hunting pressure is heavy. The beautiful Spix's Macaw from northeast Brazil is now extinct in the wild, and some of the most characteristic species, notably edible mammals like the Giant Armadillo, are now quite scarce and localized.

Nearly all flights from São Paulo or Rio de Janeiro to the Amazon stop in the capital, **Brasília**, and the landing approach is a good opportunity to study relatively undisturbed *cerrado* from a bird's eye view. A stopover here offers a chance to visit an outstanding example of protected *cerrado*, **Parque Nacional Brasília ❷**.

The park, created in 1961, is about 8 km (5 miles) to the northwest of the city, and covers 300 sq km (115 sq miles). Areas fall into six different categories of protection. After Tijuca and Iguaçu, this park has the third highest number of visitors of all Brazilian national parks, because it caters to crowds of local visitors who are interested in enjoying their weekends in the zone of intensive use, where there are picnic areas and artificial swimming pools. There are no accommodations, however – one has to spend the night in the city.

Visitors interested in the park's *zonas primitivas* have to be fairly self-reliant, an easy task in this landscape. In the dry, sunny climate with temperatures sometimes exceeding 35°C (95°F), it is advisable to bring drinking water and wear a hat for protection against the sun. Protection is also needed against the cold: in the dry season, temperatures can drop to freezing.

Only a small population of the Maned Wolf exists, while the Giant Armadillo persists in fairly high numbers; the Brazilian Tapir has an increased population, and there are fairly frequent encounters with Tufted-ear Marmosets, Capybaras, peccaries, and Giant Anteaters. Attractive large birds include the Rhea, Seriema, Red-winged Tinamou and the Spotted Nothura.   ❑

BELOW: a Greater Rhea.

Map on page 278

THE ATLANTIC FOREST

*Stretching down the Brazilian coast, the littoral
forests are havens of wildlife diversity – although, as elsewhere,
they are under threat from logging and development*

The Atlantic forest of Brazil originally covered more than 1 million sq km (385,000 sq miles) along a narrow band, 4,200 km (2,600 miles) long, hugging the coastline from the state of **Rio Grande do Norte** to the state of **Rio Grande do Sul**. Its long, narrow shape makes the Atlantic forest a heterogeneous ecosystem. To the north, it is influenced by the drier, hotter climates characteristic of Brazil's *nordeste* and to the south by the cooler climates of the temperate forests.

There is also the marked effect of its proximity to the ocean, which is reflected in the structure and composition of the vegetation. These north–south and east–west gradients interact strongly, adding significantly to the richness and diversity of the Atlantic forest.

There is a third axis that also contributes to the heterogeneity of the Atlantic forest: elevation. The coastline is flanked by a series of mountain ranges – **Serra do Mar**, **Serra da Mantiqueira**, and others – that provide marked elevational gradients and create a complex landscape, rich in isolated valleys and ridges with different microclimates and corresponding changes of the biological composition.

Sketching the forest

Any brief characterization of the Atlantic forest is inadequate and would be an over-simplification, but it is possible to make a thumbnail sketch of what this ecological "hot spot" is all about. Keep in mind, however, that the Atlantic forest is, in reality, the host to many vegetation types.

Moving from the coast inland there are mangroves, *restingas*, rainforests, high-elevation grasslands at the top of the highest mountains, and, in the most temperate areas to the south, there are – or at least were – large stands of *Araucaria* pines. Mangroves, although not unique to the Atlantic forest, are an integral part of this ecosystem. This habitat type is associated with coastal areas where rivers and oceans mix their waters, creating a brackish environment. The resulting system tends to be oxygen-poor and exposed to the ebb and flow of the tides twice a day.

These waters are known for their high productivity as well as for the fragility of the ecological communities they support. Many species of commercial fish breed and complete the early stages of their development under the mangrove roots which are also an important substrate for other commercially important species, such as oysters.

In the **Lagamar**, a region of 32,000 sq km (12,350 sq miles) between the states of São Paulo and Paraná, lie the 20 percent of remaining Atlantic forest and one of the most extensive mangrove stands on the coast. To explore the Atlantic forest, take the train from Curitiba to Marumbi to explore Marumbi Park, one of the world's largest remaining Atlantic rainforests and a UNESCO World Heritage Site and Biosphere Reserve.

Mangroves had virtually disappeared from the shores of **Ilha Comprida** and the vicinity of the city of **Iguape** as a result of a channel being opened in the first decades of the 20th century to shorten the distance that ships had to travel from the coast up the **Rio Ribeira**. The channel immediately changed the water quality of the **Mar Pequeno**, the estuary that separates Comprida Island from the mainland, killing all the mangrove stands and reducing fish catches significantly for many years. But the resilience of the mangroves can be demonstrated by their quick recovery. Together with the mangrove, the local fishing industry also recovered after the channel was closed in the late 1970s. Now, Comprida Island has regained its healthy mangrove stands and a reviving fishing industry.

PRECEDING PAGES: Atlantic rainforest and the 'Finger of God" peak in Sierra dos Orgãos.
LEFT: a perching tree frog.

Inland from the mangroves lies a very different habitat – the *restinga*. This mix of shrubs and short forest is the result of a successional process that takes place as the coastal dunes are stabilized. This complex system includes dunes, interdunal ponds, and extensive wetlands. Some of the best examples are the large wetland systems of **Rio Grande do Sul** (**Lagoa dos Patos** and **Lagoa dos Peixes**), and the easily accessible *restingas* of Rio de Janeiro's **Barra da Tijuca**.

One important characteristic of all dune systems that is also apparent in the *restingas* is the prevalence of vegetative growth as a method for consolidating new and unstable dunes; a slow process that can easily be disrupted. Consequently, any damage to the *restingas* by too many visitors and off-road vehicles significantly sets back plant succession. These disturbances can initiate a process that reverses succession, taking stabilized dunes with vegetation cover back to the stage of shifting dunes – dunes that move due to wind action.

Further inland from mangroves and *restingas* are the rainforests. Lush vegetation covers the rocky mountains that follow the coast and touch the waters, as in **Rio de Janeiro**, or that are well inland, as in the state of **Minas Gerais**. The vegetation of those mountains varies significantly with elevation. The lower strata, from sea level to 800 meters (2,620 ft) is lush and has a canopy that can reach 25 meters (82 ft) in the wettest sites, with emergent trees that can easily exceed those heights. In the drier ridges the canopy drops to 12 meters (40 ft). Further up, at 800–1,700 meters (2,620–5,580 ft), is the montane forest with epiphyte-covered trees. In the **Parque Nacional do Serra dos Orgãos**, close to 800 species of epiphytes have been collected, and it is estimated that the number of species may reach 1,000.

At the top of the highest mountain ranges are the *campos de altitude* – the high elevation grasslands which are like islands in a sea of forest. These remnants of a habitat that was more common during

BELOW: a Green-headed Tanager.

Map on page 278

the last glaciation period have a species composition that sets them apart from everything else. The level of endemism – species that occur only there – is astonishing. Near Rio de Janeiro, within the **Desengano State Park** and the Parque Nacional do Serra dos Orgãos are excellent examples of all these habitat types.

In the more temperate and humid areas of the southern state, there were once large, single-species stands of conifers like the *Araucaria* and *Podocarpus*. However, these stands have been reduced to isolated patches, due to intense harvesting. A number of parrot species, endemic to the Atlantic forest, have been associated with this habitat type, such as the Red-spectacled Parrot, Red-tailed Parrot, Vinaceous Parrot, and the Glaucous Macaw (now considered extinct). Other bird species, such as the Azure Jay, have been specialized seed dispersers of *Araucaria*. The decline of this conifer is thought to be largely responsible for the reduction in the jay populations.

This complex of habitats and ecological associations harbors some of the most diverse plant and animal life on the planet. Because of its high diversity and endemism, the Atlantic forest is considered one of the top biological hot spots on earth. In spite of having been reduced to less than 14 percent of its original size, it still contains an estimated 10,000 species of plants, with 53 percent of the tree species, 64 percent of the palms, and 74 percent of the bromeliads being endemic.

Animal life of the forest is equally diverse and endemic. Out of the 130 species of mammals, 51 occur nowhere else in the world. Primates have one of the highest levels of endemism in the Atlantic forest: 10 species of marmosets, three species of lion tamarins, as well as the Woolly Spider Monkey, are endemic. Among the birdlife, there are 30 genera and 160 endemic species found in this ecosystem.

Most of this biological diversity still remains after intensive exploitation for close to five centuries. Since the year

BELOW: a Red-necked Tanager.

1500, when Brazil was "discovered" by the Portuguese, the Atlantic forest has gone from occupying 12 percent of Brazil's territory to approximately 1.6 percent. This forest has been through many phases of resource use: from sugar cane and coffee to industrial development and logging, which is still carried out today with an intensity disproportionate to its sustainability.

In the process, close to 60 percent of Brazil's population – over 80 million people – has been concentrated in the region that was once the original Atlantic forest, maintaining tremendous pressure on the little that is left of this ecosystem.

The responsibility for the current condition of the Brazilian Atlantic forest must be shared by all. Poverty drives most of the inhabitants of the forest to over-exploit the resource base, as happens everywhere in the world. And greed drives businesses to engage in activities that destroy the last remnants of this imperiled ecosystem for the sake of short-term profits.

There are, however, communities using sustainable practices to increase the productivity of the system with a minimum environmental impact. Local fishermen in the Lagamar, for instance, are beginning to use low-technology methods to increase oyster yields by providing artificial substrates for the attachment of oyster larvae. They are gradually abandoning the non-sustainable practice of cutting mangrove roots to collect the oysters.

Other important changes in resource use practices are also being implemented by some businesses and farmers. However, in spite of all the efforts, the Atlantic forest continues to disappear at an ever-increasing rate. Strong protection is needed because there is simply not sufficient time to change old habits, based on the misconception that the riches of the forests and the seas are inexhaustible.

On the conservation side, there are major efforts being made to find integrated ways to deal with the critical problems of poverty and habitat loss. Brazilian gov-

BELOW: a Black-fronted Piping Guan.

Map
on page
278

ernment and non-governmental organizations in São Paulo and Paraná State, with help from the Nature Conservancy, the WWF, and other international organizations, have focused on the development of a major ecosystem conservation program for the Lagamar, where 20 percent of the remaining Atlantic forest is to be found.

This program includes protection of natural areas, sustainable development, research, and education. Nevertheless, programs like this cannot be successful without the leadership of conservation organizations and the strong participation of government agencies through programs like Projeto Tamar, designed to save the sea turtles.

At the government level, it is important to mention the Atlantic Forest Consortium formed by the secretariats of the environment from the Atlantic forest states. Their objective is to increase the efficiency of their programs by coordinating their efforts. Through this project the Atlantic forest component of a $167-million environmental program, jointly funded by the Brazilian government and the World Bank, will be implemented.

As visitors fly into Rio, circling **Guanabara Bay** and flying over the **Parque Nacional da Tijuca**, they must remember that they are looking at a disappearing treasure. The Atlantic forest, with all its biological diversity, is hanging on to Brazil's coastline tenaciously but will disappear eventually if nothing is done to preserve and conserve it.

The Atlantic Forest is a sad example for a worst-case scenario of the Amazon's future: a contiguous bio-geographic region can slip toward destruction by becoming fragmented into small parcels, which then collapse under continual human encroachment. The other side of the coin for the nature tourist is that nature reserves within this region, such as Tijuca, Serra dos Orgãos, Itatiaia, and Iguaçu, are small and close to human settlements and provide – in contrast to the Amazon – access, accommodations and hiking trails. ❑

BELOW: a Spot-billed Toucanet.

Map on page 278

PARKS CLOSE TO RIO DE JANEIRO

*Rio is the most common entry point to the country,
and most visitors will pass through the city. Close by are
a number of beautiful parks*

Few international airlines fly directly into cities on the Amazon, so most visitors will often have to use **Rio de Janeiro ❸** as a gateway to Manaus, Belém, or the Pantanal. Although the tropical rainforest has been overshot at this point, a stopover in Rio provides an excellent chance to see some of the few remaining examples of southeastern Brazil's Atlantic forest *(see pages 291–5)*, a habitat often classified as subtropical rainforest.

Like most tourists, the traveling naturalist will normally select a hotel in the districts of Ipanema, Leblon or Copacabana. The latter suburb, with more than one million inhabitants and a long beachfront, offers a chance to get away from the bustling city. While walking along the famous beach, try to spot the numerous beautiful seabirds. Brown Boobies shoot over the waves in search of fish, and Magnificent Frigatebirds soar with Black Vultures in mixed flocks over the city.

To get away from the crowds and the traffic of Rio, visit the botanic garden, **Jardim Botánico**, located about 5 km (3 miles) from Copacabana, in front of an impressive backdrop of steep mountains. In the east, **Corcovado** rises to 704 meters (2,300 ft), crowned by the **Cristo Redentor** statue, a landmark of Rio. To the west are the forested **Tijuca mountains**.

The botanic garden is nearly 200 years old, and contains a fine selection of old tropical trees of local and exotic origin. From the surrounding forests, Common Marmosets, Gray Tree Squirrels, and several species of toucans and parakeets, venture into the garden. Hummingbirds, called in Brazil "flower-kissers" *(Beija-Flores)*, zoom between the trees, the large Swallow-tailed Hummingbird being the most common species. Masked Water-Tyrants catch insects between water lilies

and Rufous-bellied Thrushes sing in the trees. This garden is contiguous with the Tijuca forest, but it is not easy to identify a trail into this area. A road to Tijuca passes between the botanic garden and Corcovado, and can be used for hiking, since there is little traffic, but it is not advisable to hike alone because of the city's high crime rate.

Tijuca National Park

Parque Nacional da Tijuca ❹ is an unusual national park, first because it is completely surrounded by the suburbs of Rio, and second because its forests have been planted by humans. Rio has been one of the continent's population centers since its colonization by the Portuguese in the 16th century, and the surrounding forests were the first to be logged. In the 18th and early 19th centuries, the Tijuca mountains were used to grow sugar cane and coffee, and virgin vegetation and wildlife survived at best in tiny pockets in inaccessible places. In the mid-19th century, most land was acquired by the government to create a watershed for Rio's growing needs.

Beginning in 1861, the plan was to recreate a natural landscape and to plant 100,000 native and exotic trees in the montane area of 30 sq km (11 sq miles). Over time, the planted forest was enriched by the invasion of additional native plants from remaining natural pockets, and today, nearly 140 years later, a succession towards natural old growth forest is well advanced.

Tijuca is an excellent illustration that restoration of nature is possible in the tropics, given the political will and funds, and provided that the native species have not disappeared from the planet. In Tijuca, many mammals had become locally extinct, but could be reintroduced from elsewhere. This was unfortunately not

LEFT:
a Margay Cat.

done in the case of the Lion Tamarin, a native here. It was replaced by the Common Marmoset from the Brazilian northeast, which has become one of the most common mammals in the park. Also common is the Gray Tree Squirrel. Several other mammals had either survived in the area or have been reintroduced, such as the Brown Capuchin, pacas, and Agoutis.

Tijuca is excellent for bird-watching, with sightings of some particularly colorful passerines, such as Red-necked and Green-headed Tanagers, and Pin-tailed, White-bearded and Swallow-tailed Manakins. Among larger birds are several species of toucans, seven species of parakeets, the Scaly-headed Parrot, and two species of fowl, the Rusty-margined Guan and the Spot-winged Wood-Quail.

Tijuca is split into three parts by private property, which includes some restaurants as well as a hotel. For hiking and wildlife observation, it is best to start at one of the parking lots close to the **Pico da Tijuca**, past the small but beautiful Cascatinha waterfall. It is possible to hike up the Pico da Tijuca, the park's highest mountain, at 1,021 meters (3,350 ft), following a 2.8-km (1.7-mile) trail. It is worth driving to several of the other access routes to enjoy the spectacular view of Rio and the coastline from points such as the **Vista Chinesa**.

Parks to the north

Parque Nacional da Serra dos Orgãos ❺, two hours north of Rio by car, is a pleasant one-day excursion. The 50-sq km (19-sq mile) park protects the eastern flank of an impressive mountain range stretching between **Petrópolis** and **Teresópolis**. One entrance is at the edge of this last city, which also offers accommodations.

With a patchwork of virgin and secondary vegetation, crystal-clear creeks and spectacular mountains like the **Dedo de Deus**, some of them above 2,000 meters (6,500 ft), the park caters to naturalists, hikers, mountaineers, or to people who just want to bathe in natural or artificial pools. For serious hikers, there is a

BELOW: a group of Coatis.

Map
on page
278

trail across the mountains from Teresópolis to Petrópolis, a hike of three to four days through a gneiss and granite rock landscape. These heights are exposed to freezing temperatures in winter and violent thunderstorms in summer. For shorter walks, start at the fringe of Teresópolis, or turn right off the highway at the "Parque Nacional Subsede" sign, 11 km (7 miles) before Teresópolis.

The park's fauna is much richer than that of Tijuca. There are three species of monkeys, five species of cats, the Coati and the Lesser Anteater. Among numerous bird species are tinamous and toucans.

Possibly the richest flora and fauna in this part of Brazil is in **Parque Nacional de Itatiaia** ❻. Located halfway between Rio de Janeiro and São Paulo, north of km 150 on the Presidente Duarte freeway, this park is difficult to reach without a car. Close to the town of **Itatiaia**, turn down a 14-km (9-mile) asphalt road to the park headquarters, where there is an office that provides maps and information; basic accommodations can be booked through the Administração do Parque Nacional de Itatiaia, tel: 27580-000.

With an area of 119 sq km (46 sq miles), Itatiaia protects part of the slopes and the top of a mountain plateau, the **Serra da Mantiqueira**. Within the park is **Agulhas Negras**, the third highest mountain in Brazil at 2,787 meters (9,144 ft). Itatiaia has well-preserved examples of Atlantic forest. The flora is rich at all elevations, and contains more than 100 endemic plants.

The park has a virtually full complement of original fauna. Some species are declining due to insufficient habitat or small population size; such as, the Muriqui, the largest neotropical primate, the Jaguar and the Brazilian Tapir. Other mammals are faring well: the Masked Titi, the Brown Capuchin, the Brown Howler Monkey, and the Pale-throated Three-toed Sloth. The rich birdlife (250 species) includes the Red-legged Seriema, the Solitary and the Brown Tinamou, the Pileated Parrot and the Purple-bellied Parrot. ❑

BELOW:
an Emerald
tree Frog.

IGUAÇU

*One of the most spectacular waterfalls in the world,
Iguaçu is also home to a wonderful national park, which is
a haven for many endangered species*

guaçu/Iguazu, which means "big water" in Guarani, is one of the most spectacular waterfalls in the world. The Iguaçu river created this waterfall at its turn from a southwesterly to a northwesterly course, by cutting into the deep layer of basalt which covers a large part of southern Brazil. The waterfall is the central feature of a 2,200 sq km (850 sq mile) national park of the same name, an area rich in vegetation and wildlife. Three-quarters of this protected area is in Brazil, and one-quarter in Argentina. The Iguaçu river and the waterfall mark the border between Brazil and Argentina, while these two countries meet Paraguay at the confluence of the Iguaçu and Paraná rivers, about 30 km (18 miles) northwest of the falls.

From cities in Brazil like São Paulo, the park is reached by flying into the airport halfway along the 30-km (18-mile) road between the falls and the small city of **Foz do Iguaçu**. From here, it is possible, when the bridge is open, to drive to **Puerto Iguazu** in Argentina, and on to the Argentine side of the waterfall and the forests of the national park. Alternatively, there are exciting boat trips to both sides of the falls, although no connection between the two countries exists directly at the falls. There are various hotels in town or close to the airport, or – best but usually booked out – in the **Hotel das Cataratas**, close to the falls.

About 2,500 km (1,500 miles) south of the equator, the **Parque Nacional do Iguaçu** ❼ has subtropical rather than

PRECEDING PAGES: Iguaçu falls. **BELOW:** Black Vultures *(Coragyps atratus)* beside the falls.

Map
on page
278

BELOW:
a Capybara
(Hydrochaeris
hydrochaeris).
OVERLEAF:
a spider waits
at the center
of its web.

tropical forest. There are only about 1,000 mm (40 inches) of precipitation per year, half the amount of the Amazon or the Atlantic coast, but the forest is mostly evergreen rather than deciduous, since the dry winter season is not strictly without rainfall. Much of the vegetation has small- or medium-sized leaves, but the broad-leafed plants, palms, lianas, epiphytes and the generally vigorous tree growth, gives an impression comparable to the tropical forests.

The forest of Iguaçu was contiguous with the Atlantic forest of southeastern Brazil until extensive clearing of much of the state of Paraná in the middle of the 20th century led to its isolation. Due to this previous connection, there are many links with the flora and fauna of the Atlantic forest and even the Amazon, and, because of today's isolation, Iguaçu National Park is important for the protection of many species.

Thanks to its large area, populations persist of large mammals such as the Brazilian Tapir, White-lipped and Collared Peccary, Capybara, Jaguar, and several smaller cats.

The locally endangered Southern River Otter lives in the park's waterways.

The park is a stronghold of large predatory birds like the Harpy and the Crested Eagle, seriously threatened elsewhere. Unfortunately, one of the continent's most beautiful parrots, the Glaucous Macaw, has become extinct, a fate possibly faced by several of its relatives, unless trapping is curtailed. Viable populations exist at this time of the Blue-and-Yellow Macaw, while the Vinaceous Amazon Macaw is endangered because of the reduction of its habitat of *Araucaria* groves. There is a faint possibility that one of Brazil's rarest birds, the Brazilian Merganser, still exists here.

Although there is, unfortunately, little chance of spotting these rarities, it is easy to find common but beautiful and colorful tropical birds, such as the Red-breasted Toucan, Red-capped Parrot, Red-rumped Cacique, Blue Dacnis, or Green-headed Tanager. One can also watch the tame and impressively large Common Iguanas in the meadows close to the waterfall. ❑

INSIGHT GUIDES

Travel Tips

Insight FlexiMaps

Maps in Insight Guides are tailored to complement the text. But when you're on the road you sometimes need the big picture that only a large-scale map can provide. This new range of durable Insight Fleximaps has been designed to meet just that need.

Detailed, clear cartography
makes the comprehensive route and city maps easy to follow, highlights all the major tourist sites and provides valuable motoring information plus a full index.

Informative and easy to use
with additional text and photographs covering a destination's top 10 essential sites, plus useful addresses, facts about the destination and handy tips on getting around.

Laminated finish
allows you to mark your route on the map using a non-permanent marker pen, and wipe it off. It makes the maps more durable and easier to fold than traditional maps.

The world's most popular destinations
are covered by the 125 titles in the series – and new destinations are being added all the time. They include Alaska, Amsterdam, Bangkok, Barbados, Beijing, Brussels, Dallas/Fort Worth, Florence, Hong Kong, Ireland, Madrid, New York, Orlando, Peru, Prague, Rio, Rome, San Francisco, Sydney, Thailand, Turkey, Venice, and Vienna.

INSIGHT GUIDES
The world's largest collection of visual travel guides

CONTENTS

Colombia

Getting Acquainted

GOVERNMENT

Colombia has had a "limited democracy" where two parties have shared power for several decades. Andres Pastrana won the 1998 presidential elections.

He has the difficult task of governing the country while left-wing guerilla groups, namely the Revolutionary Armed Forces of Colombia (FARC) and the National Liberation Army (ELN), are increasing their popular support and unofficially control parts of the country.

ECONOMY

Colombia has one of South America's strongest economies and is self-sufficient in energy. Agriculture is the most important source of revenue, and Colombia is the world's second largest coffee producer. Manufacturing comes a close second as Colombia produces 60 percent of the world's emeralds. It also produces 70 percent of the world's cocaine, and President Pastrana introduced the "Plan Colombia" to help fight drug problems. Unfortunately violence

and kidnapping are serious issues and foreign visitors are urged to avoid traveling on public transport at night and to seek official advice before visiting what is otherwise a most beautiful and fascinating country.

CLIMATE

As with all countries in the tropics, the climate is determined by altitude. The highlands enjoy cool, moderate weather all year round, while the lowlands and coast are hot and humid. The only seasons are wet and dry, varying by region: in Bogotá, the driest time is between December and March, and between July and August.

BUSINESS HOURS

Generally Monday–Friday, 8am–noon and 2–6pm. Banks open 8am–3pm.

Planning the Trip

TIME ZONE

Colombia is five hours behind Greenwich Mean Time.

MONEY MATTERS

The unit of currency is the Colombian *peso*.

Many *casas de cambio* change US dollars into local money. Traveler's checks are more difficult to change, although the **Banco Union Colombiano**, **Banco**

Visas

Citizens of the US, the EU, Australia and New Zealand do not need visas. You should be given a 90-day stay on arrival, but this is not always automatic – if you plan to stay more than 30 days, ask for a longer-stay visa. Everyone else needs a visa. The price varies from country to country.

Industrial de Colombia and **Banco Anglo Americano** will do so, as well as some others, depending on the region.

It is recommended you take American Express dollar travellers' cheques as sterling cheques are impossible to change. Hotels will not change travellers' cheques and offer a very poor rate of exchange for cash. Do not change money on the street: they are almost all unscrupulous hustlers.

WHAT TO WEAR

The weather in Bogotá and the highlands stays cool all year round: a warm jumper/jacket and rain protection is useful. The coast and lowlands have a tropical climate: take light summer clothes.

GETTING THERE

By Air

Being the northernmost country in South America, Colombia is many people's first stop. There are many flights from Europe and North America to Bogotá, Cartagena and Barranquilla, with onward connections to the rest of the continent. Cheap flights are available from Costa Rica in Central America.

By Road

From Ecuador, buses go to and from the border at Tulcan/Ipiales. From Venezuela, you can choose between the coastal route from Maracaibo to Santa Marta, or the highland route from Caracas and Mérida to Cucutá via San Cristobal. From Brazil, the only way to enter is via Leticia in the Amazon basin and then fly to Bogotá.

On Arrival

Bogotá's El Dorado Airport is one of the most modern in the world. Buses and taxis run to the city center. Be sure to fill out a tax exemption form on arrival or departure, so you don't have to pay the Colombian citizen exit tax on departure. Even with this form,

Staying Healthy

Drink bottled mineral water, avoid ice cubes, salads and unpeeled fruit. If you are spending time in the Amazon rainforest, do take malaria tablets with you. Travelers arriving in Leticia will be asked to show a yellow fever certificate or be inoculated on the spot.

foreigners pay $26 departure tax from Bogotá when leaving by air ($48 if you have stayed more than 60 days).

Practical Tips

MEDIA

Bogotá has several daily newspapers. Both *El Tiempo* and *El Espectador* are considered the most comprehensive in terms of news coverage.

POSTAL SERVICES

Mail is generally very reliable in Colombia. The General Post Office in Bogotá is at the Avianca Building in the center of the city, opposite the Parque Santander.

TELECOMMUNICATIONS

The Colombian Telecom offices in all big cities have international communication facilities to enable you to dial home.

TOURIST INFORMATION

Contact the **Fondo de Promoción Turistica de Colombia**, Carrera 16ª, No. 78–55, tel: 611 4330/4471 Information is also available at the office in El Dorado airport, and the Instituto Distrital de Cultura y Turismo, Carrera 8, 9-83, tel: 286-6555.

There are branches of Fondo de Promoción Turistica, known as Fondos Mixtos, in the major cities.

EMBASSIES & CONSULATES

The following embassies and consulates are in Bogotá:

British Embassy
Calle 98, No. 9-03
Tel: 218-5111; fax: 218-2460.
Canadian Embassy
Calle 76, No. 11–52
Tel: 313-1355.
United States Embassy
Cra 50, Av El Dorado
Tel: 315-0811.
Venezuelan Consulate
Calle 33, No. 6-94, piso 10
Tel: 285-2035; fax: 285-7372.

MEDICAL SERVICES

There are many private clinics which offer emergency treatment including **Clinica Marly**, Calle 50, No. 9-67 tel: 287-1020 and **Clinica del Country**, Cra 15, No. 84-13 tel: 257-3100.
The two main medical centers in Bogotá are:
Cruz Roja Internacional, Avenida 68, No. 66–31, tel: 250-661 or 231-9027; and, **Centro Medico de Salud** Carrera 10, No. 21–36, tel: 243-1381 or 282-4021.

Getting Around

FROM THE AIRPORT

Taxi services run from all the major airports to the city center at fixed rates. Local buses also run to Bogotá airport.

DOMESTIC TRAVEL

By Air
The national carrier **Avianca** can be found at Carrera 7, No. 16–36 (tel: 410-1011 or 295-4611; airport, tel: 413-8295). It flies to all parts of the country and operates a useful Air Pass, which must be bought outside Colombia in conjunction with an international ticket.

Other domestic flights are offered by SAM, ACES and SAETANA.
SAM, Carrera 10, No. 27–91, tel: 286-8402; airport, tel: 413-8868.
ACES, Carrera 10, No. 27–51, tel: 401 2237/336 0300.
Satena, Centro Taquendama, Carrera 10 y Carrer 27, tel: 286-2701; airport, tel: 413-8158.

By Road
Bus transport along the main routes is generally good, often luxurious, but deteriorates when heading into the more remote areas. Normally, the rougher the road, the poorer the bus. The alternative on main routes is a *buseta* (minibus) or *colectivo* (a shared taxi that is more expensive but much quicker).

Within the main cities, taxis are cheap and relatively reliable. A meter clocks over the price in *pesos*.

Security & Crime

The center of major cities is generally safe, but with such extreme poverty in Colombia it is worthwhile to check which outlying areas can be visited safely at night. The unsafe areas are clearly defined, usually red-light districts – ask at your hotel for details. With a little common sense, Colombia is as safe as any other country.

Although most Colombians are honest and friendly, theft occurs in the larger cities and you should take care in looking after your valuables at all times.

Be aware that some thieves pose as policemen asking to check your passport and documents in the street. A good policy to minimise risk is to hire a local guide.

There are "Tourism Police" at major airports and towns to help if you have anything stolen.

There is sporadic political violence in certain parts of the country and it is recommended to check with the Foreign Office before planning a trip across Colombia; www.fco.gov.uk/travel

TOURS

Eco-Guias, Carrera 3, No. 18–56A, Of 202, tel: 334-8042 or 284 8991, fax: 284-8991. Email: ecoguias@elsitio.net.co; www.ecoguias.com
 Recommended Bogotá-based adventure travel and eco-tourism agency, specialising in tailor-made cultural, ecological and wildlife tours across Colombia, including the Amazon, coffee fincas, Cartagena, the Bogotá region and the national parks and reserves.
 Package tours as well as travel within and beyond Colombia can be arranged with **Viajes Chapinero**, Av. 7, 124–15, tel: 612-7716; www.viajeschapinero.com
 The UK specialist tour operator for Colombia is **Ecolatina**, PO Box 395, Richmond, Surrey, TW10 7FE, tel: (020) 8549-9430; email: ecolatina@btinternet.com. Also, **Trips Worldwide**, 9 Byron Place, Clifton, Bristol, BS8 1JT; tel: (0117) 987-2626; www.tripsworldwide.co.uk

HOTELS

Accommodation in Colombia ranges from luxurious hotels at $150 a night to the most basic lodgings at around $5 a night. A useful website is Colombia Hotels www.infortur.com.co

BOGOTA

Some of the five-star hotels include:
Bogotá Royal
Avda 100, No. 8A-01
Tel: 218-9911
Fax: 218-3362
Hilton
Carrera 7, No. 32–16
Tel: 637-2400
Fax: 612-5023
Tequendama
Carrera 10, No. 26–21
Tel: 382 0300
Fax: 282-2860
La Fontana
Diagonal 127A, No. 21–10
Tel: 274-7868
Recommended by many travelers.

La Opera
Calle 10, No. 5–72
Tel: 336-2066 or 336 5285
Nice, colonial hotel in Candelaria.
 For a hotel with some atmosphere and still perfectly comfortable, though with less conveniences, try the **Hostería de la Candelaria**, Calle 9, No. 3–11, tel: 342-1727. It is in a converted colonial mansion and furnished with antiques. Rates are around US$40 single, $60 double.
 The better budget hotels are located in the Candeleria. A good option for backpackers is **Platypus**, Calle 16 No. 2–43, La Candelaria, tel: 341-2874 or 341-3104; email: platypushotel@yahoo.com
 Platypus has single, double and dormitory rooms, internet, kitchen and a friendly, helpful owner.

CARTAGENA

Cartagena-Hilton
Tel: 665-0666
Fax: 665-0661
A luxury hotel in nearby El Laguito.
Hostal Baluarte
Media Luna, No. 10–81
Tel: 664-2208
$20 single.
Santa Clara
Calle del Toro, Barrio San Diego
Tel: 664-6070
Fax: 664-8040
Luxury hotel run by Sofitel. A beautifully-converted 16th century convent with fine restaurants and excellent facilities.

In Bogotá, the **Casa Vieja** restaurants are considered to offer the best in regional food. They are at Avenida Jimenez, No. 3–73, Carrera 10, No. 26–50, and Carrera 3, No. 18–60. **Refugio Alpino**, Calle 23, No. 7–94, serves European food. **Tierra Colombina**, at Carrera 10, No. 27–27, features an evening show. Budget meals serving comidas corrientes, usually a fried piece of meat with beans and rice, are found in many of the restaurants.
 There are also many good

Colombian cooking varies by region, although in the big cities any kind of international cuisine can be enjoyed at a price.
 For Colombian dishes, a few local specialties worth trying are:
Ajico – a soup of chicken, potatoes and vegetables, common in Bogotá.
Arepa – a maize pancake.
Arroz con coco – rice cooked in coconut oil, special to the coast.
Bandeja paisa – a dish of ground beef, sausages, beans, rice, plantain and avocado.
Carne asada – grilled meat.
Cazuela de mariscos – seafood stew.
Chocolate santafareño – hot chocolate accompanied by cheese and bread.
Mondongo – tripe soup.
Puchero – broth of chicken, beef, potatoes and pork, typical of Bogotá.
Tamales – chopped pork with rice and vegetables in a maize dough.

restaurants between Carrera 5 and Calle 27 including **El Patio**, **Arcanos** and **Merlin**, and also on Parque de la 93 in the north, such as **Mamas** and **San Angelo**.
 In Cartagena, **Santa Clara** is a restored 16th-century convent, now a hotel, serving excellent food.

ALTO ANCHICAYA NATIONAL PARK

Geographic location: Pacific slope of the Department of Valle, Colombia.
Size: Not defined; protects catchment for a hydroelectric plant.
Access: By bus (TRANSUR company) along old Cali–Buenaventura road. Get off at Danubio (60 km). Report at gatehouse across river bridge and wait for a generating company bus to travel the 10 km to the work camp. If driving, leave main Cali–Buenaventura road at KM 21, turning left onto old road.

Bars & Nightclubs

● **Bogotá:** There are several good salsa bars around the intersection of Carrera 5 and Calle 27, although this is the seedy section of town. Others can be found in the Candelaria area – just wander the streets and listen for the blaring music. None gets moving until after midnight on Fridays and Saturdays.

For taped tangos from Argentina, head for **El Viejo Almacén** at Carrera 5, No. 14–23.

● **Cartagena: La Muralla,** Plaza Bolivar, on the old city wall, has salsa music. For discotheques, head for the Bocagrande district.

Approximate cost of access: Bus Cali–Danubio approximately $4.
Tour operators: None.
Open: All year, crowded during holiday periods.
Registration: Essential. Book through CVC (Central Autonoma del Cauca), Carrera 56, No. 11–36, Cali.
Accommodation: In chalet apartments. Usually free.
Food/restaurant: Meals available in the camp canteen. $8 per day.
Transport in park: Good system of roads with side tracks. Lifts can be obtained on camp vehicles.

AMACAYA NATIONAL PARK

Geographic location: Extreme south-east of Colombia, extending northwards from the Amazon River.
Size: 300,000 hectares (1,158 sq miles).
Access: By plane to Leticia (four flights a week by Avianca). Boat upstream 60 km to the Visitors Center at Matamata.
Approximate costs of access: Return air ticket Bogota–Leticia $145. Boats carrying up to 15 passengers can be hired for $150 a day. Alternatively, a public river taxi travels between Leticia and Puerto Nariño twice a week, stopping at Matamata (ask conductor). Leticia–Matamata costs $10 per person and takes two hours.
Tour operators: Many travelers make arrangements at the well-run Visitor Centre or in Leticia. Try Amaturs, c/o Hotel Anaconda, Calle 8/Carrera 25, Leticia; or Turamazonas, Parador Ticuna, Aptdo Aereo 074, tel: 7421.

Eco-Guias in Bogotá, tel; 284-8991/334-8042, www.ecoguias.com offers 3–5 day jungle tours to various Amazon lodges with specialist guides.
Open: All year, usually full during Easter Week.
Registration: Essential. Entry fee $10. Arrange visit with the Ministerio del Medio Ambiente, Carrera 11, No. 12–45, Leticia.
Accommodation: Visitors Center has dormitory accommodation for 40–50 with hammocks or folding beds, $16 full board.
Trails: Well-marked trails enter the forest. Visitors are encouraged to hire a local guide (approximately $2 per day, but ask for the current standard rate at the Visitors Center). During February to early June access into the forest may only be possible by dugout canoe (which can be hired from the park). It is possible to visit Tikuna Indian communities within easy reach of the Visitor Centre.

CHINGAZA NATIONAL PARK

Geographic location: On the border of Cundinamarca and Meta, east of Bogotá.
Size: 50,000 hectares (193 sq miles).
Access: Take the road from Bogotá to La Calera. Two kilometers after La Calera, on the road to Guasca, take a right turn onto a good, but unpaved road, traveling through a checkpoint, then past a cement factory, following the route for 35 km before entering the park.
Approximate cost of access: No

---

public transport available, although it may be possible to arrange transport with the water company which runs buses daily to the work camp at the reservoir in the park (Empresa de Acueductoy Alcantarillado de Bogotá, Calle 22c, No. 40–99, Bogota).
Tour operators: Eco-Guias, (see box) offers day hikes and overnight camping trips to various parts of the park.
Open: All year.
Registration: Essential. Contact UAE (Unidad Administriva Especial del Sistema de Parques Nacionales Naturales) at the Ministerio del Ambiente. Main office: Caja Agraria, Carrera 10, No. 20–30, Floor 8, Bogotá, tel: 283-0964, fax: 341-5331. Regional office for Amazonia, tel: 283-3009, fax: 243-3091.
Accommodation: At the New Visitors Center. Prices about $6–8 per room per night (accommodating up to 5 persons).
Food/restaurant: No restaurant.

Iquaque National Park

Geographic location: North-east of Bogota in the Department of Bocaya, close to the town of Villa de Leiva.
Size: 3,600 hectares (8,900 acres).
Access: On the road to Arcabuco, 12 km from Villa de Leiva.
Tour operators: Eco-Guias, tel: 284-8991 or 334-8042, www.ecoguias.com can arrange tours staying at nearby coffee fincas or at one of the charming colonial hotels in Villa de Leiva.
Accommodation: Visitors centre or wide choice of accommodation in Villa de Leiva. Camping allowed and safe.
Food/restaurant: Good food available at the restaurant at the tourist centre.
Transport in park: on foot or private vehicle.
Trails: Excellent trails leading up to high altitude lakes, 'paramo' and cloud forest.

Cullture

Museums
Like those in most other countries in this part of the world, most museums in Bogotá are closed on Mondays.
Archeological Museum, Carrera 6, No. 7–43. Good collection of pre-Columbian pottery housed in a beautiful 16th century colonial building.
Museo del Oro (Gold Museum), Parque de Santander, corner of Calle 16 and Carrera 6A. Open Tues–Sat 9am–4pm, Sun 9am–noon. Unique collection of pre-Columbian gold.
Museo Mercedes de Perez, Carrera 7, No. 94–17. Colonial life.
Museum of Modern Art, Calle 24, No. 6–55.
National Museum, Carrera 7, No. 28–66. In an old Panopticon prison.
Donacion Botero – Calle 13, Cra 4. New excellent modern art museum, with work by famous Colombian artist/sculptor Fernando Botero, and collection of modern art including works by Picasso and Miro. Closed Tues.

Theater and Cinema
For details of what's on, see the "Espectaculos" sections in the daily newspapers.

facilities, so food needs to be brought in. The reservoir work camp has a cafeteria and it may be possible to arrange meals there.
Transport in park: On foot or private vehicle.
Trails: Good system of paths.
Further advice: Be prepared for sudden changes in the weather.

LA PLANADA RESERVE

Geographic location: South-west of Colombia in the Department of Nariño, close to the town of Ricaurte.
Size: 3,600 hectares (14 sq miles).
Access: Take the Pasto–Tumaco road. At the village of Chucunes (a

few kilometers before Ricaurte) take a left turn. The reserve lies 7 km (4 miles) along this road. The journey time between Pasto and La Planada is approximately four hours. Buses are available for Chucunes from Pasto (try Servicio Transportes Especiales leaving at noon from Hotel Chambu). Cost is about $3 or a taxi costs about $70 return.
Tour operators: None.
Open: All year.
Registration: Essential. Book through FES (Fundacion Educación Superior), La Planada, Apartado Aereo 1562, Pasto, tel: (928) 845-933.
Accommodation: Accommodation in study-bedrooms.
Food/restaurant: Meals available in a small restaurant. Full board and lodging $30 a day.
Transport in park: On foot.
Trails: Well-marked system of paths and tracks.
Further advice: Since it can be wet, waterproofs and rubber boots are essential, although the Visitors Center does hold a stock of boots for use by visitors.

RIO TATABRO

Geographic location: On the Pacific slope of Valle, Colombia.
Size: Undefined.
Access: On the old Cali–Buenaventura road, between Agua Clara and Lianobajo.
Approximate costs of access: TRANSUR bus from Cali, about 4½ hours, approximate cost $3.
Tour operators: None.
Open: All year.
Registration: Essential. Book through Fundación Herencia Verde, Calle 4 Oeste, No. 3A–32, El Penon, Cali (in person) or AA 32802, Cali (by post), tel: 593-142.
Accommodation: Limited accommodation in bunk beds. $3 per night.
Food/restaurant: Bring food, to be prepared by caretaker's wife.
Transport in park: On foot.
Trails: Two main trails; directions can be obtained from the caretaker.

SIERRA NEVADA DE SANTA MARTA

Geographic location: The Sierra Nevada massif, situated close to the Caribbean coast on the borders of the Colombian departments Magdalena, Guajira and Cesar.
Size: 383,000 hectares (1,479 sq miles).
Access: To the San Lorenzo ridge, travel to the village of Minca from Santa Marta. The road continues through the village and rises through coffee-growing areas to the MMA buildings at San Lorenzo at 2,200 m (7,220 ft) above sea level. The drive takes about 2½ hours. Between Minca and San Lorenzo the road is unpaved and very rough, particularly in the wet season. To visit Ciudad Perdida, it is essential to organize a guide for the week-long trek.
Approximate costs of access: Regular buses (about $1) between Santa Marta and Minca but no public transport for the 35 km between Minca, and San Lorenzo. Car hire at international rates. Treks to Ciudad Perdida average $250.
Tour operators: For Ciudad Perdida, Turcol, Carrera 1, No. 22–77, Santa Marta Eco-Guias, tel: 284-8991 or 334-8042, www.ecoguias.com has regular departures from Santa Marta to Ciudad Perdida.
Open: All year.
Registration: Essential. Book through MMA, Carrera 12, No. 16D–05, tel: 203-116, fax: 204-506. Access into the high sierra is restricted. Check with: ONIC (Oganización Nacional de Indigenas de Colombia), Calle 13, No. 4–38, Bogotá, tel: 284-2168; Instituto Colombiano de Antropologia, Calle 8, No. 8–87, Bogotá, tel: 333-0535; or Fundación Pro-Sierra Nevada, Edif., Los Bancos 502, Santa Marta, tel: 214-697, fax: 214-737. For Ciudad Perdida, also essential to seek permission from Instituto de Anthropologia, Carrera 7a, No. 28-66, Santa Marta, tel: 342-5925, 334-2639.
Accommodation: MMA Visitors Center at San Lorenzo, about $3.50 per night.
Food/restaurant: All food has to

be purchased beforehand in Santa Marta.

Transport in park: On foot.

Trails: The main track along the San Lorenzo ridge provides the easiest opportunities to find Santa Marta specialities.

Further advice: Some armed activity occurs in the Sierra Nevada de Santa Marta. Seek current advice from the MMA office in Santa Marta before venturing into the area.

TAYRONA NATIONAL PARK

Geographic location: On the Caribbean coast of the Department of Magdalena, Colombia, extends for about 35 km from just east of Santa Marta.

Size: 12,000 hectares (46 sq miles).

Access: A tourist bus leaves daily at 10.30am from Hostal Miramar in Santa Marta into the park, costing $12 including entry fee. Or take a local bus which is heading for Rioacha from the market in Santa Marta. Ask to be dropped off at El Zaino (34 km/21 miles) from Santa Marta) for Canaveral, the park's administrative center. The park entrance is on the left side of the road. There is a 4-km (2-mile) walk along the road to the park center. It is also possible to take a taxi from Santa Marta straight through to the center, or hire a vehicle. The park gate closes at 5pm.

Approximate costs of access: Registration fee $10 or tourist bus from Santa Marta including fee, $12. Bus from Santa Marta to El Zaino, about $2. Taxi from Santa Marta to Canaveral, about $4. Hire cars at international rates.

Tour operators: Check with the Santa Marta tourist office, Carrera 2, No. 16–44, tel: 35-773.

Open: All year.

Registration: Essential. Register at park entrance. If closed, check with MMA in Santa Marta.

Accommodation: At Canaveral, in "Ecohabs" (thatched-roofed 4- to 6-person cabins, with en suite bathrooms) for $80–100 per

Shopping

For Colombian handicrafts, the best place is **Artesanias de Colombia** in the old San Diego church, Carrera 10, No. 26–50.

Near the Museo del Oreo there is an excellent handicraft market and there are several handicraft shops on Cra 15 near Calle 75 such as Artesenia El Balay.

Emeralds can be bought in the *joyerias* in Bogotá's **Centro Internacional** or at **La Casa de la Esmeralda**, Calle 30, No. 16–18.

The best antique shop in Bogotá is on the **Plaza Bolívar**, next to the cathedral. Pre-Columbian pottery is sold in the Centro Internacional. Colombian leather goods are one of its lesser-known bargains, available in shops around the city.

person. Also there is a campsite approx $40 per tent but you need to take all your own supplies. At Arrecifes, hammocks $3 or tents $5. The high season periods are December and January, Easter week and July. During these periods the park gets very crowded.

Food/restaurant: At Canaveral, two restaurants serving three meals a day. Expect to spend between $6 and $9 a day.

Transport in park: The only road to Canaveral is the access road from El Zaino.

Trails: There is a well-marked system of trails, with maps available. These include a walk from Canaveral to Arrecifes (where accommodation is possible in hammocks, but food will have to be carried in) and thence to Pueblito (an archeological site).

Further advice: Access to Tayrona National Park is also possible at Palangana (closer to Santa Marta) and those with hire cars may wish to explore this and other parts of the park. Accommodation is, however, centered on Canaveral, which provides an excellent base, particularly for short-stay visitors.

Venezuela

Getting Acquainted

GOVERNMENT

Venezuela achieved independence in 1821 and is a centralized federal republic formed by 22 States, a Federal District, two Federal Territories and 72 Federal Departments (corresponding to Venezuela's numerous islands in the Caribbean Sea). It has enjoyed an uninterrupted period of democratic freedom since January 23, 1958. (The period from 1821 to 1945 was largely dominated by the rule of military dictators.)

The government system is a representative democracy with one authority for each of the three branches of Public Power – Legislative, Executive and Judicial. Executive power rests solely with the President designated by direct, popular and secret ballot for a six-year non-renewable term. With the Council of Ministers, the president is answerable to the legislature – a unicameral parliament. While ex-presidents of the Republic automatically become life members of the Senate, its other members are elected, two from each State and Federal District plus 55 senators who represent proportionally the country's minority political parties.

PEOPLE

More than eighty percent of Venezuela's 23.2 million population lives in the urban areas of its many middle-sized cities, giving the country an average population density of 50 inhabitants per square mile.

CLIMATE

Although Venezuela predominantly has a tropical climate with an average temperature of 27°C (80°F), four well differentiated climatic zones are represented within its boundaries: hot, mild, cool and cold.

In the Andes region the highest mountains are covered with permanent snow. Like everywhere else in the tropics, temperature depends greatly on altitude above sea level, and temperate climates can be found among beautiful mountain landscapes away from the coast. Hot temperatures prevail throughout the year in the lowlands, mainly along the coast. Between December and February there is a slight lowering of temperature and in zones like the Caracas valley, and up to 1,981 m (6500 ft) above sea level, the climate during this period is temperate and similar to France or Spain during April or the beginning of May.

The rainy season in Venezuela is, for want of a better word, called winter, generally starting in mid-May and lasting until the end of October. But showers may fall in December or January. In the capital, Caracas, the January average temperature is 18.6°C (65°F) rising to 21°C (70°F) in July.

BUSINESS HOURS

Business hours are 8am–noon and 2–6pm, although some stores stay open until 8pm. As in all Latin countries, Venezuelans tend to enjoy extended lunch hours.

Useful Facts

● **Weights and measures:** The metric system of measurement is used throughout Venezuela.
● **Electricity:** Electric power in the country is 110 volts (60 cycles fluctuating) for domestic and personal appliances.
● **Time zone:** Venezuelan time is four hours behind Greenwich Mean Time.

Planning the Trip

VISAS

Entry is by passport and visa or passport and tourist card.

When arriving by air, tourist cards can be issued to citizens of the United States, Canada, Japan and Western European countries (except Spain and Portugal).

When arriving by land, a multiple-entry visa must be obtained from a consulate before arrival, which is often a lengthy process and will cost around $22. It is a good idea to carry your passport with you at all times (be sure to have photocopies of all documents guarded in a separate location in case of loss) as the police mount spot checks and anyone without ID is detained.

MONEY MATTERS

Major credit card companies have offices in Caracas, including:
American Express, c/o Turisol in Centro Comercial Tamanaco shopping mall, tel: 959-3050.
Visa, tel: 8001-2169.
Diners, tel: 202-1716.
Mastercard, tel: 8001-2902.

The Banco Union or Banco de Venezuela can be used for Visa transactions; Banco Mercantil or Banco de Venezuela for Mastercard transactions. Since 1997 it has become difficult to change dollars or travellers' cheques even in *casas de cambio*, who may insist upon proof of purchase to process the latter. Only accept US$ if changing money in hotels.

HEALTH

Health conditions in Venezuela are good. Water in all major urban areas is chlorinated but it is better to drink bottled water. Medical care is good. Inoculation against typhoid and yellow fever is advisable and you should have protection against malaria if you plan to visit the Orinoco basin and

other swampy or forest regions in the interior. It is always good to take some form of remedy for stomach upsets, and, always have a handy roll of toilet paper.

WHAT TO WEAR

Tropical cotton clothing in normal city colors is best for Caracas, while in Maracaibo and the hot, humid coastal and low-lying areas, washable tropical clothing is best. In western Venezuela, in the higher Andes, a light overcoat and warm jacket are handy. Khaki bush clothing is practical for a visit to the oil fields, but remember it is the local custom that men should wear long trousers except at the beach. Women should wear slacks or cotton dresses, with an extra wrap for cooler evenings as well as in air-conditioned restaurants and cinemas.

GETTING THERE

By Air

Venezuela has six international airports and 282 airdromes, of which 250 are private. These are used mostly by small planes and helicopters. The Metropolitan area of Caracas is served by the Simon Bolívar International Airport for international flights and the adjacent Maiquetia which is the main domestic terminal – both are on the coast 30 km (18 miles) from the city centre.

On Arrival

Tourist authorities warn that taxi services at the main airports are infamous for over-charging hapless tourists. Corpoturismo, the Venezuelan State Tourist Authority, has made efforts to control excesses. Take an official taxi in the taxi line ($20–25 to center; official tariffs are posted by the taxi stand). Ask for assistance at the airport information desk if in doubt and always take a licensed taxi.

Public Holidays

January 1: New Year's Day
April 19: Anniversary of the National Declaration of Independence
May 1: Labor Day
June 24: The Feast day of San Juan Batista
July 5: Anniversary of the Signing of the Venezuelan National Independence Act
July 24: Anniversary of the birth of the Liberator, Simon Bolívar
October 12: Columbus Day (Anniversary of the Discovery of America)
December 25: Christmas Day

GETTING AROUND

By Road

There are 64,516 km of roads, about 29,032 km of which are paved and there are many freeways.

Buses to nearby destinations leave from Nuevo Circo bus station in Caracas city center. There is a new terminal at La Bandera for western destinations. Buses to the east leave from the new Terminal de Oriente on the eastern outskirts of the city. Bus travel varies a lot in quality with most companies liable to run bone-shaking wrecks of buses on the same routes as luxury coaches.

Practical Tips

MEDIA

The Daily Journal, founded in 1945 by Jules Waldman, is the country's only English-language daily newspaper and is favored by newcomers to Caracas as well as international businessmen and diplomats. Also in English is the *Guardian Weekly* (Latin American edition), available as a supplement to *Economia Hoy*.

The main Spanish-language papers published in Caracas with a national distribution are *El Nacional*, *El Universal* and *El Diario*.

POSTAL SERVICES

The Venezuelan postal service IPOSTEL is extremely slow and inefficient, although much better at mailing abroad than locally, and efforts are being made to speed it up. As a result, motorcycle courier services abound in Caracas, and other courier services give a very much better service than IPOSTEL to almost everywhere in the country.

EMBASSIES & CONSULATES

Australia: Quinta Yolanda, Avenida Luis Roche, between 6 and 7 Transversal, Altamira, tel: (0212) 283-3090.
Austria: Edif. Torre Las Mercedes, Piso 4, Ofic. 408, Avenida La Estancia, Chuao, tel: (0212) 913-863.
Brazil: Centro Gerencial Mohedano, Piso 6, Avenida Los Chaguaramos and Avenida Mohedano, Urb. La Castellano, tel: (0212) 261-4481 or 261-5506 or 261-6529.
Canada: Torre Europa, Piso 7, Avenida Francisco de Miranda, Urg. Campo, Alegre, tel: (0212) 951-6166 or 951-6306.
Colombia: Edif. Consulado de Colombia, Calle Guiacaipuro, El Rosal, tel: (02) 951-3631 or 951-6692.
France: Edif. Embajada de Francia, Calle Madrid and Av. La Trinidad, Las Mercedes, tel: (0212) 910333/324.
Italy: Edif. Atrium, Pent House, Calle Soracaima between Av. Tamanaco and Venezuela, El Rosal, tel: (02) 952-7311 or 952-8939.
Japan: Av San Juan Bosco, entre 8 y 9 Transversal, Altamira, tel: (0212) 261-8333.
United Kingdom: Torre Las Mercedes, Piso 3, Avenida La Estancia, Urb. Chuao, tel: (0212) 993-4111.
United States: Calle 5 with Calle Suapure, Colinas de Valle Arriba, tel: (0212) 997-2011 or fax: 997-0843.

SECURITY & CRIME

The crime rate in Venezuela, particularly in Caracas, is high. It is recommended that you take precautions by not wearing jewelry or carrying money in such a way that it can be snatched. You should not walk alone in narrow streets in downtown Caracas after dark and it is strongly recommended that you do not travel by car at night, particularly in the countryside where, should you have an accident or a breakdown, the risk of robbery and worse crimes is much greater.

TOURIST INFORMATION

CORPOTURISMO (Corporacion de Turismo de Venezuela) is the Venezuelan State Tourist Authority with overall responsibility for tourism. Although they claim to have a hotel booking or reservation service, it is strongly recommended that you make arrangements via a local travel agency or tour organizer. For tourist information you may like to contact CORPOTURISMO at: Torre Oeste, Parque Central, (metro Bellas Artes), Av. Lecuna, Caracas, tel: (0212) 507-8815; freephone 800 43-328 or www.indecu.gov.ve

Medical Services

In an emergency, contact any of the following hospitals in Caracas (code 0212):
Centro Medico de Caracas, tel: 509-9111
Policlinica Metropolitana, tel: 908-0100, 908-0140
Hospital de Clinicas, tel: 574-2011

Where to Stay

The following are recommended hotels in Caracas (code 0212):
Avila
Avenida Washington, San Bernandino
Tel: 515-128
Fax: 523-029

Complaints

The Venezuelan Hotel Association, ANAHOVEN, tel: (02) 574-3994 or 574-7172, liaises with the Venezuelan State Tourist Authority, CORPOTURISMO, to supervise hotel standards and to deal with all customer enquiries and complaints.

Caracas Hilton
Avenida Libertador and Sur 25
Tel: 503-5000
Fax: 503-5003
Hotel Tampa
Avenida Francisco Solano (metro Plaza Venezuela)
Tel: 762-3771
Fax: 762-0112
Las Americas
Calle Los Cerritos, Bello Monte
Tel: 951-7387
Fax: 951-1717
Lincoln Suites
Avenida Francisco Solano, between San Jeronimos and Los Jabillos
Tel: 761-2727
Fax: 762-5503
Tamanaco Intercontinental
Avenida Principal Las Mercedes
Tel: 792-4522
Fax: 208-7116

Eating Out

WHAT TO EAT

Venezuelan cuisine is very varied because of the diverse cultural influences the country has been subjected to over four centuries. At Christmas and national celebrations, the *hallaca* is paramount as Venezuela's national dish – it's a stew of chicken, pork, beef and spices used as a filling to a pie-like dough of maize, which is then wrapped in banana leaves and cooked in boiling water. Another

Restaurant Guide

A useful guide to restaurants, available in most bookshops, is the *Guia Gastronómica de Venezuela.*

favorite is *pabellon*, which combines rice, black beans, shredded beef and *tajadas* (sliced and fried ripe plantains).

In the Andes there is *pisca,* a rich and tasty soup, as well as local dishes based on trout and sausage. Coro is famous for its *tarkari de chivo* (made from goat), marinated fish and goat milk preserves.

Zulia State has delicious coconut-based specialties like *conejo en coco* (rabbit cooked in coconut milk) and a selection of sweets and candies. The Eastern region is widely known for its tasty seafood specialties, like *consomé de chipichipi* (small clams broth); cream of *guacucos* (middle-sized clams) and *empanadas de cazón* (small shark pie).

A typical and very popular Venezuelan dish is *mondongo* (a soup-like stew which uses specially processed tripe as a main ingredient). The *arepa* is traditional Venezuelan bread made from maize and served either fried or baked.

WHERE TO EAT

Eating out is extremely cheap by US and European standards and there are literally hundreds of restaurants catering to all tastes. From the *tascas* of Candelaria, where the best Spanish cooking in the whole of Venezuela is said to be available, to the elegant and exorbitant French restaurants of Las Mercedes, there are enough to satisfy every whim.

The cheapest food is at *Fuentes de Soda* and cafés. Food in bars may cost 50 percent more, and there is no need to feel reticent about asking for prices before you order, as the price of a beer can often be three times higher than elsewhere.

A recommended list of restaurants in Caracas follows.
In the Las Mercedes district try:
Il Cielo
Avenida La Trinidad
Tel: 993-4062
Italian cuisine.
Le Gourmet
Hotel Tamanaco
Tel: 208-7242
French.

Taiko
Avenida La Trinidad
Tel: 993-5647
Japanese.

In the La Sabana district, go to Avenida Francisco Solano for a variety of restaurants and cafés. The Altamira district has many good restaurants too.

Parks & Reserves

ECOTOURISM AGENCIES

Geodyssey
116 Tollington Park, London N4 3RB, tel: (020) 7281-7788, fax: (020) 7281-7878, www.geodyssey.co.uk
Independent travel company specializing in mainland Venezuela with a strong eco-tourism base and working closely with the local communities; offers classic tours to all regions, including birding, butterfly tours, treks and diving trips.
Last Frontiers
Fleet Marston Farm, Aylesbury, Bucks HP18 0PZ, tel: (01296) 658-650, fax: (01296) 658-651, www.lastfrontiers.co.uk
Small independent South American specialist with in-depth knowledge of Venezuela. Also offer painting and photo trips, riding and wildlife holidays.
Lost World Adventure Tours
Avenida Abraham Lincoln, Caracas 1050, Edif. 3–H, Piso 6, Oficina 62, tel: (0212) 761-7538.
US office: tel: 1-800-810-5021, fax: 914-273-6370, www.lostworldadventures.com
Orinoco Tours
Boulevard de Sabana (metro Plaza Venezuela), Edif. Galerias Bolivar, Piso 7, Oficina 75–A, tel: 761-7712, fax: 761-6801, email: orinoco@sa.omnes.net www.he.net/ven/orinoco

---

AMAZONAS

Except for a few kilometres of road near the state capital of Puerto Ayacucho, travel is almost exclusively by boat or small charter plane in this beautiful and largely untouched region. Easily reached by air from Caracas, Puerto Ayacucho is the gateway to the region where local operators offer three-day trips up the Sipapo and Autana rivers and local Piaroa and Guariba settlements. It is not recommended to travel alone, and much of Southern Amazonas is heavily restricted to protect the Yanomami Indians and the Alto Orinoco-Casiquiare Biosphere Reserve.

Lodge Orinoquia

c/o Orinoco Tours, tel: (0212) 761-7712, fax: 761-6801, www.he.net./ven/Orinoco

Only 20 minutes away from Puerto Ayacucho and set between the famous Atures and Maypures rapids, considered by Humboldt to be the 8th Wonder of the World, Lodge Orinoquaia is a small rustic lodge with 5 rooms, each with private bath. Good base to visit indigenous villages and Tobogan de la Selva Park as well as take boat-trips to watch birds, otters and maybe dolphins.

Camani Lodge

Tel: (48) 24 4865 or 21 4553, www.canami.com

On the banks of Ventuari river, 2 hours by boat from San Juan de Manapiare or directly by charter flight from Caracas. 13 attractive 'churuatas' with private bath and hot water. Recommended for boat and canoe trips, sport fishing, mountain bikes, observation tower and access to Tencua Falls and Cano de Piedra, where there are beaches and waterfalls.

Junglaven

Tel: (02) 993-2617, email: kitti@cantv.net, or contact, Geodyssey in the UK tel: (020) 7281-7788, www.geodyssey.co.uk

Reached by private charter from Puerto Ayacucho, this superb sport fishing lodge and birding camp has access to a wide range of habitats and some 280 species of birds. 10 cabins with private bathrooms, bar and game room. It's possible to explore the surrounding jungle, river and waterfalls or even stay overnight at a nearby indigenous settlement.

CANAIMA NATIONAL PARK AND THE GRAN SABANA

One of the six largest parks in the world, characterised by its spectacular table mountains, open savannah and stunning waterfalls; there are many exciting options ranging from overflights of the Angel Falls to trips by motorized canoe and treks into Conan Doyle's Lost World of Mount Roraima. Many travelers choose to visit the area in June–Nov after the heavy rains when, in addition to Angel Falls, there are many swollen rivers spilling over the edges of escarpments and table mountains to create dramatic cascades.

Tour Operators

Most tour operators offer itineraries that include a visit to Angel Falls. It is possible to take a day-trip from Caracas or Ciudad Bolivar but many travelers prefer to spend a few nights in this idyllic region.

Geodyssey

Tel (UK): (020) 7281-7788, fax: (020) 97281-7878, www.geodyssey.co.uk

La Gran Sabana Tours

www.venezuelatuya.com

Lost World Adventures

Edif 3-H, Piso 6, Oficina 62, Av Abraham Lincoln, Caracas 1050 Tel: (761) 7538, www.lostadventures.com

The Canaima Camp Lodge

The area's largest hotel is owned by Hoturvensa, a division of the Venezuelan airline Avensa, and is set at the edge of the Canaima Lagoon. Currently only clients staying at the lodge can book seats on their daily direct flight by DC-3 from Caracas to Canaima.

The standard Hoturvensa package offers good value for money and includes one night's accommodation in one of the 35 thatched bungalows, a panoramic sightseeing flight over the Angel Falls (weather permitting) and a short excursion by motorized dugout boat on the Canaima Lagoon to Hacha Falls. From Jun–Oct, when river levels are usually high, boat trips are available from Canaima to a lookout near the foot of Angel Falls, with two nights spent sleeping in hammocks in a permanent shelter in the jungle. Book through a tour operator or Avensa direct tel: (2) 562 3022, fax: (2) 564 7936.

ORINOCO DELTA

The tourism potential in the huge delta of the Orinoco, where the river makes its final journey to the Atlantic through narrow channels around lowlands of palm forest and mangroves, has only recently been exploited, and now it is possible to take river trips from Tucupita or stay at one of the few rustic lodges. In the Delta there is an amazing diversity of flora and fauna, including macaws, toucans, water birds, howler and capuchin monkeys, river otters, cayman, freshwater dolphins, turtles and piranha, and also the opportunity to observe the river-life of the indigenous Waroa people, famous for their boat-building and their stilted villages.

Boca de Tigre Lodge

c/o Boca de Tigre Tours, tel: (91) 41 704; email: bocatigre@telcel.net.ve

Located 2½ hours by boat from Bujas Port in the north-western part of the delta, this Guyanese-style lodge has 30 rooms with private facilities, convivial public area and an impressive selection of tours. Nearby, the lodge has built a school for the local Warao children which lodge guests are welcome to visit.

Maraisa Lodge

San Francisco de Guayo, tel/fax: (87) 21 6660/0553.

7½ hours by motor-boat from Tucupita and close to the Atlantic, Maraisa Lodge has stilted guest cabins with walkways to the lodge, each with private bathroom; it's sited on the opposite bank to part of San Francisco de Guayo, with wonderful river views from its verandah. Trips can be made in canoes with the Warao to observe the plentiful local wildlife and visit indigenous communities.

Tour Operators

Geodyssey, tel: (020) 7281-7788,
fax: (020) 7281-7878,
www.geodyssey.co.uk
Jakera Tours, www.jakera.com
Tucupita Expeditions,
www.orinocodelta.com

THE LLANOS

Extending between the Andes and
the Orinoco, these vast flat cattle-
raising plains have an annual
pattern of flood and drought which
makes it one of the most exciting
places to view wildlife in South
America. Apart from more than 300
species of birds including, scarlet
ibis, hoatzins and spoonbills, it is
possible to spot jaguars, ocelots,
anteaters, honey-bears, capybaras,
alligators and freshwater dolphin.

The best, and almost only,
places to stay are the wildlife *Hatos*
– large cattle ranches which
pioneered eco-tourism by banning
hunting, setting up wildlife
conservation programmes and
offering simple accommodation and
excellent wildlife-watching
opportunities to scientists and
travelers. In most *hatos*, the high
season coincides with the dry
season Nov–April. Access is by air
to San Fernando or Barinas
followed by a road transfer of 2 to 3
hours.

Hato El Frio

Tel: (47) 81793, fax: (47) 81223.
One of the largest ranches in the
region and highly recommended to
first-time visitors to the Llanos.
Ten basic double rooms with fans
and private bathrooms.

Hato El Cedral

Booked through Turismo Aventura,
Caracas, tel: (212) 951-1143.
Formerly owned by the Rockefellers,
this ranch has good facilities,
including a small swimming-pool
and air-conditioned accommodation.
An excellent choice at any time of
the year.

Hato El Pinero

Tel: (212) 916965,
www.branger.com/pinero
A traditional and rustic lodge which
banned hunting 40 years ago, this
hato has well-established birding
trails and a wide range of habitats,
but it's 7 hours' drive from Caracas
and lacks the grand wildlife
spectacle of the low Llanos.

LODGES

**Avensa Lodge/Canaima/Angel
Falls** are served daily by Avensa
Airlines from Caracas. Turn to a
travel agent or contact Avensa
directly. Avensa Lodge is nicely
situated at a lagoon; however,
it is normally not mentioned
that visiting Angel Falls is an
additional two-day trip (six hours
by boat one way), and from
December to May often not
possible because of low water.

Hortuvensa run a Canaima two-
day camp, costing $430 including
accommodation and flight. They can
be contacted through Avensa, tel: 58-
2-562-3022, fax: 58-2-564-7936.

Culture

As a result of different customs,
rites, religions and musical
expressions handed down through
generations, Venezuelan folklore is
both colorful and varied. In the
warmer areas in particular, local
musical instruments have African
origins – the *Diablos Danzantes*
("Dancing Devils") of Yare and the
San Juan Dance are characteristic
expressions with a lineage from the
West African slave trade.

The *Joropo* is regarded as the
national dance of Venezuela. It
comes from the Llanos (Prairies)
and is a lively form of music
generally interpreted using the harp,
cuatro and maracas.

Caracas has some fine museums
and exhibition centers, which
include: Arte Hoy, Avenida El
Empalme and Urb. El Bosque. The
following museums are
recommended for a visit:

Casa Natal del Libertador, Plaza
San Jacinto A Traposos, Centro,
tel: (0212) 545-7693.
A faithful colonial reconstruction on
the original site where the Liberator,
Simon Bolívar, was born.

Furnishings and memorabilia are of
that period in Venezuela's history.
Tues–Fri 9am–noon and
2.30–5.30pm, Sun and holidays
10am–5pm.

Cuadra Bolívar, Avenida Sr 2 entre
Esq. Barcenas y Las Piedras,
tel: (0212) 483-3971.
A reconstruction on the original
site of the Bolívar family's country
home "El Palmar", with gardens
and patios, colonial furniture, a
restored kitchen, portraits and
books of the period. Tues–Sat
9am–1pm and 2.30–5pm, Sun
and holidays 9am–5pm.

Galeria de Arte Nacional, Plaza
Morelos, Urb. Los Caobos,
tel: (0212) 571-3519 or 572-1070.

Jardín Botánico, Calle Salvador
Allende, Ciudad Universitaria, UCV,
Tues-Sun 8.30am–5pm.

Casa de Arturo Michelena, Esq.
Urapal No. 82 (a block south of La
Pastora Church), Urb. La Pastora,
tel: (0212) 825-853.
Former residence of the famous
19th-century Venezuelan painter
Arturo Michelena. It is a stony, old-
style house containing the painter's
personal belongings and some
unfinished canvasses. Daily except
Mon and Fri, 9am–noon and
3–5pm.

Museo de Arte Colonial, Quinta
Anauco, Avenida Panteon (at the
Cota Mil exit), San Bernardino,
tel: (0212) 518-517.
This former residence of the
Marqués del Toro, War of
Independence hero, has been
faithfully restored with a collection
of colonial furniture, household
objects, paintings, sculptures etc.,
surrounded by beautiful gardens.
Tues–Sat 9am–12pm and
2–4.30pm, Sun and holidays
10am–5.30pm.

Museo de Caracas, Palacio
Municipal (Concejo Municipal),
Plaza Bolívar, Esq. Las Monjas,
tel: (0212) 545-6706 or 545-
8688. Features wood-carved
miniatures by Raul Santacan
depicting scenes of life in
Caracas, from colonial times to
the beginning of the 20th century,
and the life works of Venezuela's
internationally acclaimed impres-

sionist painter Emilio Boggio (1857–1920). Mon–Fri 9am–noon and 2.30–4.30pm, Sat, Sun and holidays 10.30am–1.30pm.
Panteón Nacional, Plaza del Panteón, Avenida Panteón y Avenida Norte, Centro, tel: (0212) 821-518. Contains the tomb of Simon Bolívar and memorials to other military heroes of the War of Independence. Tues–Sun 9am–12pm and 2.30–5pm.

Shopping

Venezuela's craftsmen enjoy a prestigious position because of the variety and quality of their workmanship. For example, outstanding Quibor ceramics have pre-Hispanic origins, molded using styles and techniques handed down through the generations. Baskets, hammocks, hats and other products made from vegetable fibers are to be found in the towns and villages along the eastern coast.

Beautifully woven square and round ponchos, colorful blankets and caps are sold by the Andean people, while craftsmen of the Llanos (Prairies) sell four-string guitar-like musical instruments called *cuatros* as well as harps and mandolins. Craftsmen of San Franscisco de Yare make devil's masks for their Festival of the Dancing Devils, while hand-carved furniture and objects made from goat skins are samples from Coro.

SHOPPING AREAS

In Caracas, the **Sabana Grande Boulevard** is an excellent commercial street with hundreds of boutiques, jewelry stores, bazaars and stores. There are busy bars and coffee shops.

Also visit **Chacaito**, **CCCT**, **Paseo Las Mercedes**, **Concresa**, **Plaza Las Americas** – all have supermarkets, department stores, cafés, restaurants, beauty parlors and just about every imaginable ware on display.

The Guianas

Getting Acquainted

FRENCH GUIANA

Area: 85,500 sq km (33,000 sq miles)
Main city: Cayenne
Population: 150,000
Status: an overseas *départment* of France.

GUYANA

Area: 215,000 sq km (83,000 sq miles)
Capital: Georgetown
Official language: English
Population: 874,000

SURINAME

Area: 163,820 sq km (63,250 sq miles)
Capital: Paramaribo
Languages: official language is Dutch, English, Hindi, Javanese and Chinese are also spoken.
Population: 452,000

Climate

The temperature remains warm with equatorial humidity all year round, but it's not excessively hot; average temperatures are between 24°C (75°F) and 31°C (88°F). The best time to visit is July/August–November/December, between the two wet seasons, which are from May through June and from December through January. Average annual rainfall is 230 cm (91 inches).

What to Wear

As in all humid tropical climates, by far the best fabric to wear is light cotton. Loose clothing that covers arms and legs will be comfortable and protect you from the sun and mosquitos. Shorts and swimming costumes should not be worn away from the beach.

Tourist Information

Tourism in the Guyanas, with the exception of French Guiana (an overseas *département* of France) is still undeveloped, and travel facts and listings are hard to come by. For further background and a wide selection of nature and adventure tours in the Guyanas, Wilderness Tours have a very comprehensive website at: www.wilderness-explorers.com Another useful website is: www.exploreguyana.com

For tourist information on French Guiana, contact the French Guiana Tourist Board at 1 rue Clapeyron, 75008 Paris, tel: (01) 4294 1515; fax: (01) 4294 1465.

For travel information in Suriname and Guyana, check out www.surinfo.org and www.tourisme-guyane.gf

Iwokrama

The Iwokrama International Centre for Rain Forest Conservation and Development is a 360,000 hectare (890,000 acre) area of virgin rainforest set aside for the study of conservation and sustainable development of the rainforest.

Here visitors regularly see jaguars and there are probably 1,500–2,000 different plant species, 120 snakes, lizards and frogs and 105 mammals, plus more than 450 species of birds. Their website is very informative and can be found at: www.iwokrama.org

Forest Cover

In 1995 Guyana had 185,800 sq km (71,738 sq miles) of forest, and Suriname 147,200 sq km (56,834 sq miles) – 94.4 and 94.8 percent of their respective total areas.

Trinidad & Tobago

THE PLACE

Trinidad

Situation: Of all the Caribbean islands, Trinidad is furthest south, just 11 km (7 miles) from the coast of Venezuela, by the Gulf of Paria.

Area and geography: Roughly triangular in shape, the island is about 80 km (50 miles) long and almost 64 km (40 miles) wide, with three mountain ranges traversing the interior. These areas remain heavily forested and rich in wildlife. The east and west coasts tend to be swampy; the most popular beaches are in the north, over the mountains from Port of Spain, the capital. The south is very different, with its famous pitch lake, oilfields and mud volcanoes.

Tobago

Situation: Tobago lies 32 km (20 miles) northeast of Trinidad.

Area and geography: Fish shaped Tobago is 41 km (25 miles) long and 11 km (6 miles) wide. A central range of mountains forms a spine through the Tobago Forest Reserve in the north. Coral reefs surround much of the southern tip of the island. Several smaller islands nearby are wildlife sanctuaries.

TIME ZONE

Trinidad and Tobago operates one hour ahead of Eastern Standard Time, or the same as Eastern Daylight Savings Time and four hours behind Greenwich Mean Time.

ELECTRICITY

Trinidad and Tobago operate on 110 volts and 60 cycles of current.

CLIMATE

The weather is almost always perfect. Because of the trade winds, the sun's heat is mitigated by sea breezes. Average temperatures are 28°C (84°F) during the day and 23°C (74°F) at night.

There are two seasons: the dry from January to May, and the wet from June to December respectively. All this means – except in late October and November, when hurricanes further north can create stormy weather – is that from June to December it rains briefly in the afternoon. June is the wettest month; February and March the driest. It never rains during carnival.

GOVERNMENT & ECONOMY

Trinidad and Tobago became fully independent from Britain in 1962, and declared the Republic into being in 1976, when the President replaced the British Monarch as head of state. The government is a parliamentary democracy, headed by the President, and governed by Parliament and the Prime Minister. Tobago's legislative body, the House of Assembly, sets domestic policy for the island.

Since gaining political independence, Trinidad and Tobago's leaders have also worked to achieve economic independence and stability. This goal seemed well within their grasp during the late 1970s, when prices for exported Trinidad and Tobago oil were at an all-time high. But much has changed. Oil prices declined precipitously, unemployment claimed more of the workforce, and a populace that had grown accustomed to prosperity began to experience harder times.

Beset by economic problems, the government has turned more to tourism, a source of hard currency

Public Holidays

- Good Friday
- Easter Monday
- Whit Monday – 8th Monday after Easter
- Corpus Christi
- Labour Day – June 19
- Emancipation Day – August 1
- Independence Day – August 31
- Eid-Ul-Fitr – as decreed
- Divali – as decreed
- Republic Day – September 24
- Christmas Day – December 25
- Boxing Day – December 26
- **Carnival:** This raucous and unforgettable national party takes place on the Monday and (Shrove) Tuesday before Ash Wednesday. Though not a legally decreed holiday, it's safe to assume that a very relaxed attitude will prevail.

that has remained uncultivated for many years, and encouraged light industries to diversify the country's economic base.

PEOPLE

As a nation, the population of Trinidad and Tobago is divided between roughly equal numbers of African and East Indian descendants, and much smaller percentages of Chinese, Europeans and Syrians. This diversity is mainly confined to Trinidad, however; in Tobago, over 90 percent of the people are of African descent.

BUSINESS HOURS

Monday through Friday, most shops and offices are open from 8am–4pm or 4.30pm. Some food stores stay open until 6pm, except on Thursdays when food and liquor stores close at noon. Saturday closing time is at noon or 1pm, except for food and liquor stores, which are open all day. Large malls like the Long Circular stay open until quite late; others are open late on Fridays and Saturdays.

All banks stay open from 8am–2pm Monday–Thursday, and

9am–1pm and 3–5pm Friday. Most of the major commercial banks have offices on Independence Square, as well as various branches.

Planning the Trip

VISAS & PASSPORTS

All visitors must have a return or ongoing ticket and a valid passport for entry into Trinidad and Tobago. Citizens of the United Kingdom, United States and Canada do not need visas unless they are doing business in Trinidad and Tobago. Visitors from certain European countries do need visas, however, so check with authorities in your own country before leaving.

MONEY MATTERS

The TT dollar (TT$) is the basic unit of currency in Trinidad and Tobago. Since many TT$ prices are the same as they were before devaluation, Trinidad and Tobago is a more reasonably priced destination than it once was, especially if you buy goods and services geared to locals or West Indian travelers.

Currently, the TT$ is classified as a "restricted currency," a result of policies aimed at curbing inflation. For visitors, this means you should reconvert all the TT$ you have not spent before you leave Trinidad and Tobago, as you probably won't be able to do so in another country. Also, be sure to get a receipt for all cash and travelers' checks you change into TT$. You must show the receipt in order to change your money back. There is no limit to the amount of foreign currency you can bring in, but TT$1,200 is the maximum you will be able to reconvert on departure.

The National Commercial Bank of Trinidad and Tobago has a foreign exchange branch at Piarco Airport (tel: 664-5281/5322), which changes British pounds, American and Canadian dollars and all Caribbean currencies for a small fee. In Tobago, the only bank

outside Scarborough is the Republic Bank at Crown Point Airport. Bring some US dollars or pounds sterling to change at the airport for initial expenses like a taxi to your hotel.

Major credit cards can be used at the larger hotels, restaurants and car rental firms throughout the islands, and at stores that cater mainly to visitors. However, many of the smaller hotels, guesthouses, restaurants and shops do not accept them, so it's wise to have an international brand of travelers' checks and a good supply of cash on hand.

HEALTH

Vaccinations for smallpox and yellow fever are not required for entry into Trinidad and Tobago unless you have just passed through an infected area.

Throughout the Republic, the water is safe to drink and even street food is generally tasty and fresh for most of the year. However, during Carnival it's probably best to avoid the seafood and meat specials sold from numerous shanties around the Savannah. Often the food is cooked at the vendor's home in the morning, and stored without refrigeration throughout the day.

Tourist Information

For advice before your trip, consult one of the TIDCO (Tourism and Industrial Development Company of Trinidad and Tobago Ltd) offices in New York, Miami, Toronto or London.

The TIDCO information Office at Piarco Airport (tel: 664-5196) is open every day except Christmas from 6am–midnight. Here you will find friendly and efficient answers to questions about accommodations, transportation and events of interest to visitors. There is also an information office at Crown Point Airport, Tobago, tel: 639-0509. TIDCO staff also meet all cruise ships and arrange special activities like tours and

The National AIDS Program of Trinidad and Tobago also issues a timely reminder for Carnival and all year round: "AIDS is preventable; don't give it to yourself." Avoid casual sex, any procedures which pierce the skin, and remember that alcohol abuse can affect your judgment. The sun shines with dangerous intensity all year round. Wear ample sunscreen, build tanning time slowly up from 15 minutes in the early morning and late afternoon, and bring a hat or buy one.

WHAT TO WEAR

Casual clothes in lightweight fabrics are most comfortable for touring the countryside. In town, most men wear dark trousers and a light jacket, or more formal business suits. Women wear dresses or skirts and blouses for work and shopping. Note that skimpy clothing, particularly on women, will attract attention and occasionally provoke comment.

Even at the best restaurants there's not much call for evening wear. Casual clothes with a sense of style in lush colors and lavish fabrics prevail in fashionable Trinidad. On the beaches, anything

calypso performances for the passengers.

TIDCO addresses in Trinidad and Tobago are:

In Trinidad: 10–14 Philipps Street, Port of Spain, tel: 623-1932/4 or 623-INFO, fax: 623-3848.

In Tobago: Unit 26, IDC Mall Sangster's Hill, Scarborough, tel: 639-4333, fax: 639-4514, www.tidco.co.tt

In London: c/o Morris Kevan International Ltd, Mitre House, 66 Abbey Road, Bush Hill Park, Enfield, Middx, EN1 2RQ, tel: (020) 8350-1000, fax: (020) 8350-1011; email: mki@ttg.co.uk; www.VisitTNT.com

goes. But, especially in Tobago, keep in mind that this is a small-town society, so rules of propriety should be observed. Wear bathing suits on the beach only.

GETTING THERE

By Air

Almost touching the coast of South America, Piarco Airport in Trinidad is a good six hours from New York by plane. Tobago's Crown Point Airport is just 12 minutes flying time further, on one of the six daily flights of national carrier, BWIA.

Regular airfares to Trinidad run relatively high. There are fewer air-plus-hotel packages than for some of the more heavily advertized and tourisy Caribbean countries, but several airlines do offer them occasionally.

BWIA offers flights from Miami to Piarco Airport daily, from London and New York five days a week, Baltimore–Washington three times a week, Toronto twice, Stockholm and Frankfurt once a week.

American Airlines has a daily nonstop service from San Juan, linked to cities throughout the United States. There are also scheduled flights from London on British Airways. British visitors using Tobago as a base may find it cheaper to book a package holiday through a tour operator, flying on Caledonian Airways, which has very competitive fares. Air Canada's direct flights leave from Toronto; some are nonstop, depending on the day of the week. KLM flies to and from Amsterdam once a week. There are also connections to other Caribbean islands and South America via BWIA, LIAT (Leeward Island Air Transport), LAV (Linea Aeropostal Venezolana), and Guyana Airways.

By Sea

Trinidad and Tobago are developing their assets as a cruise destination through large scale expansion of Scarborough Harbor and other improvements. This, combined with a boom in the cruise industry, means new options for travelers and more cruise lines with one or both islands on their routes. Check with a travel agent before booking.

The main problem with taking the sea route is the difficulty in extending your stay beyond the few hours the ship is docked. Most cruise ship tickets are sold for complete voyages and you'll probably want to avoid the logistical maneuvering involved in a change.

Ships from the following cruise lines stop at Trinidad and/or Tobago:

Cunard Cruise Lines, tel: 800-327-9501 or 528-6273
Epirotiki Lines Inc. (World Renaissance), tel: 800-221-2470
Princess Cruises, tel: 800-421-0522
Windjammer Cruises, tel: 800-245-6338 or 327-2601

ON ARRIVAL

On the airplane or ship, just before you arrive, you will be asked to fill out an immigration form. Immigration officials collect one copy as you enter the country. Save the duplicate; you will need it for departure. Also, be prepared to show your return ticket and give the address where you will be staying in Trinidad and Tobago. If you do not have a Trinidad and Tobago address, Tourism Development Authority staff meet all flights and can provide information on accommodations.

WATER TRANSPORTATION

The daily (except Saturdays) sea-ferry between the Port of Spain terminal and Scarborough takes 5 hours, and food and drink are available on board. Cabins are available for TT$50/60 and double cabins cost approx TT$160. Sailings from Port of Spain can be rough and depart at 2pm, returning from Scarborough at 11pm, though schedules are liable to change. Tickets are sold only at the Port of Spain and Scarborough offices; passenger ticket sales close two hours before sailing time. For more information, tel: 625-4906 in Port of Spain, or 639-2417 in Tobago.

PUBLIC TRANSPORTATION

By Bus

The Public Service Transport Service (PTSC) runs both buses and maxi-taxis. Buses go to every part of both islands, and are quite cheap. But because they are often hot and crowded, most tourists, and many locals, prefer the maxi-taxi which costs slightly more.

By Public Taxi

Maxi-taxis: These color-coded mini-buses follow particular routes on both islands. In Tobago, ask directions to the nearest stop.
Route taxis: These cars carry up to five passengers along set routes from fixed stands to various destinations. Sometimes drivers will digress slightly for visitors if the car isn't crowded. Unless they render special services, drivers do not expect tips. In Scarborough, Tobago, route taxi stands are located across from the bus terminal and central shopping plaza.

PRIVATE TRANSPORTATION

Hire Taxis

Like route taxis, hire taxis have "H" on their licence plates, but they are essentially private taxis, carrying only you and your companions exactly where you want to go. Hire taxis do not have meters and are rather expensive. Always agree on fares in advance. Also note that fares in Tobago double after 9pm. Hire taxis wait near most large hotels.

Driving

If you have a valid driver's license from the United States, Canada, France, United Kingdom, Germany or the Bahamas, you may drive in Trinidad and Tobago for a period of up to three months – but only the

Holiday villas beyond indulgence.

BALEARICS ~ CARIBBEAN ~ FRANCE ~ GREECE ~ ITALY ~ MAURITIUS
MOROCCO ~ PORTUGAL ~ SCOTLAND ~ SPAIN

If you enjoy the really good things in life, we offer the highest quality holiday villas with the utmost privacy, style and true luxury. You'll find each with maid service and most have swimming pools.

For 18 years, we've gone to great lengths to select the very best villas at all of our locations around the world.

Contact us for a brochure on the destination of your choice and experience what most only dream of.

INTERNATIONAL
CHAPTERS

Toll Free: 1 866 493 8340
International Chapters, 47-51 St. John's Wood High Street, London NW8 7NJ. Telephone: +44(0)20 7722 0722
email: info@villa-rentals.com www.villa-rentals.com

66 I was first drawn to the Insight Guides by the excellent "Nepal" volume. I can think of no book which so effectively captures the essence of a country. Out of these pages leaped the Nepal I know – the captivating charm of a people and their culture. I've since discovered and enjoyed the entire Insight Guide series. Each volume deals with a country in the same sensitive depth, which is nowhere more evident than in the superb photography. **99**

Sir Edmund Hillary

class of vehicle specified on your license. (China, South Vietnam and South Africa are excluded.) If your stay exceeds this limit, you must apply to the Licensing Department on Wrightston Road in Port of Spain.

While driving stay to the left, and at all times carry both your driver's license and a document (such as a passport) certifying the date of your arrival in Trinidad and Tobago.

Hitchhiking

When driving in the country, you might see hitchhikers, particularly in Tobago. Everyone from schoolchildren to grandmothers hitch (though visitors shouldn't) and it's only friendly to "give a drop" to children, the elderly, and mothers with babies. Exercise more caution with young men.

ON DEPARTURE

If you are leaving by plane, plan to arrive at the airport two hours before your flight is due to depart. Last minute schedule changes are not unheard of, so reconfirm your reservation 24 hours in advance, and make sure the flight is on time before leaving for the airport. Allow extra time of at least an hour to get to Piarco if you are traveling in a taxi or private car from Port of Spain during rush hour. Between about 6.30am and 8.45am, and 3.30pm and 5.45pm, traffic is most horrendous and will definitely cause frustrating delays.

Note that there are some strict departure rules for business people: find out what regulations apply to you at least a week before your planned departure, as they may necessitate a visit to various government offices that are not always open.

On leaving Trinidad and Tobago, you will be required to pay a US$15 departure tax. You will also be asked to surrender the carbon copy of the Immigration Card you filled in on arrival. Beyond the customs and immigration checkpoint, the airport departure area has duty-free shops, refreshment stands, and telephones for overseas calls.

Where to Stay

CHOOSING A HOTEL

Trinidad has a limited number of hotels, compared to the Caribbean's resort islands, and nothing that could be called a resort, American-style. Many business travelers stay at the Hilton or the Holiday Inn, which have facilities for business people. The Kapok Hotel and Hotel Normandie also have large business clienteles and popular restaurants. Facing the Savannah within walking distance of downtown, the Queen's Park Hotel attracts value conscious Europeans and West Indians. As beautiful as the beaches are here, visitors come for music, culture, business, and Carnival, then retire to Tobago for relaxation.

The Trinidad and Tobago Hotel and Tourism Association is happy to provide brochures on member hotels, and information about facilities and services and can be contacted on (868) 623-1932 or at www.tidco.tt

HOTEL LISTINGS

Trinidad
Charconia Inn
106 Saddle Road, Maraval
Tel/fax: 628-8603. Moderate.
Grande Riviere
Mount Plaisir Estate,
Hosana Street, Grande Riviere
Tel: 670-8381.
www.mtplaisir.com
Moderate.
Hilton International Trinidad
P.O. Box 442, Port of Spain
Tel: 624-3111
Fax: 624-485.
www.hilton.com
Expensive.
Holiday Inn
P.O. Box 442, Port of Spain
Tel: 625-3366
Fax: 625-4166.
Expensive.
Hotel Normandie
P.O. Box 851, Nook Avenue,
St. Ann's, Port of Spain
Tel: 624-1181/4
Fax: 624-0108.

Reservations

Reservations will make your travels here run smoother. But you can probably find a room without advance reservations in Port of Spain or Tobago, though you may not get your first choice. The exception to this is Carnival season when reserving a room far in advance is essential, especially if you want to stay near the Savannah.

Email: normandie@wow.net
Moderate.
Kapok Hotel
16–18 Cotton Street, Port of Spain
Tel: 622-6441/4
Fax: 622-9677.
www.kapok.co.tt
Moderate.
ASA Wright Nature Centre
Blanchisseuse Road, Arima
Tel: 667-4655.
Email: asawright@tstt.net.tt
www.asawright.org
This bird-watching paradise, is located at 360m (1,180 ft) above sea-level, with incredible views of the rainforest. Up to 40 bird species can be seen from the verandah alone and there are daily tours with expert guides.
Expensive.

Tobago
Arnos Vale
P.O. Box 208, Scarborough
Tel: 639-2881
Fax: 639-4629.
Expensive.
Blue Waters Inn
Batteaux Bay, Speyside
Tel: 660-4341
Fax: 660-5195.
Email: bwitobago@trinidad.net
Moderate.
Cocorico Inn
Plymouth
Tel: 639-2961
Fax: 639-6565.
Moderate.
Grafton Hotel
Black Rock Beach Resort
Tel: 639-0191
Fax: 639-0030.
www.grandehotel.com

Jimmy's Holiday Resort
Crown Point
Tel: 639-8292
Fax: 639-3100.
Cheap to moderate.
Man-O-War Cottages
Charlotte
Tel: 639-4327
Fax: 660-4328.
Cheap.
Mount Irvine Bay Resort
P.O. Box 222, Scarborough
Tel: 639-8800.
Expensive.
Tropikist Beach Hotel
Crown Point
Tel: 639-8512
Fax: 628-1110.
Cheap.

GUEST HOUSES

Trinidad
Alicia's House
7 Coblentz Gardens, St. Ann's,
Port of Spain
Tel: 623-2802.
Email: 2getaway@tstt.net.tt

Car Rental

Car rental in Trinidad and Tobago
is handled by numerous local
fleets and can be pre-booked
before departure. Many take
credit cards for both rental fees
and the substantial deposit (up
to TT$ 1,000) that is often
required. (Caution: remember
that there is a limit to the
amount of TT$ you can reconvert
on departure. If you expect the
refund of a large TT$ deposit
when you turn in your car, make
sure it won't put you over the
limit, or allow time to spend it.)
     Day rates for cars with manual
transmission begin at about
US$30; weekly rates start at
TT$725. The difficulty with car
rental is that business clients
engage many of the cars for long
periods, so a short-term rental
can be hard to find. Plan to shop
around well in advance and
consult a travel agent for
suggestions.

Monique's Guest House
114 Saddle Road, Maraval,
Port of Spain
Tel: 628-3334.
moniques@carib-link.net
Zollna House
12 Ramlogan Development,
LaSeiva, Maraval
Tel: 628-3731.
www.lagunamar.com

Tobago
Coral Reef Guest House
Milford Rd. Scarborough
Tel: 639-2536.
Islanders Inn
Main Road
Speyside
Tel/fax: 639-6575.
The Speyside Inn
Windward Road,
Speyside
Tel: 660-4852
www.caribinfo.com/speysideinn

BED & BREAKFAST

The Bed and Breakfast Association
of Trinidad and Tobago publishes a
leaflet listing a number of private
homes with one or more rooms to
spare. All have been inspected and
many have extra features like
television and access to kitchens.
Breakfast is included in rates which
range from US$17 to $40, slightly
higher during Carnival. TIDCO offices
have information or write to the
Association at Diego Martin P.O.
Box 3231, Diego Martin, Republic
of Trinidad and Tobago; or phone
Miss Grace Steele, tel: (809) 637-
9329, or Mrs Barbara Zollna,
tel: (809) 628-3731/660-4341.

Eating Out

RESTAURANTS & CAFÉS

Travelers from the budget
conscious student to the business
executive can find food to suit their
tastes and pockets, though food
prices all the while are relatively
high. The more expensive hotels
tend to serve a kind of bland
continental cuisine, but some offer
special Creole meals.

Drinking Notes

Mixed drink specialties in Trinidad
and Tobago revolve around rum,
and it's not hard to find rum in a
punch that will infuse any time of
year with a bit of Carnival. Trinidad
is also home to Angostura Bitters,
a secret concoction that has been
adding zest to mixed drinks for
generations. Two popular brands
of beer are Stag and Carib. And
don't forget to sample local soft
drink favorites like sorrel, ginger
beer and mauby.

Restaurants run the gamut from
roti shops frequented by working
people to sophisticated restaurants
housed in restored Victorian
mansions, serving the best French,
Creole, Indian, and Asian cuisines.
Roadside stands proliferate, as do
American-style fast food outlets.

Reserves and Sanctuaries

**El Tucuche Reserve or Northern
Sanctuary, Maracas**
Area: 936 hectares (2,313 acres)
El Tucuche is the second highest
peak at 936 m (3,072 ft). Forest with
interesting flora and fauna such as
Giant Bromeliad and the Golden Tree
frog. Birds include the Orange-Billed
Nightingale Thrush abound. Hiking
trails include the Ortinola Estate Trail.
Guides required.
**Valencia Wildlife Sanctuary,
Valencia**
Area: 2,785 hectares (6,881 acres)
At least 50 species of birds,
including antbirds and tanagers.
Deer, wild pigs, agouti, tatoo and
iguana roam this area. Easy access
by road and guides can be hired.
Caroni Bird sanctuary, Caroni
Area: 136 hectares (337 acres)
The third largest swamp in Trinidad.
At least 138 species of birds
present including the Scarlet Ibis
and many species of fauna and fish.
Spectacular views of returning birds
at dusk. Guided boat tours daily.
Tours start at 4pm and can be
booked through Winston Nanan,
tel: (868) 645-1305.

Bush Bush Wildlife Reserve, Nariva Swamp
Area: 1,554 hectares (3,840 acres). Fresh water swamp with hard wood forest. The Red Howler Monkey and the Weeping Capuchin are among more than 57 species of mammals that are found here. It is also possible to find The Savannah Hawk and the Red breasted blackbird, reptiles, fish and manatee. Prior permission needed and access by boat only.

Pointe à Pierre Wild Fowl Trust
Area: 26 hectares (64 acres) The Trust breeds endangered species of waterfowl and birds and reintroduces them into natural wildlife areas. A library and small museum containing unique Amerindian artifacts are located at the Trust, which is easily accessed by car. Reservations required, tel: (868) 637-5145.

Attractions

TOUR OPERATORS & NATURE ORGANIZATIONS

Trinidad

ASA **Wright Nature Center and Lodge**, Arima, tel: (868) 667-4655, fax: (868) 667-0493.

Avifauna Tours, Diego Martin, tel: 633-5614
Roger Neckles is a famous bird

Trips to Other Islands

If your main stop is Trinidad, don't leave without spending some time in Tobago – or vice versa. Or you can include in your schedule a trip further afield:
● BWIA has one-day packages to Tobago at a cost of TT$130, including transfers to the beach at Pigeon Point and a trip to Buccoo Reef. Tickets are sold only at BWIA offices and must be bought at least 24 hours in advance.
● **Nealco Air Services** (tel: 664-5416, Piarco; or 625-3426, Port of Spain) can arrange plane and helicopter sightseeing trips within Trinidad and Tobago and to other islands.

photographer and his tours are suitable for amateur birders as well as experienced ornithologists.

Chaguaramas Development Authority: guided tours, tel: (868) 634-4364 or 634-4349. Arrange waterfall and walking tours.

The Forestry Division, Long Circular Road, Port of Spain, Trinidad, tel: (868) 622-4521 or 622-7476 or 622-7256 or 622-3217. Contact for details of turtle-watching (May–Sep).

Hikers 20, c/o Gia Gaspard-Taylor, Apt. 3, 10 Hunter Street, Woodbrook, Port of Spain, Trinidad, tel: (868) 622-7731.

Trinidad and Tobago Field Naturalists' Club, c/o Secretary, Rosemary Hernandez, No. 1 Errol Park Road, St. Ann's, Port of Spain, Trinidad, tel: (868) 625-3386 or 645-2132, evenings or weekends.

Winston Nanan, 38 Bamboo Grove Settlement No. 1, Uriah Butler Highway, Valsayn Post Office, Trinidad, tel: (868) 645-1305.

Cruises
Jolly Roger, Point Gourde Road, (just beyond the Small Boats Jetty), tel: 634-4334. Wednesday Calypso Cruise; Friday Heart to Heart Cruise. Boarding time 8.30pm, cruise 9pm–midnight.

Tobago

Educatours, Carnbee, tel: (868) 639-7422.
Nature Tours, P.O. Box 348, Scarborough, tel: (868) 639-4276.
Pioneer Tours, Man-O-War Bay Cottages, Charlotteville, tel: (868) 660-4327.

FESTIVALS

March and October: **Flower and orchid shows**
May and June: **Drama Festival**
July: **Heritage Festival** (Tobago)
October: **Natural History Festival**
October to December: **Indian festivals** – Phagwah, Divali Nagar, Eid ul-Fitr and Hosay; and the **Best Village Folk Festival**
Every two years: **Carnival** and **Steelband Music Festival**.

Ecuador

Getting Acquainted

GOVERNMENT & ECONOMY

There are 22 Ecuadorian provinces, including the Galápagos Islands. Ecuador has been under democratic rule since 1979. The economy is based on oil, shrimp farming, bananas, roses, and to a lesser extent coffee and cocoa.

CLIMATE

Each of Ecuador's regions has a distinct climate. The coast is hottest and wettest from January to April. The mountains are driest, and therefore clearest, between June and September, and in December. The Oriente usually has a wetter season between June and August, followed by a drier season until December. As on the coast, however, a torrential downpour is never out of the question. The best months to visit the Galápagos Islands are March, April and November.

The mountains soar and the temperature drops: Quito, at an altitude of 2,850 m (9,350 ft), is cold when the sun disappears. The other regions are invariably warm and often steaming hot. Due to the country's equatorial climate, seasonal temperature variation is minimal.

Planning the Trip

VISAS

For stays of up to 90 days, visas are not required for most citizens of Europe, North and South America,

Australia and New Zealand, as long as there is at least six months until the expiry date on the passport. Stay permits (known as "T3") are issued at the border or point of arrival, usually for 30 days, which can be easily extended. Officially, tourists are only allowed to stay in Ecuador 90 days in any 365-day period, although in practice extensions are often given. For visa extensions, go to Avenida Amazonas 2639, open from 8am–noon and from 3–6pm. The Immigration Office in Quito is at Independencia y Amazonas 3149, tel: 454 122.

MONEY MATTERS

Travellers should bring US dollars as this became the official currency of Ecuador in 2000. The government has introduced a series of local coins of 1, 5, 10, 25 and 50 cents that are the same size and colour as American coins which are also accepted. Travelers' checks are also acceptable. Diners, Mastercard and Visa credit cards are widely accepted for meals and purchases, though a commission is often charged.

GETTING THERE

By Air
Flights from the United States to Ecuador are available with American Airlines, Continental or Lacsa. Flights from Europe are available with Avianca, Iberia, KLM or Lufthansa. Mariscal Sucre Airport in Quito is not far from the city center. Taxis cost around US$2. There is a US$25 airport and police departure tax.

By Land
One of the most popular ways for travelers to enter Ecuador from Peru is at the Tumbes-Huaquillas border post, changing buses at the frontier. Buses are frequent both ways. To Colombia, almost everyone takes the Quito-Ibarra, Tulcán-Pasto road.

Practical Tips

MEDIA

The main national newspapers in Ecuador are *El Comercio, Hoy, El Universo* and *Extra.*

POSTAL SERVICES

The central post office *(Correo Central)* in Quito is at Eloy Alfaro 354 y 9 de Octubre in New Quito. Address *poste restante* to this office, otherwise it will go to the post office in the old town at Espejo (between Guayaquil and Venezuela). Members of the South American Explorers Club can receive their mail at the Club House, Washington 311 y Plaza.

TELECOMMUNICATIONS

Andinatel is at 10 de Agosto and Av. Colon. Calls cost US$5 per minute to the United States, US$7 per minute to Europe. Phonecards can be bought all over town for cellular public phone boxes from which you can make international calls. There are many cybercafés in Quito and you can get a list from the Internet Café Guide at www.netcafeguide.com

EMBASSIES

US Embassy: Avenida 12 de Octubre y Avenida Patria, Quito, tel: 562-890 (560-307 after hours).
UK Embassy: Gonzales Suarez 111 y 12 de Octubre, tel: 560-670 or 560-309.

SHOPPING

Otavalan goods are the best buys in Ecuador. If you cannot make it to Otavalo, there are stores in Quito, Guayaquil, Cuenca and Baños where prices are only slightly higher than in the market. Cuenca is perhaps the best place to buy quality gold and silver jewelry, and is – along with Montecristi – a center of Panama hat production.

Police Matters

The main police office dealing with tourist matters is on Calle Mantúfar. Go to the police office on Roca y Juan Leon Mera if you need to report a robbery; this must be done within 48 hours.

Shops and stalls on Avenida Amazonas in Quito sell everything that Ecuador has to offer.

GETTING AROUND

Domestic Travel
If a road exists, some form of public transport will run along it. In the Oriente, motorized dugout canoes ply the main rivers.

Flight prices within mainland Ecuador are as follows:
Quito–Guayaquil – $55
Quito–Cuenca – $55.
Quito–Galápagos Islands – $385.80 return.

Travel Club
South American Explorers Club:
Washington 311 y Plaza, Quito
Tel: 225-228
Fax: 225-228
Email: explorer@ecnet.ec
www.samexplo.org
A non-profit making information resource center for travel in South America. Annual membership $40 single, $70 for a couple. Resources include trip reports written by members, library, message boards, luggage storage, mail service and travel advice. Quarterly magazine for members ($7 postage overseas).

Where to Stay

QUITO

Café Cultura
Reina Victoria and Robles
Tel/fax: 224-271
Good standard of accommodation and atmospheric café.
Chalet Suisse
Reina Victoria 312 and Calama
Tel: 562-700, fax: 563-966
First class.

Hilton Colón
Av. Amazonas and Patria
Tel: 560-666, fax: 563-903
Luxury hotel within easy walking
distance of many shops, bars and
cafés.
Posada del Maple
Rodriques 148 and 6 de Diciembre
Tel: 544-507
Budget accommodation.

GUAYAQUIL

The best area to stay is by the river,
where the air is cleaner and cooler.
Hilton Colón Guayaquil
Av Francisco de Orellana 111 Cdla.
Kennedy Norte
Tel: 689-000
Luxury hotel and the largest in the
city.
Oro Verde
Garcia Moreno and Av. 9 de Octubre
Tel: 327-999, fax: 329-350
Luxury.
Ramada
Malecon and Orellana
Tel: 565-555, fax: 563-036
First class.

Eating Out

QUITO

Numerous places serve cheap set
meals, and Chinese restaurants,
known as *chifas*, are good value.
The streets surrounding Avenida
Amazonas in the new town have
many good "international" restau-
rants. These include the **Columbus
Steak House** and the **El Cebiche**,
after the national dish of this name.
The Hotel Colón offers a sumptuous
buffet daily at lunchtime, and the
pavement cafés on Amazonas are
good for snacks, sunshine and
street life. The following restaurants
are also recommended:
La Bodeguita de Cuba
Reina Victoria 1721 y La Pinta
Great food and music.
La Choza
12 de Octubre 1821 y Cordero
Excellent Ecuadorian food.
Magic Bean
Foch 681 y Juan Leon Mera
Good food and atmosphere.

Mare Nostrum
Foch 172 y Tamayo
Very good seafood. Try the *cebiche*
(marinated fish and vegetable dish).

GUAYAQUIL

In Guayaquil, the most pleasant
places to eat are on the Malecón, a
2 km (1 mile) waterfront walk. For
cheaper and often more fiery
seafood meals, the area just north
of Parque del Centenario is good.
Calle Escobedo has two popular
street cafés serving breakfast. The
following restaurants are also
recommended:
Lo Nuestro
Victor Emilio Estrada 903 &
Higueras
Tel: 386-396
Excellent Ecuadorian food.
La Trattoria da Enrico
Balsamos 504 between Ebanos &
Las Monjas
Tel: 387-079
Crabs, pasta and Mediterranean
food.
La Parrilla del Ñato
Victor Emilio Estrada 1219 &
Laureles
Tel: 387-098
The most famous steak house in
Guayaquil.
El Café del Parque
Circunvalación 112 & V.E Estrada
Tel: 611-137
Cozy European ambiance and
excellent food.

Parks & Reserves

INEFAN is the official agency
responsible for national parks and
reserves and it produces useful
books and leaflets; tel in Quito is:
548-924 or (3) 963779 in Riobamba.
The South America Explorers Club
(see page 324) also give advice on
planning treks and hiring guides.

SANGAY NATIONAL PARK

Geographic location: South and
east of the capital Quito, from the
eastern *cordillera* to the Oriente.
Approximate area bounded by

Baños in the north to Guamote in
the south and the confluence of the
Rios Palora and Sangay in the east.
Size: 270,000 hectares (1,042 sq
miles).

Culebrillas Section
Access: For Sangay volcano and
Culebrillas (cloud forest and
páramo) or El Placer (cloud forest
and montane forest) best access is
from Alao (mountain village near
Riobamba).
Approximate costs of access: Taxi
from Riobamba to Alao
(approximately $20) or get a lift on
onion lorry. Guides, porters and
horses can be hired in the village at
$10 each. Park entry $20.
Tour operators: Sierra Nevada,
tel: 224-717, fax: 554-936, and,
Expediciones Andinas at
Argentinas 3860 and Zambrano,
Riobamba organize all-inclusive
hikes to the volcanoes. **Tropic
Ecological Adventures**, tel: 225-
907, fax: 560-756, email:
tropic@uio.sat.net work closely
with the indigenous communities.
Also recommended are
Andismo Tours,
www.andismo.com/exped
Open: All year
Registration: Park entry permit
($20). Check in at the park guard
station at Alao.
Accommodation: Large range of
hotels and hostels in Riobamba.
Camping may be possible at the
guard station in Alao. Suitable
camp site in Culebrillas valley and
at La Playa, at base of volcano.
Food/restaurant: Purchase basic
supplies in Riobamba, fresh vegetables available in Alao.
Transport in park: Horses and
mules can be hired in Alao, but
general means of access is on foot.
Trails: Hire guides to show the way.

Purshi Section
Access: Guamote to Atillo, then
Laguna Negra by road and then trail
through cloud forest to the Oriente at
Macas. It is possible to get a bus
from Macas to the village of Nueve
de Octubre and then trek to Purshi
which is the official entrance of
Sangay Parque Nacional. Hire guides

and porters (information from Park Headquarters in Riobamba).
Open: All year.
Registration: Pay at guard station at Atillo ($10).
Accommodation: Camp at Laguna Negra on edge of the park and along trail.
Food/restaurant: Purchase supplies in Riobamba or Guamote.
Transport in park: On foot.
Trails: Guides are required. Main trail from Laguna Negra to Macas, may take several days to walk

PODOCARPUS NATIONAL PARK

Geographic location: Southern Ecuador, between Loja and Vilcabamba and east toward Zamora.
Size: 146,280 hectares (565 sq miles).
Access: Western section (*páramo*, Podocarpus forest, elfin forest): car, taxi or bus from Loja, 15 km along road to Vilcabamba. Six-kilometer (4-mile) walk from road to Guard Station. Guides and horses for hire in Vilcabamba. Permit costs $10. Northern section (subtropical): car or taxi from Zamora. Park guards and guides at the ranger station on the Rio Bombusgaro. Permit costs $20.
Approximate costs of access: Inclusive taxi fee Loja to Guard Station costs $10.
Open: All year.
Registration: Pay at Guard Station.
Accommodation: Variety of hotels and hostels in Loja. Alternatively, use Cabanas Madre Tierra, tel: 673-123 or 580-269, in Vilcabamba or sleep at the Guard Station. There is also camping near the refuge and budget hotels in Zamora.
Food/restaurant: Bring your own.
Transport in park: On foot.
Trails: Extensive well-marked trail system from Guard Stations in both sections or INEFAN in Loja, tel: 571 534 or in Zamora.

ECUADORIAN ORIENTE

Geographic location: The Ecuadorian Oriente comprises the

upper Amazon basin east of the Andes; it is a place of endless virgin rainforest cut by fast-flowing rivers and is inhabited by more than 450 species of birds and a fascinating variety of mammals, primates, fish and flora, as well as the indigenous communities of the Siona, Cofan, Quechua, Quijo and Huaorani peoples. Travel in the Ecuadorian Amazon is easier than in neighbouring countries, as distances are shorter and transportation quite straightforward. However, there has been a large influx of people into the region near Lago Agrio since the discovery of oil in 1967, which has sent wildlife deeper into the forests. A large part of the Northern Oriente is taken up by the remote Yasuni UNESCO Biosphere Reserve and the Cuyabeno Wildlife Reserve and there are several excellent lodges close to the town of Coca. Macas is the gateway to the Southern Oriente and is only a short flight from Quito or Cuenca.
Size: Not applicable.
Access: Flight from Quito to Coca (Francisco de Orellana). River journey along Rio Napo and tributaries can be arranged independently or as part of an inclusive tour. Yasuni Reserve is approx 4 hours by boat from Coca.
Approximate costs of access: Inclusive tours are highly variable in price depending on venue and duration.
Tour operators and lodges: Many of the best tour operators in Quito offer jungle lodges and tours as part of an Ecuador itinerary. Along the River Napo, 3 hours from Coca, is the award-winning La Selva Jungle, which has interesting jungle excursions, observation tower, butterfly farm and a birding list of over 550 species. La Selva also has an Amazon Light Brigade luxury safari for those wanting a more intense jungle experience and both can be booked through La Selva Lodge, 6 de Diciembre 2816, Quito, tel: 550-995, fax: 567-297 or www.discoveramazonia.com /Ecuador/laselva
   Also the slightly cheaper Sacha

Lodge, Lizardo Garcia 613 and Reina Victoria, Quito, tel: 566-090, fax: 508-872 or www.sachalodge.com
Yachana Lodge, 2 hours by canoe from Coca, is part of the Funedesin Foundation Montana Project. All profits help 30 indigenous communities. Highly recommended. Tel: (2) 237 133, fax: (2) 220-362; email: info@yachana.com and www.yachana.com
Open: All year.
Registration: Not necessary unless also intending to visit the Yasuni National Park, or the Cuyabeno Reserve Area.
Transport in park: River transport and trails in vicinity of lodges.

CUYABEÑO RESERVE

Geographic location: North of Rio Aguarico in province of Sucumbios, northeastern corner of Ecuador.
Size: 603,380 hectares (2,330 sq miles).
Access: Flight from Quito to Lago Agrio (or bus). Bus from Lago Agrio to Chiritza (one hour), then down the Aguarico by canoe. Or continue on the road to Puerto Cuyabeno (two more hours) and then transfer to canoe on the Cuyabeño River.
Open: All year.
Registration: $20. Tour company or guide will take care of registration process.
Accommodation: Nativo cabin can be booked through Native Life, Calle Foch 167 and Amazonas in Quito, tel: 505-158, fax: 229-077, www.natlife.ec
Cuyabeno Lake Lodge can be booked through Neotropic Turis, tel: 521-212, fax: 554-902 or email: info@neopicturis.com, or www.ecuadorexplorer.com/neotropic Currently the popular Flotel Orellana, operated by Metropolitan Touring, is being relocated and refurbished. Check up-to-date information on www.ecuadorable.com
Transport in park: Canoe and motorized canoe. The Lago Agrio–Tipisca road cuts through park (crossing Rio Cuyabeño).
Trails: A guide is essential.

Attractions

TOURS

Otavalo

Zulaytours offers good value, all-day tours of the villages surrounding Otavalo. The history and culture of the peoples is outlined, and a visit to a pre-Inca cemetery is included. As each village specializes in a particular craft, the tour visits various homes to observe the different methods of production. Most Ecuadorian tour operators and specialists offer stays at haciendas in the region.

The Galápagos Islands

Metropolitan Touring offers 3, 4 and 7 night cruises on its 90 passenger MV SANTA CRUZ and 40 passenger yacht ISABELLA II. Day tours are also available with Delfin Yacht, based at Hotel Delfin on Isla Santa Cruz. Alternatively, Quasar Nautica operate small exclusive yachts as well as the 48 passenger Eclipse which can be booked through Quasar Nautica in Quito, tel: (2) 446-996, fax: (2) 436-625, www.quasarnautica.com or with Penelope Kellie World Wide Yacht Charter and Tours who represent Quasar Nautica in the UK, tel: (01962) 779-317, fax: (01962) 779-458.

EXCURSIONS FROM QUITO

Among Ecuador's snow-capped peaks, Chimborazo, Cotopaxi, El Altar and Tungurahua are certainly the most challenging.

Twenty kilometers (12 miles) north of Quito is Mitad del Mundo, where a monument and museum mark the equator.

On a clear day, a train or bus ride along the Avenue of the volcanoes from Quito to Riobamba is breathtaking. There are several cloud forest areas that can be visited in 1–3 days from Quito, notably Mindo and Maquipicuna Reserve.

NATURE TRIPS

Emerald Forest
Amazonas 1033 and Pinto, Quito, tel: 526-403, fax: 568-664.
Camping trips to Rio Tiguino, Huaorani Reserve and Parque Nacional Yasuni.
Flotel Orellana (Metropolitan Touring)
tel: 464-780, fax: 464-702
Operate out of Lago Agrio. Luxury tour down Rio Aguarico aboard Flotel Orellana. Four days, $520.
Huaorani Safari
Calama 380 and Juan Leon Mera, Quito, tel: 552-505, fax: 220-426.
Camping trips to Rio Shirapuno, Huaorani Reserve.
Kempery Tours
Pinto 539 and Amazonas, Quito, tel: 226-583, fax: 226-715.
Tours to Rio Tiguino, Huaorani Reserve.
Native Life
J Pinto 446 and Amazonas, tel: 505-158, fax: 229-077.
Email: natlife@natlife.com.ec
Camping trips out of Lago Agrio to Cuyabeno area.

Nightlife

BARS & DISCOS

Nightlife in Ecuador often involves frequenting bars with astonishingly loud music.
Quito
In Quito, the new town has its share of discos, as well as an authentic English pub, the Reina Victoria, on the street of the same name. The best *peña* is the Taberna Quiteña on Avenida Amazonas. Or try Bodegón del Cuba for live Cuban music.
Guayaquil
In Guayaquil, countless all-night discos play the latest American and *latino* tunes, and some good Colombian *salsa*. They are not places for the faint-hearted.
Otavalo
Otavalo has a couple of *peñas* which get rather lively on Friday and Saturday nights. In addition to excellent local music, they regularly feature groups from Colombia, Peru and Chile.

Peru

Getting Acquainted

THE PLACE

Language: The official languages of Peru are Spanish and Quechua, although almost all school children study English. Near Puno and Lake Titicaca, Aymará is spoken. In the Amazon basin, several tribal dialects are spoken.
Time zone: Lima time coincides with Eastern Standard Time in the United States (five hours behind Greenwich Mean Time). For official Peruvian time, call 652-800 in Lima.
Electricity: Appliances in Peru run on 220 volts, 60 cycles.
Weights and measures: The metric system is used for all units of measurements.

The People

The population of 24 million in Peru are spread across the arid coast, the jungle, Peru's fertile valleys and the Andes mountains. Because of successive immigration, Indian and *mestizo* citizens live side by side with Europeans and Chinese and Japanese. Peru's coastal cities to the south include a cohesive black population.

GOVERNMENT & ECONOMY

A country which has sporadically fallen into the hands of military governments, Peru since 1979 has been governed by a democratic government under a constitution. Presidential elections are held every five years and municipal elections every three years. The country is divided into 24 departments and the province of Callao – Peru's main port.

Fishing, mining and tourism play important roles in the economy, although the government has gradually revitalized the farming sector. Petroleum is extracted from the Amazon jungle area; these supplies currently account for about half of all domestic demand.

Peru's economy is growing fast, but poverty and unemployment are still serious problems. Terrorist activity was suppressed in the early 1990s under the government of Alberto Fujimori.

CLIMATE

On the coast, the temperatures average between 14°C and 27°C (58°F and 80°F), while in the highlands – or *sierra* – it is normally cold, sunny and dry for much of the year, with temperatures ranging from 9°C to 18°C (48°F to 65°F). The rainy season in the *sierra* is from December to May. The jungle is hot and humid with temperatures ranging from 25°C to 28°C (77°F to 82°F).

Lima alone suffers from the climatic condition the Peruvians call *garua* – a damp mist that covers the city without respite for the winter.

BUSINESS HOURS

Offices normally operate from Monday–Friday 8.30am–6pm with a break from 12.30–3pm; or some open continuously from 9am–5pm. Banking hours are from 9am–12.30pm and from 3–6pm from Monday–Friday all year round.

ENTRY REGULATIONS

If you intend visiting the Amazon basin area, a yellow fever vaccination certificate is required.

Visas

Visitors from all European countries, the US and Canada do not need a visa. Just get a tourist card on arrival, which is usually valid for 90 days, and hold onto this piece of paper until you leave the country. On departure, a $25 exit tax must be paid in US dollars or the equivalent in soles. There is a $4 airport tax on domestic flights. To extend the length of stay, go to the Dirección General de Migraciones, España 700, Breña, Lima. A 30-day extension is available upon payment of $20 and presentation of a return travel ticket.

Customs

It is illegal to take archeological artefacts out of the country. Special permits are needed to bring professional movie or video equipment into Peru. In Cuzco, a special tax is assessed for professional photographers passing their equipment through that airport.

MONEY MATTERS

In 1991, the nuevo sol was introduced as Peru's currency. The sol is divided into 100 céntimos. Coinage is in denominations of 10, 20 and 50 céntimos, and 1, 2 and 5 sols; with 10, 20, 50 and 100-sol notes.

Banks, money exchange houses, hotels and travel agencies are authorized to exchange money, either from traveler's checks, cash or, in some cases, money orders. The exchange rate fluctuates daily. Always check for forged notes.

Most international credit cards are accepted, including Diners Club, Visa, Mastercard and American Express at hotels, restaurants and stores. These last four companies all have offices in Lima.

WHAT TO BRING

Tourists should bring with them any medicines or cosmetics they use; tampons and contraceptives are difficult to find outside Lima. Since Peruvians are smaller in stature than North Americans or Europeans, it may be difficult to find large-sized clothing.

TOUR OPERATORS

There are many excellent tour operators in the UK that offer interesting itineraries in Peru. Call the Latin American Travel Association on (020) 8715-2913 for a free guide to Latin America and list of members.

In Lima there is a wide range of companies that can put together tailor-made and group programmes including:

Amazonas Explorer
Tel: (084) 225284, www.amazonas-explorer.com Recommended for adventure expeditions in Peru and Bolivia, including white-water rafting, canoeing and trekking.

Lima Tours
Jr Belen 1040, Lima, tel: (1) 424-5110, fax: (1) 426-3383, email: inbound@limatours.com.pe Highly recommended for quality tours across the whole of Peru.

InkaNatura
Av Manuel Banon, 461 San Isidro, Lima, tel: (1) 440-2022, fax: (1) 422-9225, www.inkanatura.com One of the leading Peruvian wildlife and archeological tour companies, specializing in natural history tours to Manu Wildlife Centre, Sandoval Lake Lodge and a unique tour into Machiguenga territory, staying at the new Machiguenga Centre, the only 100 percent indigenous-owned rainforest lodge in the Amazon.

GETTING THERE

By Air

Flights from the US are operated by: American Airlines, Continental Airlines, Delta Airlines and Aerocontinente. From Europe flights are available with: KLM (from Amsterdam); Lufthansa (from Frankfurt); Avianca (via Bogota), and Iberia (from Madrid).

By Land

There is a bus service between Lima and the major cities although most buses are not luxury vehicles. There are train services in the south to and from Arequipa,

Puno and Cuzco and Machu Picchu.

The Pan American Highway still links Chile and Ecuador, and Colombia and Venezuela to the north, but was damaged by the El Niño hurricane in 1998. Car rental is available, but the distances between main attractions and poor road conditions – especially in mountain areas – make air or bus travel preferable.

Practical Tips

TOURIST OFFICES

There are tourist information booths in Parque Kennedy, Miraflores, at the airport, and at the Centro Cultural Ricardo Palma, Avenida Larco, Miraflores. Information is also available from:
Promperu
Edificio Mitinci, Piso 13, Calle 1 Corpac, San Isidro,
tel: 224-3279, www.promperu.org
South American Explorers Club
Avenida Portugal 146
Breña, Lima, tel: (51-1) 425-0142, fax: (51-1) 425-0142.
Email: montague@amauta.rcp.net.pe
See page 324 for details of membership benefits.

Useful websites include: www.PeruTravelNet.com for general information and useful weblinks. www.andeantravelweb.com.peru for travel advice, destination information and lists of operators in the Peruvian Association of Adventure Tourism and Ecotourism.

EMBASSIES

US Embassy: Avenida La Encalada, Block 17, Monterrico, Lima, tel: 434-3000 or 434-3032.
UK Embassy: Edif. Pacifico-Washington, Plaza Washington, Avenida Arequipa, Block 5, Lima, tel: 433-5032.

SECURITY & CRIME

Pickpockets and thieves – including senior citizens and children – have become more and more common in Lima and Cuzco. It is recommended that tourists do not wear costly jewelry and that their watches, if worn, be covered by a shirt or sweater sleeve. Thieves have become amazingly adept at slitting open shoulder bags, camera cases and knapsacks; keep an eye on your belongings. Officials also warn against dealing with anyone calling your hotel room or approaching you in the hotel lobby or on the street, allegedly representing a travel agency or specialty shop. Carry your passport at all times.

A special security service for tourists has been created by the Civil Guard. These tourism police are recognizable by the white braid they wear across the shoulder of their uniforms and can be found all over Lima, especially in the downtown area. In Cuzco, all police have tourism police training. The tourist police office in Lima is at Museo de la Nación, J. Prado 2465, 5th Floor, San Borja, tel: 476-9896.

Also, check with the Foreign Office website www.fco.gov.uk/travel or the South American Explorers Club for up-to-date security advice.

MEDICAL SERVICES

Most major hotels have a doctor on call. Three clinics in Lima and its suburbs have 24-hour emergency service and usually an English-speaking person on duty. They are the Clinica Anglo-Americana, Avenida Salazar in San Isidro, tel: 221-3656; the Clinica International, Washington 1475, San Isidro, tel: 433-4295, and the Clinica San Borja, on Avenida Guardia Civil 337, San Borja, tel: 475-3141.

Getting Around

FROM THE AIRPORT

There is a bus service, Transhotel (Logrono 132, Miraflores, tel: 448 2179) which runs to and from Jorge Chavez Airport, Lima. Make reservations beforehand. A bit more

Maps

Maps and useful leaflets are available from PROMPERU, Peru's national tourism office, free of charge. Larger city maps are sold at newspaper kiosks, bookstores and the Instituto Geográfico Nacional at Avenida Aramburu 1190 in the suburb of Surquillo, tel: 475-9960. Some maps are available from the South American Explorers Club, tel: 425-0142, www.samexplo.org

expensive but reliable are Lima Taxis (tel: 476-2121). A normal taxi will charge about $20–25 to or from the airport (US$40 to Miraflores) and it's much safer than trying to flag one down in Avda Faucett, outside the terminal. Always agree the cost of the trip before getting in the taxi as there are no taximeters in Peru.

You can also rent cars at the airport, where Avis, Budget, Hertz and National have offices open 24 hours a day. An international driver's license is needed and is valid for 30 days. For additional days, it is necessary to obtain authorization from the Touring and Automobile Club of Peru, located at Cesar Vallejo 699 in the suburb of Lince, tel: 440-3270, fax: 422-5947, or visit www.hys.com.pe/tacp

DOMESTIC AIRLINES

Domestic airlines Aero Continente, LanPeru and TACA serve most cities in Peru.
Aero Continente is at Avda José Pardo 651, Miraflores, tel: 242-4260, fax: 241-8098.
LanPeru is based at Avda José Pardo 269, Miraflores, tel: 446 6995.
TACA is based at Avda Cmdte, Espinar 331, Miraflores, tel: 213-7000.

PUBLIC TRANSPORT

Lima has a multitude of city bus lines, although most are

overcrowded, slow and not recommended for tourists. Cab fares are generally so inexpensive by international standards that they are preferred.

For travel outside the city, cars may be rented, or transportation can be arranged through travel agencies; buses are available to just about every part of the country.

There is a train service between Arequipa, Puno and Cuzco. Since trains are the most economical means of transportation, they are also sometimes the most crowded. Make sure you have a guaranteed seat and a tourist (first-class) ticket. In 1999, Orient-Express and Peru Rail Corp took over the Peruvian southern railway system and are making huge improvements between Cuzco-Machu Picchu, Cuzco-Puno and Arequipa-Puno. Visit www.perurail.com or book as part of an itinerary with a tour operator.

Where to Stay

HOTELS

Lima
Country Club Lima Hotel
Los Eucaliptos Cuadra 5, San Isidro
Tel: 211-9000, fax: 421-7220
Luxury hotel with just 75 suites.
Member of the Leading Hotels of the World group.
Crillon
La Colmena 589
Tel: 428-3290, fax: 432-5920
Hotel Bolívar
Plaza San Martin
Tel: 427-2305, fax: 433-8626
The doyen of Peruvian hotels.
Miraflores Cesar
Corner of Avenida La Paz and Diez Canseco, Miraflores
Tel: 444-1212, fax: 444-4440
Pueblo Inn
KM 11 of the Central Highway
Tel/fax: 494-1616
An exclusive hotel complex.
Sans Souci Hotel
Avenida Arequipa 2670, San Isidro
Tel: 422-6035, fax: 441-7824
Sheraton Lima Hotel & Towers
Paseo de la Republica 170

Tel: 315-5000, fax: 426-5920
Large deluxe hotel in central area.

Cuzco
El Dorado Inn
Sol 395
Tel: 233-112, fax: 240-993
Email: doratur@telser.com.pe
Good value for money.
Golden Tulip Libertador
San Agustin 400
Tel/fax: 231-175
A fine traditional hotel with good restaurant.
Hotel Monasterio Orient Express
Calle Palacio 136
Plazoleta Nazarenas
Tel: 241-777, fax: 231-711
www.monasterio.orient-express.com
Restored 16th century monastery with beautiful rooms, courtyards and cloisters.

Iquitos (Northern Amazon)
Hotel Amazonas Plaza
Abelardo Quinonez, Km 2
Iquitos
Tel: 23-5731/23-1091
Reservations in Lima: tel: 444-1199.
Iquitos' top (5-star) hotel, 3 km (2 miles) out of town.
Acosta I
Esq. Huallaga & Araujo
Tel: 23-5974
Hotel Victoria Regia
Ricardo Palma 252
Tel: 231-983, fax: 232-4999
Swimming pool.
Hostal La Pascana
Pevas 133
Tel: 231-418
Recommended budget hostel.

LODGES

Iquitos
Amazon Lodge
Iquitos: Raymondi 382
Tel: 237-142
In Lima, tel: 221-3341;
fax: 221-0974
Explorama Lodge and Explornapo Camp
c/o Explorama Tours
P.O. Box 446, Iquitos
Tel: (094) 252-526 or 252-530,
fax: (094) 252-533.
Email: amazon@explorama.com

www.explorama.com
In US, tel: 1-800-233-6764,
fax: 581-0844.
Explorama run 5 highly recommended sites comprising Explorama Inn, Explorama Lodge, ExplorNapo Camp, ACEER scientific research stations and Explor Tambos. Explorama Inn is only 40 km (25 miles) from Iquitos and has comfortable accommodation and good activities; with food it costs approx $175 per person for two days/one night; Explorama Lodge costs $250 per person for three days/two nights; Explornapo Camp costs $1,000 per person for five days/four nights, with excellent canopy walkway. Well-organized, comfortable, with fairly structured itineraries. Clients staying at ACEER spend the first and last nights at Explorama. Explor Tambos is 2 hours from ExplorNapo and is more basic but with excellent opportunities to see rare wildlife.
Yacumama Lodge
Avenida Benavides 212, Ofic. 1203, Miraflores, Lima
Tel/fax: 241-022
www.gorp.com/yacumama
Excellent lodge on Rio Yarapa.
Yarapa River Camp
Reservations: Amazonia Expeditions, Napo 150
Tel: (094) 222-049
Or for something a bit different, contact **Amazon Tours & Cruises** who offer adventure cruises, rainforest and nature expeditions and private boat charters in the Peruvian Amazon. Tours include trips upriver to view wildlife and national reserves and downriver to visit river villages and indigenous communities as well as longer cruises from Iquitos to Manaus.
Tel: (94) 233-931, fax: (94) 231-265, www.amazontours.net

Madre de Dios (Southern Amazon)
Manu Biosphere Reserve
Manu Nature Tours
Avenida Pardo 1046, Cuzco
Tel: (084) 252-721,
fax: (084) 234-793
Email: mnt@amauta.rcp.net.pe
www.manuperu.com

Cultural Zone
Amazonia Lodge
Alto Madre de Dios River,
Cultural Zone
Tel: 231-370
Family-run converted hacienda.
Manu Expeditions and Treks
Av Pardo 895, Cuzco
Tel: (084) 226-671,
fax: (084) 236-706
www.ManuExpeditions.com
Owned by the ornithologist Barry
Walker, Manu Expeditions operate
4–9 day tours in the reserves,
featuring camps at the Manu
Wildlife Centre, which they jointly
own with the Silva Sure
Conservation Group, a non-profit
making organization involved in
rainforest conservation projects.
The centre is in its own private
reserve that backs onto the Manu
Biosphere Reserve on the Madre de
Dios River, 90 minutes from Boca
Manu in pristine rainforest.
Pantiacolla Lodge
c/o Pantiacolla Tours
Plateros 360, Cuzco
Tel: 238-323,
fax: 233-727
www.pantiacolla.com
Itahuania; situated 30 minutes
downriver from Shintuya.
Over 500 species of birds have
been seen from this lodge.

Tambopata-Candamo Reserve
Explorers Inn
c/o Peruvian Safaris
Garcilaso de la Vega 1334
Casilla 10088, Lima
Tel: 433-7963,
fax: 432-8866
Rainforest Expeditions
Arambru 166-4B, Lima 18
Tel: (1)421-8347
Fax: (1) 421-8183
www.perunature.com
Award-winning Rainforest
Expeditions owns two eco-lodges
which give visitors unique
opportunities to experience a
pristine rainforest alongside working
scientists, while injecting money
into the local community.
Tambopata Research Centre is
located in the heart of the reserve,
500 metres from the world's
largest macaw clay lick. 5 hours

away by boat is the comfortable 24
bedroom Posada Amazonas, co-
owned by the local Ese'eja Native
Community of Tambopata. The
posada is easily accessible from
Cusco and is close to oxbow lakes
and small parrot and parakeet clay
licks.
Sandoval Lake Lodge
c/o Inkanatura
www.inkanatura.com or
see page 328.
A 25 minute motor canoe ride
down the River Madre de Dios
from Puerto Maldonado is followed
by a 2 mile walk or rickshaw ride
through secondary forest to the
beautiful lake where the lodge is
located. Good opportunities to
see Giant Otters, monkeys and
macaws.
Tambopata Jungle Lodge
c/o PAT/Peruvian Andean Tours
Peruvian Andean Treks Ltd
Avenida Pardo 705, P.O. Box 454,
Cuzco-Perú
Tel: (084) 225-701,
fax: (084) 238-911
www.cbc.org.pe/pat/patinde2.htm

Eating Out

Peru's *criolla* cuisine evolved
through the blending of native and
European cultures. *A la criolla* is the
term used to describe slightly
spiced dishes such as *sopa a la
criolla*, a wholesome soup
containing beef, noodles, milk and
vegetables.

Drinking Notes

In towns, many Peruvians drink the
soft drink *chicha morada*, made
with purple maize (different from
the *chicha de jora*, the traditional
home-made alcoholic brew known
throughout the Andes).
The lime green Inca Cola is more
popular than it's northern
namesake, as well as Orange
Crush, Sprite and Seven-up. The
jugos, juices, are a delightful
alternative to sodas and there are
many fruits available. Instant
Nescafé is often served up even in
good restaurants, although real

Throughout the extensive coastal
region, seafood plays a dominant
role in the Creole diet. The most
famous Peruvian dish, *ceviche*, is
raw fish or shrimp marinated in
lemon juice and traditionally accompanied by corn and sweet potato.
Other South American countries
have their own version of *ceviche*
but many foreigners consider Peru's
to be the best.
Corvina is sea bass, most simply
cooked *a la plancha*, while scallops
(*conchitas*) and mussels (*choros*)
might be served *a lo mancho*, in a
shellfish sauce. *Chupe de
camarones* is a thick and tasty
soup, of salt or freshwater shrimp.
A popular appetizer in Peru is
palta a la jardinera, avocado
stuffed with a cold vegetable
salad or *a la reyna*, chicken salad.
Choclo is corn on the cob, often
sold by street vendors during
lunchtime. Other Peruvian "fast
food" includes *anticuchos*, shish
kebabs of marinated beef heart;
and *picarones*, sweet lumps of
deep fried batter served with
molasses. For *almuerzo* or lunch,
the main meal of the day, one of
four courses might be *lomo
saltado*, a stir-fried beef dish, or
aji de gallina, chicken in a creamy
spiced sauce.
The most traditional of Andean
foods is *cuy*, guinea pig, which is
roasted and served with a peanut
sauce. Another speciality of the
Sierra is *pachamanca*, an

coffee can be found at a price. Tea
drinkers would be advised to order
tea without milk, to avoid receiving
some peculiar concoctions.
The inexpensive beers are of
high quality. Try *Cusqueña, Cristal*
or *Arequipena*. Peruvian wines
can't compete with Chilean
excellence, but for a price,
Tabernero, Tacama, Ocucaje and
Vista Alegre are the reliable
names. The Peruvian *pisco sour*
is, however, a strong contender.
Try the famous, potent *Catedral* at
the Gran Hotel Bolívar in Lima.

assortment of meats and vegetables cooked over heated stones in pits within the ground. Succulent freshwater trout is plentiful in the mountain rivers.

Peruvian sweets might be *suspiro* or *manjar blanco*, both made from sweetened condensed milk, or the ever popular ice-cream and cakes. There are many weird and wonderful fruits available in Lima, notably *chirimoya,* custard apple, *lucuma,* a nut-like fruit, delicious with ice-cream, and *tuna,* which is actually the flesh from a type of cactus.

WHERE TO EAT

Lima

There is a huge choice of restaurants in the city area, particularly in the Miraflores and San Isidro districts. Below are just a few recommendations:

International
Carlin
La Paz 646, Miraflores
Tel: 444-4134
Noon–4pm and 7pm–midnight daily. A cozy international style restaurant which is very popular with foreign residents and tourists.
El Alamo
Corner of La Paz and Diez Canseco, Miraflores
12.30–3pm and 7pm–midnight Mon–Sat. A cozy spot for imported wines and cheeses. Specializing in fondues.
Pabellon de Caza
Alonzo de Molina 1100, Monterrico
Tel: 437-9533
Noon–3pm and 7pm–1am Mon–Sat. Brunch: Sun 10.30am–4pm. Located only a short stroll from the Gold Museum, containing a luxuriant garden and elegant decor.
Criolla
Las Trece Monedas
(The Thirteen Coins)
Jr. Ancash 536, Lima
Tel: 427-6547
Noon–4pm and 7–11.30pm Mon–Satu. Formal presentation within a beautiful 18th-century colonial mansion, including courtyard and antique coach.

Manos Morenos
Avenida Pedro de Osma 409, Barranco
Noon–3.30pm and 6pm–midnight Tues–Sun. The best *criollo* fare Lima has to offer, specializing in *anticuchos* (shish kebab of marinated beef heart).

French
L'Eau Vive
Ucayali 370, Lima
Tel: 427-5612
Noon–2.45pm and 8.15pm–10.15pm, Mon–Sat. Fine provincial dishes prepared and served by an order of French nuns. The inner courtyard is one of Lima's most pleasurable settings – the perfect place to take a break from sightseeing. (*Ave Maria* is sung nightly at 10pm and dinner guests are invited to join in.)

Italian
La Trattoria
Manuel Bonilla 106, Miraflores
Tel: 446-7002
1–3.30pm and 8pm–midnight Mon–Sat. Authentic Italian.
Valentino
Manuel Banon 215, San Isidro
Tel: 441-6174
Noon–3pm and 7.30pm–midnight Mon–Sat. Great Italian food at Lima's best international restaurant.

Japanese
Matsuei
Manuel Banon
Tel: 422-4323.
Excellent quality Japanese food.

Pizza
La Pizzeria
Diagonal 322, Miraflores
Tel: 46-7793
9.30am–midnight daily. Night-time haunt of trendy Miraflores crowd.

Seafood
La Costa Verde
Barranquito Beach, Barranco
Tel: 477-2424
Noon–midnight daily. Romantic at night under a thatched roof and right on the beach.

La Rosa Nautica
Espigon No. 4, Coasta Verde, Miraflores.
Tel: 447-0057
12.30pm–2am daily. Lima's most famous seafood restaurant located at the end of an ocean boardwalk.

Cafés
Café Café
Martir Olaya 2250, Miraflores
Tel: 445-1165
Good atmosphere.
La Tiendecita Blanca – Café Suisse
Avenida Larco 111, Miraflores
Tel: 45-9797
Boasts 30 different kinds of chocolates, excellent ice-cream and desserts. Popular with shoppers but also serves lunch and dinner.

Cuzco

While in Cuzco, sample the succulent pink trout, prepared in many of the Peruvian and international style restaurants. *Cuzquenos* have also mastered the art of pizza baking, and there is always a new establishment to be found within the plaza's portals.

Cafés around the plaza serve *mate de coca* (chocolate mud cake) and hot milk with rum on chilly Cuzco nights. Good restaurants will provide Andean music. Cuzco's two best hotels, the **Hotel Monasterio Orient Express** and the **Golden Tulip Libertador**, have excellent restaurants, with high prices set in US dollars.

El Ayllu
Portal de Carnes 203
(beside Cathedral)
Good place to visit for breakfasts of ham and egg, toast and fruit juice, or homemade fruit yoghurt, cakes, teas and good coffee. Continuous opera Muzak and cosmopolitan clientele give this place its unique character.
El Truco
Plaz Regocijo 247
Catering to most of the large tour groups, this nightly dinner and show is the most elaborate in Cuzco and very good value.
Mesón de los Espaderos
Espaderos 105

A quaint wooden balcony provides perfect dining views of the Plaza. The mixed grill, *asado*, is excellent and huge.

Pizzeria La Mamma
Portal Escribanos 177,
Plaza Regocijo
Just one of the many restaurants which serves mouth-watering garlic bread and fresh mixed salads as well as pizzas.

Quinta Zarate
Calle Tortera Paccha
One of the many *quintas*, or inns, in the suburbs. The specialty here is roast *cuy* – guinea pig.

Trattoria Adriano
Sol 105, corner of Mantas
Friendly service and excellent pink trout. Try it simply grilled and with a squeeze of lemon.

Parks & Reserves

MACHU PICCHU SANCTUARY

Geographic location: Eastern *cordillera* of Peru, 100 km (62 miles) northwest of Cuzco.
Size: 32,594 hectares (126 sq miles) including Archeological Park.
Access: There are three classes of tourist train from Cuzco to Aguas Calientes: Autovagon, Pullman and Expresso; check www.perurail.com for more details. At Aguas Calientes a tourist bus meets the trains to take visitors up to the hidden city. Inclusive day-tours are available from Cuzco which include all rail travel, bus transfers and entry fee to the monument. Alternatively, by rail from Cuzco to the station at KM 88, which is about 22 km (13 miles) beyond Ollantaytambo station, then join the Inca trail for a 33 km (20 mile), three/five-day hike. Camping overnight – equipment can be hired in Cuzco. For personal security, camping is not recommended for groups of less than four persons.
Approximate costs of access: Inclusive day trip by train and bus $110. Entrance ticket for independent travelers $11.
Tour operators: Any travel agent in Lima or Cuzco can arrange day-trip or overnight stay.

Open: All year.
Registration: Only required if climbing Huayna Picchu (the mountain overlooking the site). Register at hut on the trail to the monument.
Accommodation: Orient Express Hotels have recently refurbished Machu Picchu Sanctuary Lodge which is the only hotel within the ruins, giving its guest access to the site when all the day-trippers have gone back to Cuzco. www.orient-expresshotels.com Cheaper hotels are available in Aguas Calientes near the railway station. No camping available at the monument. Official camp site is at Puente Ruinas.
Food/restaurant: Food is available at most railway station stops. A self-service restaurant can be found at the monument.
Trails: Various tourist trails in the vicinity of the monument, although not necessarily designed with nature observation in mind.
Further advice: Ornithologists would find it worthwhile visiting the Cross Keys Bar in the Plaza de Armas, Cuzco. The owner, Barry Walker, is one of Peru's top ornithologists. Machu Picchu is a highlight of any Peru itinerary and it is highly recommended to pre-book accommodation, especially at the Machu Picchu Sanctuary Lodge.

TAMBOPATA-CANDAMO RESERVED ZONE

Geographic location: South of Puerto Maldonado – east of Cuzco. Area enclosed by the Rio Inambari (tributary of the Rio Madre de Dios) and the Rio Heath which forms the western border of Bolivia.
Size: 1,479,000 hectares (5,710 sq miles), including 5,500 hectares (21 sq miles) of Tambopata. Adjacent to Heath Pampas National Sanctuary of 101,109 hectares (390 sq miles).
Access: Flight from Lima or Cuzco to Puerto Maldonado. Hire river transport for journey up the Rio Tambopata.
Approximate costs of access: Hire of river transport from Puerto

Maldonado to Tres Chimbadas approximately $70. Hire guides and buy food from the Esse'eja community along the river.
Tour operators: Inkanatura (Lake Sandoval Lodge), Peruvian Safaris (Explorers Inn) and PAT/Peruvian Andean Tours (Tambopata Jungle Lodge). $200 for 3 days/2 nights. *See page 331* for contact details.
Open: All year.
Registration: Not necessary.
Accommodation: Many hotels of all price ranges in Puerto Maldonado, but the town itself has little to offer and it is recommended to pre-book one of the lodges through an operator as outlined above. At least two or three nights' stay recommended to justify the journey and get the best wildlife-viewing experience.

Camping at Tres Chimbasas and Macaw Lick Camping may be available through Peruvian Andean Tours as a special arrangement.
Transport in park: Motorized dugout canoe and walking.
Trails: Extensive trail system at Explorers Inn with expert scientist guides. Trail system in vicinity of other lodges still being developed, using local people as guides.
Further advice: The Tambopata Reserve is one of the oldest in the Madre de Dios region and has a high reputation amongst naturalists. The Tambopata-Candamo Reserve was declared in 1990. Because of the limited tourist activity to date, good opportunities may exist for nature observation.

In the UK send a SAE to Tambopata Reserve Society (TReeS), J Forrest, 64 Belsize Park, London NW3 4EH, for conservation information.
It is also possible to visit the **Santuario Nacional Pampas del Heath**, which comprises 102,109 hectares (394 sq miles) adjacent to the Tambopata-Candamo Reserved Zone, near the border with Bolivia.
Contact **Conservation International** in the US: 2501 M Street NW, Suite 200, Washington DC 20037. www.conservation.org
Tel: 1-202-974-9729

In Peru, tel: 1-202-429-5660, fax: 1-202-887-0192.

MANU BIOSPHERE RESERVE

Geographic location: North of Cuzco. Eastern slopes of the Andes protecting the entire Manu watershed. Northern limit defined as high ground between the Rio De las Piedras and Manu basins, north of Fitzcarrald; western limit is the Paucartambo mountains which divides the watersheds of the Urubamba and Alto Madre de Dios.
Size: Biosphere Reserve of total extent 1,881,200 hectares (7,263 sq miles), most which is not usually accessible. Tourism limited to the Zona Reservada (257,000 hectares/353 sq miles), and Zona Cultura (91,394 hectares/992 sq miles).
Access: Flight from Lima to Cuzco. Cuzco to Atalaya on the Alto Madre de Dios takes 14 hours. Boat from Atlaya to Manu Lodge, 1½ days travel. Or fly Cuzco–Boca Manu, boat from Boca Manu up Manu River to Manu Lodge 1½ hours.
Tour operators: Manu Nature Tours, Avenida Pardo, 1046, Cuzco, organize inclusive tours visiting the Manu Biosphere Reserve and operate Manu Lodge, the only lodge in the lowland forest of the Zona Reservada. Approximate prices $800–1,000 (depending on number of persons) for seven-day/six-night tour, involving land and boat transport in and chartered flight out from Boca Manu to Cuzco. Naturalists stay at the cloud forest lodge en route to Atalaya.

Manu Expeditions, Avenida Pardo, Block 7, Cuzco, arrange more adventurous trips, camping along the Rio Manu. Also Pantiacolla Tours, Plateros 360, Cuzco.
Open: All year, but rainy season from November to April may restrict access. Driest months are July to September.
Registration: Tour operator will take care of registering if going into Reserve Zone; independent visitors not permitted in this zone. No permit needed for Cultural Zone.
Accommodation: Apart from Manu Lodge in the Reserve Zone, it is possible to stay at Amazonia Lodge, Erika Lodge, Pantiacolla Lodge, Manu Wildlife Centre or Manu Cloud Forest Lodge in the Cultural Zone – *See pages 330–1 for contact details.*
Food/restaurant: Not applicable.
Transport in Park: River transport.
Trails: Trails in vicinity of lodges. Twenty-seven kilometers of trails (17 miles) at Manu Lodge. Take excursions to Cocha Otorongo and Salvador to view Giant Otters.

PACAYA SAMIRIA NATIONAL RESERVE

Geographic location: Department of Loreto, northeastern Peru.
Size: More than 2,080,000 hectares (8,031 sq miles).
Access: Northern approach: from Iquitos, six hours, or by plane, one hour. Southern approach: Lima to Tarapoto by plane, then bus to Yuimaguas (five hours), then 12-hour boat ride to Lagunas.
Accommodation: Camping.
Tour operators: Hire local guides from Lagunas or Pro Naturaleza (Fundación Peruana para la Conservación de la Naturaleza). Contact Pro Naturaleza at Parque Blume 106 con General Córdova 518, Miraflores – Lima 18. Email: fpcn@mail.cosapidata. com.pe Check with INRENA, Calle 17,355 Urb.El Palomar, San Isidro, Lima, tel: (1) 224-32982 regarding the entry fee (US$10) and possible permit required to enter the reserve. www.orbnet.com.pe/pronatura/pacaya10.htm
Further advice: Best time to go is June to November.

The Peruvian dry season, May through September, offers the best views and finest weather. For safety reasons and for greater enjoyment, parties of three/four or more should hike together. Groups are easily formed in Cuzco, and the tourist office on the Plaza de Armas provides a notice board for this purpose. In Huaráz, the Casa de Guias, tel: 721 811, just off the main street, offers the same service. Promperu (www.peruonline.net) issue an excellent free leaflet called The Traveler's Guide to Trekking and Mountaineering in Peru, outlining the most popular areas and trails plus useful tips and a directory of relevant institutions.

Trekking Guidebook

A useful guide is *Backpacking and Trekking in Peru and Bolivia* by Hilary Bradt, Bradt Publications.

Equipment

A backpack, sturdy hiking boots, sleeping bags, insulated pad, tent and stove are the major necessities for trekking. This equipment can be hired at a number of highly visible adventure travel agencies. The cost is minimal, but quality often suffers. Inspect hired equipment carefully before departure. It can get very cold at night, and by day, the Andean sun burns quickly. Bring a good sunblock, not available in Peru, and have a hat handy.

For the less adventurous, porters can be hired for the Inca Trail. In the Cordillera Blanca, *arrieros*, or mule-drivers, are quite inexpensive and readily available.

Food

Food should be brought from a major town as little is available in smaller villages. Freeze-dried food is not available, but dried fruits, cheeses, fruits, packaged soups, and tinned fish can be acquired in a number of *supermercados*, *bodegas*, and open-air markets. Drinking water should be treated with iodine, and instant drink mixes like *Tang* can be added to offset the unpleasant taste.

High altitude

The effects of high altitude can be significantly diminished by following a few simple precautions. Alcohol, overeating and physical exertion should be avoided for the first day or two. Drinking ample liquids helps the system adjust quickly. The sugar in hard candy stimulates the metabolism, and aspirin eases the headache. *Mate de coca*, tea brewed from coca leaves, is said to be the best overall remedy.

Culture

LIMA

Museo de la Nación
Javier Prado Este 2466, San Borja, tel: 476-6577. 9am–5pm Tues–Fri; 10am–5pm Sat and Sun. Entrance fee. Lima is justly proud of its newest museum, a large modern structure containing many floors of meticulously prepared exhibitions. They include impressive models of new and established archeological sites, as well as a spacious mannequin showroom displaying Peru's magnificent folkloric costumes. Worth at least one long visit.

Museo de Oro del Peru
(Gold Museum)
Alonso de Molina 1110, Monterrico, tel: 345-1291. 11.30am–7pm daily. Entrance fee. There are actually two private museums at this outer suburb address. Apart from Peru's unique collection of Pre-Inca and Inca gold, there is an astonishingly complete Arms Museum which includes a number of highly decorative uniforms.

Nightlife

Lima is full of colorful *peñas* where folk music and dance performances go non-stop until 2am or 3am. Go about 9pm to get a good seat. Try **Hatuchay**, Trujillo 228, tel: 247-779; **Las Brisas del Titicaca**, off Plaza Bolognesi; or **La Estación de Barranco**, Pedro de Osma 112, Barranco.

Museo Rafael Larco Herrera
Bolívar 1515, Pueblo Libre, tel: 461-1312. 9am–6pm Mon–Sat; 9am–1pm Sun. Entrance fee. Private collection of over 400,000 well-preserved ceramics, pre-Columbian art and artifacts. Famous erotic *huacos* ceramics from the Moche culture are housed in a separate room beside the gift shop.

Museo Nacional de Antropologia y Arqueologia
Plaza Bolivar, Pueblo Libre, tel: 463-5070. 9am–6pm daily. Entrance fee. Highly comprehensive exhibitions depicting the prehistory of Peru from the earliest archeological sites to the arrival of the Spaniards. All the major Peruvian cultures are well represented.

Museo Nacional de la República
(National Museum of History)
Plaza Bolívar (next door to the Archeological Museum), Pueblo Libre, tel: 463-2009. 9am–6pm daily. Entrance fee. Both San Martín and Bolívar once lived in this building, now containing the furnishings and artefacts from colonial, republican and independent Peru. A must for those interested in the Peruvian revolution.

Museo Pedro de Osma
Pedro de Osma 421, Barranco, tel: 467-0915. 11am–1pm and 4–6pm. Admission by appointment only. Treasured collection of viceregal painting, sculpture and silver.

Museo de Arte
Paseo de Colón 125, Lima, tel: 423-4732. 10am–5pm Tues–Sun. Entrance fee. Grand building containing four centuries of Peruvian artworks, including modern painting. Also houses pre-Columbian artifacts and colonial furniture.

Amano Museum
Retiro 160 (off 11th of Angamos), Miraflores, tel: 441-2909. Admission by appointment only, guided tours in small groups at 2pm, 3pm, 4pm and 5pm Mon–Fri. Particularly beautiful is the textile collection from the lesser known Chancay culture.

Museo del Banco Central de Reserva
Ucayali 291, Lima, tel: 427-6250. 10am–4pm Mon–Sat; 10am–1pm Sun. A welcome relief from the hoards of money changers in this part of town, the bank offers a collection of looted pre-Columbian artefacts recently returned to Peru, paintings, from the viceroyalty to the present, and numismatics (the study of coins).

Museum of the Inquisition
Junin 548 (right side of the Plaza Bolívar as you face the Congress), Lima. 9am–7pm Mon–Fri; 9am–4.30pm Sat. Admission free. Explicit representations of torture methods in the dungeon of this, the headquarters of the Inquisition for all Spanish America from 1570–1820.

Museo de Historía Natural Javier Prado
Arenales 1256, Jesús Maria, tel; 471-0117. 9am–6pm Mon–Sat. Collection of Peruvian flora and fauna.

Numismatic Museum
Banco Wiese, 2nd Floor, Cuzco 245, Lima, tel: 427-5060 ext. 553. 9am–1pm Mon–Fri. Admission free. Peruvian coins from Colonial times to the present.

CUZCO

A bewildering number of museums and historic sights can be found in Cuzco. Independent travelers can visit them separately with a US$10 Visitor's Ticket, which is available from several of the major attractions. Opening hours are listed on the ticket.

IQUITOS

The Iron House
Putumayo & Raymondi (corner of the Plaza de Armas). Supposedly designed by Eiffel and imported piece-by-piece during rubber boom days.

Bolivia

Getting Acquainted

GOVERNMENT

Bolivia may have suffered an average of one coup per year since independence from Spain in 1825, but, it has had civilian rule since 1982. The President is elected by popular vote every four years under the 1967 constitution.

The country is divided into nine departments, each controlled by a Delegate appointed by the President. Bolivia has La Paz for a de facto capital, as the seat of the Government and Congress, with the legal capital being the small city of Sucre, where the Supreme Court sits.

ECONOMY

Bolivia is the poorest South American republic, with two out of three people living beneath the poverty line. Over half the population subsist on agriculture. Until the collapse of prices in 1985, tin was the major export earner for the country. Now Bolivia mines quantities of gold, silver and zinc. Natural gas is exported. The biggest money-spinner in Bolivia is coca growing, which the Government hopes to curb – it employs thousands of peasants and contributes a significant sum to the official economy.

PEOPLE

The 8.7 million Bolivians are mostly crowded onto the bleak highlands – some 70 percent of the population lives here, mostly around La Paz, Oruro and Lake Titicaca. About two-thirds of the total population are indigenous peoples, many speaking only Quechua or Aymará. Most of the rest are *mestizos*, or mixed Spanish and local descent, locally referred to as *cholos*. About one percent is of African heritage, and a fraction of European and Japanese descent.

CLIMATE

In Bolivia, the average temperature in the highlands – the most popular area to visit – is 10°C (50°F). Rain falls heavily from November to March here, but May to November is very dry. Naturally, the higher mountains are much colder, while the Amazon basin is hot and wet all year around. The average temperatures in La Paz are between 6°C and 21°C (42°F and 70°F) during summer and 0°C and 17°C (32°F and 63°F) in winter.

BUSINESS HOURS

Normally 9am–noon and 2pm–6pm. Banks only open until 4.30pm. Government offices close on Saturday.

Planning the Trip

VISAS

Citizens of the US, most European countries and New Zealand do not need visas. At the border, the guards will grant entry for either 30 or 90 days. The Immigration Office for visa extensions is at Avenida Gonzalvez 240, La Paz. On departure, a $25 airport tax is levied.

MONEY MATTERS

Bolivia's currency is the *boliviano*, divided into *centavos*. The exchange rate has been relatively constant, although this could change.

Banks, hotels and street money changers all offer similar rates of exchange for foreign currency – the black market is more or less dead.

Take money in US dollars and change in larger cities.

Credit cards are not always accepted outside major hotels and restaurants. American Express and Visa are the most widely recognized. ATMS are widely available in La Paz for Visa and Mastercard cash advances. Travellers' cheques are not widely accepted outside La Paz.

WHAT TO WEAR

The Bolivian highlands can be bitterly cold at night all year. Bring warm clothes and take a jumper even when the day is warm – the temperature is likely to drop dramatically when the sun sets. Note that winter is the dry season and it rarely snows in La Paz or Potosí. Winter nights do not go much below freezing in the capital, but can be bitterly cold higher up in the Andes mountains.

Practical Tips

GETTING THERE

By Air

The national Bolivian airline, Lloyd Aero Boliviano, flies direct from the US, Miami, to La Paz, as does American Airlines. From Europe, flights to La Paz are via Miami, Lima or Buenos Aires and there are flights to Bolivia from all the other South American capitals as well as Rio de Janeiro, Manaus, São Paolo and Cuzco.

Lloyd Aero Boliviano, Avenida Camacho 1460, tel: 367-701 or 0800 3001.

Grupo Taca, Paseo del Prado 1479, tel: 313111.

By Land

The majority of travelers enter Bolivia via Lake Titicaca in Peru. Many

Electricity

For visitors who bring electrical appliances, the electricity in Bolivia runs at 220 volts, 50 cycles.

Media

In La Paz, you can choose from *Presencia, El Diario, Hoy* and *Ultima Hora.* Other cities have local newspapers. An English-language paper, *The Bolivian Times*, comes out on Friday.

companies offer minibus connections between Puno and La Paz.

Crillon Tours offers a luxury hydrofoil service both ways across the lake, allowing a stopover on the Island of the Sun, a meal in Copacabana and drinks on the water, as well as a visit to their new Andean Roots complex, which gives an excellent introduction to Bolivia. The charge is $160 each way. Crillon Tours is at Avenida Camacho 1223 in La Paz, tel: 374-566, fax: 391-039.
Email: titicaca@wara.bolnet.bo
www.titicaca.com
They also have an agent in Puno.

Transturin has an alternative catamaran cruise service across the lake, with an optional overnight stop. As the catamarans are slower than the hydrofoils they are more spacious and there is more to do during the journey. The cruise includes a visit to the Sun Island and a visit to the Inti Wata Complex and Inti Wata Reed Vessel, exclusive to Transturin clients. Costs approx $160–$213.
Transturin, Calle Alfredo Ascarrunz 2518, La Paz, tel: 422-222, fax: 411-922.
Email: info@visit-bolivia.com
www.turismo-bolivia.com
It is also possible to travel from Argentina by land, crossing the frontier at La Quiaca and continuing by bus or train – a slow but fascinating journey.

From Brazil, the jokingly-named "Train of Death" comes up from Corumba in the jungle to Santa Cruz. It's not particularly comfortable but favored by more adventurous travelers. There are bus links to Arica and Iquique in Chile from Aruro and La Paz respectively.

POSTAL SERVICES

The main post office in La Paz is at Avenida Mcal Santa Cruz y Oruro, open Monday–Saturday from 8am–10pm, and Sunday from 9am–noon. Poste Restante keeps letters for three months. The outbound mail is quite efficiently posted to the rest of the world. DHL is at Avenida Mcal Santa Cruz 1297.

TELECOMMUNICATIONS

Telephone calls can now be made by satellite to the United States and Europe, although an operator is required. You can either call from your hotel or the ENTEL office in La Paz, Edificio Libertad, Calle Potosí. Or buy phonecards – there are public card telephones by the ENTEL office, on Prado and around the city.

MEDICAL SERVICES

Major hotels have doctors on call. The Clínica Alemana in La Paz, 6 de Agosto 2821, tel: 323 023 or the Clínica del Sur, Avda Fernando Siles y C. 7 Obrajes, tel: 784 –002 are competent and well-run to handle any problems that you may encounter.

EMBASSIES & CONSULATES

Peruvian Consulate: 6 de Agosto 2190 and Guachalla, Edif. Alianza, tel: 353-550. Open 9.30am–1pm.
US Embassy: Avenida Arce 2780, tel: 350-120 or 430-251, fax: 433 854.
UK Consulate: Avenida Arce 2732-2754, tel: 433-424, fax: 431-073. Also represents Australian and New Zealand citizens.

Getting Around

FROM THE AIRPORT

To get from El Alto to La Paz, you can hire a taxi for $6–8 which will take approx 30 mins, or take a minibus for about $1 which runs to and from the airport to Av Mariscal Santa Cruz.

PUBLIC TRANSPORT

The bus network in Bolivia is well-developed, although the quality of service depends on the conditions of the roads (of which only some 2 percent are paved). Luxurious services run between La Paz, Cochabamba and Santa Cruz; quite rough services to Potosí.

A wide choice of services is available to Coroico and Copacabana. The future for the Bolivian rail network is uncertain and the only reliable routes are from Santa Cruz to Puerto Quijarro on the Brazilian border and from Oruro via Uyuni to the Argentine border. To get off the beaten track, trucks are the standard transport in the Andes.

Most tour companies also offer services to major cities and attractions. Within La Paz, taxis are probably the best way to get around. They can be hired privately to go anywhere in the city for $2.50 or caught as *colectivos*, sharing with passengers along a fixed route for about $0.50.

DOMESTIC AIR TRAVEL

Domestic airline Lloyd Aero Boliviano flies between all major cities. Fares are reasonable and can be worth-while to avoid long bus journeys, especially from La Paz to Potosí (this route is by plane to Sucre then a four-hour bus journey to Potosí).

Where to Stay

LA PAZ

Hotel Gloria
Potosi 909
Tel: 407-070, fax: 406-622
Email: gloriatr@4ceibo.entelnet.bo
$60 a double.
Hotel Presidente
Potosi 920
Tel: 406-666, fax: 407-240
Email: hpresi@caoba.entelnet.bo
$120 a double.

Radisson Plaza
Avenida Arce
Tel: 441-111, fax: 440-593
Email: radissonbolivia@usa.net
$150 a double.
Residencial Rosario
Illampu 704
Tel: 451-658, fax: 451-991
Email: turisbus@wara.bolnet.bo
$30 a double – one of the best
budget hotels. Also runs local tours
and bus services to Puno and Chile.

LAKE TITICACA

Inca Utama Hotel and Spa
Bookable through Crillon Tours (see
page 336).
Located on the lakeside at
Huatajata, this deluxe hotel has a
spa based on natural remedies with
a range of astronomical, mystical,
ecological and natural health
programmes.
Posada del Inca
Also bookable through Crillon Tours,
this restored hacienda offers rustic
eco-friendly accommodation.

POTOSÍ

Hotel Claudia
Avenida Maestro 322
Tel: 22242
US$30 a double.
Hotel Colonial
Calle Hoyos 8
Tel: 24809
US$30 a double.

SUCRE

Hotel Real Audiencia
Tel: 64-60823
US$30 a double
Hostal de Su Merced
Tel: 64-42706, fax: 69-12078
Beautifully-restored colonial
building.

What to Eat

Bolivian food varies by region. Lunch
is the main meal of the day and
many restaurants offer an *almuerzo
completo* or three-course fixed menu.

Drinking Notes

The usual tea, coffee and soft
drinks are all available in cafés
and restaurants, as is the *mate
de coca* in tea bags. For coffee,
head for the Café La Paz, corner
of Camancho and the Prado.
 Bolivian beer is good, brewed
under German supervision. The
favorite drink of Bolivians is
chicha, the potent maize liquor
produced near Cochabamba.
Keep in mind that altitude
intensifies the effects of alcoholic
beverages – the intoxication and
the hangover.

Traditional foods of the highlands are
starchy, with potatoes, bread and
rice. Their meats are highly spiced.
Trout from Lake Titicaca is excellent.

WHERE TO EAT

Many restaurants in La Paz serve
international cuisine, mostly along
Avenida 16 de Julio (the Prado)
and Avenida 6 de Agosto. Below
are a few recommended
restaurants.
Casa del Corregidor
Calle Murillo 1040
Tel: 353-633
International and local dishes.
Los Escudos
Edif. Club de la Paz
Tel: 322-028
With a dance show.
Pronto
Jauregui 2248 (off Calle Guachalla)
Tel: 355-869
Italian.
Tambo Colonial
Hotel Residencial Rosario
Calle Illampu 704
Tel: 316-516
Local cuisine; live Bolivian music.

Parks and Reserves

The rainforests, swamps and
savannahs of the Bolivian Amazon
lowlands comprise nearly two-thirds
of the country's territory; a wildlife
paradise, with over 50 percent of
the country's birds and animals. In

an attempt to protect Bolivia's great
bio-diversity from the ruthless
destruction of the rainforest, and to
attract eco-tourism and wildlife
expeditions, several National Parks
and new eco-lodges and reserves
have been created with another ten
under consideration, many of them
in the Amazon basin. One of the
main gateways to explore Bolivia's
Amazon Wildlife is Rurrenabaque,
only a short flight from La Paz and
easily arranged through any good
Bolivian operator or specialist.

BENI BIOSPHERE
RESERVE

Geographic location: Northern
lowlands of Bolivia, Amazon basin.
Size: 334,200 hectares (1,290 sq
miles).
Created: 1982.
Access: Flight to Trinidad with LAB
(tel: 367-701) or SAVE
(tel: 0886-2267), then bus to El
Porvenir, one hour east of San Borja.
Some flights to San Borja with SAVE
and TAM. Bus from La Paz to
Rurrenabaque via Caranavi Mon–Sat,
18 hours or from Trinidad via San
Borja.
Cost of access: La Paz–Trinidad
one-way flight, $57.
Tour operators: Agencia Fluvial,
Hotel Tuichi, Calle Avaroa and
Rurrenabaque.
 Also Fremen Tours, C Pedro
Salazar 537, Plaza Avaroa, La Paz,
tel: 416-336, fax: 417-327,
www.andes-amazonia.com They operate
speedboat trips along the Manore
and Ibore rivers and their own Flotel
Reina de Enin, as well as tours
across the country.

Excursions

Most tour companies offer
services to major cities. For
information on trekking and adventure tours and a list of guides, call
Club Andino Boliviano, Calle
Mexico, tel: 324-692. Also contact
Turisbus, c/o Hotel Residencial
Rosario, tel: 316-156. Email:
turisbus@wara.bolnet.bo

When to go: Best time to go is in the dry season between May and October.
Accommodation: El Porvenir Station – arrange through Beni Biological Station, Bolivian Academy of Natural Sciences, P.O. Box 5829, tel/fax: 352-071. Costs $6 entrance to station, $6 bed.
Eating: $6 for three meals.

MADIDI NATIONAL PARK

Geographic location: Andean slopes of the upper Amazon basin.
Size: 1, 895,740 hectares (7,319 sq miles).
Created: 1995.
Access: La Paz to Rurrenabaque. Four flights a week to Rurrenabaque with TAM, stay there overnight. From Rurrenabaque, 5–6 hour boat ride up the Beni and Tuichi River followed by a half-hour hike to Chalalan Lodge. Or bus to Rurrenabaque (destination Riberalta and Guyaramerin) Tues, Thurs, Sat and Sun; via Caranavi Mon–Sat, 18 hours, $10.
Cost of access: $110 return or bus $10.
Tour operators: Agencia Fluvial in Rurranabaque and Chalalán Lodge, Chalalán, www.ecotour.org Also the main La Paz operators including Crillon Tours and Transturin featured under *Practical Tips: By Land (see page 336).*
When to go: May to October. Not recommended during the rainy season, November to April, due to difficulty of access.
Accommodation: Chalalan Eco-Lodge, c/o America Tours in La Paz, tel: 374-204, fax: 328-584, www.america.ecotours.com or Conservation International at www.ecotour.org A beautiful, low-impact eco-lodge and the showcase for Bolivian green tourism. Room for 14 people with showers, a library, and activities including 25 km (15 miles) of trails, canoeing, night hikes and evening talks about indigenous culture and lifestyle. Package includes transfers, a night in Rurren-abaque, meals and river transport.
Food/restaurant: Meals available at Lodge.

NOEL KEMPFF NATIONAL PARK

Geographic location: Northeastern corner of Department of Santa Cruz, on the Brazilian border.
Size: 1,523,000 hectares (5,880 sq miles).
Created: 1979.
Access: Bus from Santa Cruz to San Ignacio, then bus to La Florida (or flight), then 40-km (25-mile) walk to Los Fierros in park, or private vehicle along road open to loggers April through December. Or charter plane to Los Fierros; contact Armonia (eco-org), P.O. Box 3081, Santa Cruz, tel: 522-915, fax: 324-971. For access to Flor de Oro, contact FAN. the Fundación Amigos de la Naturaleza in Santa Cruz, tel: 352-4921, fax: 533-389, email: fan@fan.rds.org.bo Entrance fee: $30 per day. Alternatively, good local operators in La Paz can tailor-make an itinerary to include charter flights to and from Flor de Oros but roads and airstrips frequently change so get up-to-date information from International Expeditions Incorporated in the US, tel: (1) 800 633 4734 or email: nature@ietravel.com
Tour operator: Uimpex Travel, Calle Rene Moreno 226, Santa Cruz, tel: 336-001, fax: 330-785. Magri Turismo in La Paz tel: 434 74 or email: magri.emete@megalink.com and Fremen Tours at www.andes-amazonia.com have both been recommended for tours to the region.
When to go: May to October road access easier, November to April river travel easier.
Accommodation: Campamiento Los Fierros – 100 spaces in dormitaries with cooking facilities, $25 a night, camping cheaper.
Eco-Lodge at Flor de Oro – about $100 per person including meals. Book through FAN or Uimpex in Santa Cruz. Camping only $10 a night.
Further advice: Be sure to contact the FAN office (shared with Nature Conservancy and US Aid) before you go. It is in the village of La Nueve, 7 km on road west of Santa

Cruz: Fundación Amigos de la Naturaleza (FAN), Cadilla 2241, Santa Cruz.
Tel: (03) 524-921, fax: (03) 533-389.
Email: fan@fan.rds.org.bo

AMBORÓ NATIONAL PARK

Geographic location: Confluence of Amazon basin, Andes and El Chaco.
Size: 637,000 hectares (2,459 sq miles).
Created: 1973.
Access: Bus from Santa Cruz to Buena Vista, then bus (one in morning only) from Buena Vista along Rio Suturi.
Tour operators: Uimpex Travel, Calle Rene Moreno 226, Santa Cruz, tel: 336-001, fax: 330-785.
Rosario Tours, Calle Arenales 193, Santa Cruz, tel/fax: 369-656, email: aventura@tucan.cnb.net
When to go: May to October
Accommodation: Basic huts $2 a day, on Río Macunucu and other rivers Ask BID for information: office south of plaza in Buena Vista, tel: (0932) 2032. Alternatively, visit Amboró on day-trips, using Buena Vista as a base. Amboró Eco-resort can be contacted at bloch@bibosi. scz.entelnet.bo

BLANCO AND NEGRO RIVERS WILDLIFE RESERVE

Geographic location: Northeastern Bolivia, Amazon basin.
Size: 1,400,000 hectares (5,405 sq miles).
Created: 1990.
Access: Flight from Santa Cruz to Perseverancia.
Tour operators: Amazonas Adventure Tours, Avenida San Martin 756, Barrio Equipetrol, Reserva de Vida Silvestre Rios Blanco y Negro, tel: 422-760, fax: 422-748.
When to go: Dry season, March through October.
Accommodation: Lodge at Perseverancias.

Culture

MUSEUMS

Museum of Ethnography and Folklore
Corner Ingavi and Sanjines, La Paz. Exhibits relating to indigenous culture, open Mon–Fri 9am–12.30pm, 3–7pm; Sat–Sun 9am–1pm. Admission free.

Museo Tiahuanaco
Calle Tiwanaku below the Prado, La Paz. Relics from Lake Titicaca and Tiahuanaco arts and crafts. Admission fee is US$1.20. Open daily.

National Art Museum
Corner of Comercio and Socabaya, La Paz, in restored colonial building. Open Tues–Fri 9am–12.30pm,3–7pm, Sat 10am–1pm. Admission fee US$0.50.

NIGHTLIFE

La Paz has several *peñas*. Try the most popular, Peña Naira, on Sagarnaga just above Plaza San Francisco, open every night except Sunday ($5 cover charge). Others are at the Casa del Corregidor and Los Escudos restaurants. Bars are not the most pleasant places to hang out in La Paz, usually being reserved for inebriation rather than relaxation. Café Montmartre, the Forum and Café Azul are recommended in the popular Sopcachi area.

SHOPPING

The **artisan's market** is along Calle Sagarnaga, between Mariscal Santa Cruz and Isaac Tamayo. You can buy ponchos, vests, jackets and mufflers of llama and alpaca wool, and extra-ordinary tapestries. Many stalls also sell various Andean musical instruments and small sculptures. A little haggling is expected, but don't overdo it – many of these handicrafts take months to make. Prices vary depending on the quality, but are a fraction of what they'd be if bought in the US or Europe.

Brazil

Getting Acquainted

GOVERNMENT

Brazil is a federal republic with 26 states and one Federal District, each with its own state legislature. Since the federal government exercises enormous control over the economy, the political autonomy of the states is restricted. The overwhelming majority of government tax receipts are collected by the

Time Zones

Despite the fact that Brazil covers such a vast area, over 50 percent of the country is in the same time zone and it is in this area, which includes the entire coastline, that most of the major cities are located. The western extension of this zone is a north-south line from the mouth of the Amazon River, going west to include the northern state of Amapá, east around the states of Mato Grosso and Mato Grosso do Sul and back west to include the south. This time zone, where Rio de Janeiro, São Paulo, Belém and Brasilia are located, is three hours behind Greenwich Mean Time (GMT).

Another large zone, encompassing the Pantanal states of Mato Grosso and Mato Grosso do Sul, and most of Brazil's north, is four hours behind GMT. The far western state of Acre and the westernmost part of Amazonas state are in a time zone five hours behind GMT.

federal government and then distributed to the states and cities. The head of government is the president with executive powers and, in fact, exercises more control over the nation than the American president does over the US. The legislative branch of the federal government is composed of a Congress divided into a lower house, the Chamber of Deputies, and an upper house, the Senate.

In February 1987 the Federal Congress was sworn in as a National Constitutional Assembly to draft a new federal constitution for Brazil; this opened the way for new direct presidential elections every 4 years. Since being elected in 1994 and again in 1999, President Cardoso has opened up the Brazilian economy and attracted substantial investment from overseas, although the country faced a setback in 1999 when the real halved its value against the US dollar. However, Brazil has avoided the expected economic recession and has entered the new century with inflation largely under control.

ECONOMY

Despite past political problems, Brazil has – on the back of much environmental destruction – shown excellent economic growth rates for most of the past 30 years. Although growth faltered with the high inflation of the 1980s, and the 1990 austerity package took the country into recession, Brazil has held on to its position as the economic leader among developing nations.

The government has recently begun privatizing many of the giant state-owned companies that controlled much of the economy.

As finance minister, Fernando Henrique Cardoso (subsequently president) introduced the Plano Real in 1994. The new currency, the real, held its own for nearly five years against the US dollar but, at the start of 1999, pressure from speculators caused a devaluation of nearly 50 percent. The currency settled at around US$1 to R$2 at the start of 2000.

CLIMATE

In the jungle region, the climate is humid equatorial, characterized by high temperatures and humidity, with heavy rainfall all year round. The eastern Atlantic coast from Rio Grande do Norte to the state of São Paulo has a humid tropical climate, also hot with slightly more rainfall than the north, and with summer and winter seasons. Most of Brazil's interior has a semi-humid tropical climate, with a hot, rainy summer from December through March and a drier, cooler winter (June to August).

Mountainous areas in the southeast have a high-altitude tropical climate, similar to the semi-humid tropical climate, but rainy and dry seasons are more pronounced and temperatures are cooler, averaging from 18°C to 23°C (64°F to 73°F).

Part of the interior of the northeast has a tropical semi-arid climate – hot with sparse rainfall. Most of the rain falls during three months, usually March to May, but sometimes the season is shorter and in some years there is no rainfall at all. Average temperature is 24°C to 27°C (75°F to 80°F).

BUSINESS HOURS

Business hours for offices in most cities are 9am–6pm Monday through Friday. Lunch "hours" used to last hours but now normally are taken for an hour from noon–2pm in the city. Banks open from 10am–4.30pm Monday through Friday. The *casas de câmbio* currency exchanges operate usually from 9am–5pm or 5.30pm. Most stores are open 9am–6.30pm or 7pm and Saturday mornings until 12.30pm or 1pm. They may stay open even longer in some of the large malls. Post

Health

Brazil does not normally require any health or inoculation certificates for entry, nor will you be required to have one to enter another country from Brazil. Proof of a yellow fever vaccination is only necessary if you are arriving into Brazil from Peru or Colombia. It is a good idea also to have inoculations against hepatitis and typhoid, and to protect yourself against malaria in jungle areas. Although there is no vaccine against malaria, there are drugs that will provide immunity while you are taking them. Consult your local public health service and be sure to get a certificate for any vaccination.

Don't drink tap water in Brazil. Although water in the cities is treated and is sometimes quite heavily chlorinated, people filter water in their homes. Any hotel or restaurant will have inexpensive bottled mineral water, both carbonated (*com gas* – "with gas") and uncarbonated (*sem gas* – "without gas"). If you are out in the hot sun, make an effort to drink extra fluids.

offices are open from 8am–6pm Monday to Friday and from 8am–noon on Saturdays.

HOLIDAYS

January
1 New Year's Day (national holiday); Good Lord Jesus of the Seafarers (four-day celebration in Salvador, starting with a boat parade)
6 Epiphany (regional celebrations, mostly in the Northeast)
3rd Sunday: Festa do Bonfim (one of the largest celebrations in Salvador)

February/March/April
2 February: Iemanjá Festival in Salvador (the Afro-Brazilian goddess of the sea who, in syncretism with Catholism, corresponds with Virgin Mary)

February/March: Carnival (national holiday; celebrated all over Brazil on the four days leading up to Ash Wednesday. Most spectacular in Rio, Salvador and Recife/Olinda)
March/April: Easter (Good Friday is a national holiday; Colonial Ouro Preto puts on a colorful procession; passion play staged at Nova Jerusalem)
21 April: Tiradentes Day (national holiday in honor of the martyred hero of Brazil's independence – celebrations in his native Minas Gerais, especially Ouro Preto)

May
1 Labor Day (national holiday)
May/June: Corpus Christi (national holiday)

June/July
June 15–30: Amazon Folk Festival (held in Manaus)
June and early July: Festas Juninas (series of street festivals held in honor of Saint Anthony [13 June – the equivalent of St Valentine's Day] St John [24 June] and St Peter [29 June] featuring bonfires, dancing and mock marriages)
2nd half June and early July: Bumba-Meu-Boi (processions and street dancing in Maranhão)

September
7 Independence Day (national holiday)

October
Oktoberfest in Blumenau (put on by descendants of German immigrants)
12 Nossa Senhora de Aparecida (national holiday honoring Brazil's patron saint)

November
2 All Souls Day (national holiday)
15 Proclamation of the Republic (national holiday, also election day)

December
25 Christmas (national holiday)
31 New Year's Eve (all along the Brazilian coast, and most famously on Rio de Janeiro beaches, gifts are offered to Iemanjá)

Planning the Trip

ENTRY REGULATIONS

Visas
Western European citizens (except the Finnish) get a 90-day stay on arrival in Brazil. Citizens of the US, Canada and other countries must get a visa beforehand.

Customs
You will be given a declaration form to fill out in the airplane before arrival, which will be stamped by immigration and must be returned upon departure.

MONEY MATTERS

The introduction of the real in 1994 has kept inflation at a more manageable rate than in the past. You can only reconvert leftover *reais* at the Banco do Brazil at the airport on your departure, as long as you show the receipt for the initial exchange, so it is best to budget carefully and keep a supply of small dollar bills which are often accepted in tourist areas. Travelers' checks are easy to change at banks in major cities, and at exchange houses and travel agencies but normally only into reals. Credit cards widely used are American Express, Diners Club, Visa and Mastercard. Dollar cash is often accepted in tourist areas.

WHAT TO WEAR

Brazilians are very fashion-conscious but actually quite casual dressers. What you bring along, of course, will depend on where you will be visiting and your holiday schedule. São Paulo tends to be more dressy; small inland towns are more conservative. If you are going to a jungle lodge, you will want sturdy clothing and perhaps boots. However, if you come on business, a suit and tie for men, and suits, skirts or dresses for women are the office standard.

If you come during Carnival, remember that it will be very hot to begin with and you will probably be in a crowd and dancing nonstop. Anything colorful is appropriate.

Practical Tips

TRAVEL AGENTS

A variety of individual and group tours to Brazil are available; travelers may wish to plan their whole itinerary before departure by booking an all-inclusive tour package with transportation, food and lodgings, excursions and sightseeing all provided by one of the many South American mainstream and specialist tour operators. There are also special wildlife interest tours that feature boat trips and jungle lodges in the Amazon, or fishing and wildlife tours to the Pantanal. Call the Latin American Travel Association on (020) 8715-2913 or www.lata.org for a factual guide to the region and a list of members, including the Wildlife Worldwide website www.wildlifeworldwide.com

If you aren't on a tour where everything is planned, check at your hotel or a local travel agency to find out about readily available tours – however in the peak season there may be a wait, especially during Carnival, and it might be best financially advantageous and safer to make arrangements and flight reservations before departure.

Belém
Amazon Travel Service Ltda
Rua dos Mundurucus, 1826
Tel: (091) 241-1099/1020/1739
Ciatur Turismo Ltda
Avenida Presidente Vargas 645
Tel: (091) 224-1993,
fax: (091) 244-8544
Gran Pará Turismo Ltda
Avenida Presidente Vargas, 882
Loja 8
Tel: (091) 224-2111,
fax: (091) 241-5531
Lusotur Viagens e Turismo Ltda
Avenida Braz de Aguiar, 471
Tel: (091) 241–2000/2120,
fax: (091) 223-5054
Mururé Viagens e Turismo Ltda

Addresses

To understand the addresses, here's what the Portuguese words mean:

Alameda (Al.)	lane
Andar	floor, story
Avenida (Av.)	avenue
Casa	house
Centro	the central downtown business district also frequently referred to as the cidade or "the city"
Conjunto (Cj.)	a suite of rooms or a group of buildings
Estrada (Estr.)	road or highway
Fazenda	ranch, also a lodge
Largo (Lgo.)	square or plaza
Lote	lot
Praça (Pça.)	square or plaza
Praia	beach
Rio	river
Rodovia (Rod.)	highway
Rua (R.)	street
Sala	room

• Ordinal numbers are written with ° or a degree sign after the numeral, so that 3° andar means 3rd floor. BR followed by a number refers to one of the federal interstate highways, for example BR 101. Telex/telephone numbers are given with the area code for long-distance dialling in parentheses. Ramal means telephone extension.

Avenida Presidente Vargas, 134 sala-A
Tel: (091) 241-0891/3434,
fax: (091) 241-2082
Neytur
Rua Carlos Gomes, 300
Tel: (091) 241-0777,
fax: (091) 241-5669

Santarém (Pará)
Amazon Tour, Rua Turiano Meira, 1084
Tel: (091) 522-1098/2620
Email: swa@grego.com.br
www.amazonriver.com
Amazonia Turismo
Rua Adriano Pimentel 44
Tel: (091) 522-5820

Coruá-Una
Rua 15 de Novembro, 185-C
Tel: (091) 522-6303/7421
Tapam Turismo
Travessa 15 de Agosto 127 A
Tel: (091) 522-3037/1946/2234

Manaus (Amazonas)
Amazon Explorers
Rua Nhamunda, 21 – Praça
Auxiliadora
Tel: (092) 633-3319/633-2075,
fax: (092) 234-5753
Anaconda Turismo
Rua Dr. Almino, 36 – Centro
Tel/fax: (092) 233-7642/232-9492
ATA Turismo
Rua Eduardo Ribeira, 649 – Ed.
Palácio do Comércio, Térreo (Matriz)
Tel: (092) 622-4996,
fax: (092) 622-1088
Fontur
Est. da Ponta Negra, s/n – Km 18
(Tropical Hotel)
Tel: (092) 658-5000/656-2807,
fax: (092) 656-2167
Jangal Turismo da Amazônia
Rua Floriano Peixoto, 182 – 1°
andar – Centro
Tel: (092) 232-5884,
fax: (092) 232-4843
Naturis Safaris
Rua Santa Quitéria, 15 – Presidente
Vargas
Tel: (092) 622-4144/4146,
fax: (092) 622-1420
Regiatur
Avenida 7 de Setembro,
788 – Centro (Matriz)
Tel: (092) 234-6900/0519,
fax: 232-9838.
Rio Amazonas Tours
Rua Silva Ramos 42
Tel: (092) 234-7308,
fax: (092) 233-5615
Selvatur
Praça Adalberto Valle, s/n – Centro
Tel: (092) 622-2577/2578/1173,
fax: (092) 622-2177
Swallows and Amazons
R Quintino Bocaiuva 189, andar
1,Sala 13
Tel/fax: (092) 622-1246
www.swallowsandamazons.com
Viverde
Rua Emílio Moreira, 1769 – 2°
andar – Sala 4, Praça 14 de Janeiro
Tel/fax: (092) 622-4289
Cell phone: (092) 9833206

Electricity

Electric voltage is not
standardized throughout Brazil,
but most cities have a 127-volt
current, as is the case of Rio de
Janeiro and São Paulo, Belém,
Belo Horizonte, Corumbá and
Cuiabá, Curitiba, Foz do Iguaçu,
Porto Alegre and Salvador. The
electric current usage is 220
volts in Brasilia, Florianópolis,
Fortaleza, Recife and São Luis.
Manaus uses 110-volt electricity.

Email: amazon@viverde.com.br
www.viverde.com.br/home.html

GETTING THERE

By Air
A total of 28 airlines offer
international service to and from
Brazil with a variety of routes.
Although most incoming flights
head for São Paolo, depending on
where you are coming from, there
are also direct flights to Rio de
Janeiro and Brasilia, Salvador and
Recife on the northeastern coast,
and to the northern cities of Belém
and Manaus on the Amazon.
  There is an international airport
departure tax of $36 and a tax of
$7–10 for domestic flight
departures, depending on the local
airport.

By Road
There are bus services between a
few of the larger Brazilian cities and
major cities in neighboring South
American countries, including
Asunción (Paraguay), Buenos Aires
(Argentina), Montevideo (Uruguay)
and Santiago (Chile). While
undoubtedly a good way to see a lot
of the countryside, the distances
are great and the journey can last a
few days.

MEDIA

The *Miami Herald*, the Latin America
edition of the *International Herald
Tribune* and the *Wall Street Journal*

are available on many newsstands
in the big cities, as are such news
magazines as *Time* and *Newsweek*.

POSTAL SERVICES

Post offices generally are open
from 8am–6pm Mon–Fri,
8am–noon on Saturday, and are
closed on Sunday and holidays. In
large cities, some branch offices
stay open until later. (The post
office in the Rio de Janeiro
International Airport is open 24
hours a day.) Post offices are
usually designated with a sign
reading *correios* or sometimes ECT
(for *Empresa de Correios e
Telégrafos* = Postal and Telegraph
Company).

TELECOMMUNICATIONS

Pay phones in Brazil use
telephone cards which are sold
at bars, newsstands or shops:
these are usually located near
to phones. The sidewalk *telefone
publico* is also called an *orelhão*
(big ear) because of the protective
shell which takes the place of a
booth – orange for local or
collect calls, blue for direct-dial,
long-distance calls within Brazil.
You can also call from a *posto
telefônico*, a telephone company
station (at most bus stations
and airports), where you can
either use a phone and pay the
cashier afterward, make a credit
card or collect call, or buy a
telephone card.
  For making international calls, it
is advisable to buy a 100 unit card
at least; calls are rated either
cheap or normal, depending on the
time of day.
  To make a collect call from a
private telephone, dial 9 followed by
the number and say your name;
when using a public phone dial
107. Access codes are:
BT Direct: 000080-44
ATT: 00080-10
Sprint: 00080-16
MCI: 00080-12
Worldcom: 00080-11

GOVERNMENT TOURISM OFFICES

Belém
Paratur
Feira do Artesanato, Praça
Kennedy, s/n, Belém

Manaus
Empresa Amazonense de Tourismo
– Emamtur
Av 7 de Setembro 1546
Tel: (092) 633-2850,
fax: (092) 233-9973

Rio de Janeiro
Riotur
Rua da Assembleia, 10, 8-9°
andares, Rio de Janeiro
Tel: (021) 297-7117/242-8000,

Trains

Except for crowded urban
commuter railways, trains are not
a major form of transportation in
Brazil and rail links are not
extensive. There are a few train
trips, which are tourist
attractions in themselves, either
because they are so scenic or
because they run on antique
steam-powered equipment.

In the southern state of
Paraná, the 110-km (68-mile)
Curitiba–Paranaguá railroad is
famous for spectacular mountain
scenery.

The train to Corumbá, in the
state of Mato Grosso do Sul, near
the Bolivian border, crosses the
southern tip of the Pantanal
marshlands. There are train links
all the way to São Paulo, over
1,400 km (870 miles) away – a
long ride. The most scenic part is
the 400-km (248-mile) stretch
between Campo Grande and
Corumbá.

In the Amazon region, you can
ride on what is left of the historic
Madeira–Mamoré Railway: 27-km
(17 miles) of track between
Porto Velho and Cachoeira de
Teotônio in the state of
Rondônia. The Madeira-Mamoré
runs on Sundays only and strictly
as a tourist attraction.

fax: (021) 531-1872
www.rio.rj.gov.br/riotur

São Paulo
Secretaria Esportes e Turismo
Estado de São Paulo
Coordenadoria de Turismo, Rua São
Bento, 380, 6th floor, 01010 São
Paulo – SP
Tel: (011) 239-0892/239-0094

EMBASSIES & CONSULATES

United States: Rua Recife, 1010 –
Adrianópolis, tel/fax: (092) 234-
4546
United Kingdom: Rua Poraquê, 240
– Distrito Industrial, tel: (092) 237-
7869, fax: (092) 237-6437
Venezuela: Rua Ferreira Pena,
179 – Centro, tel: 233-6004, fax:
(092) 233-0481

TIPPING

Most restaurants will usually add a
10 percent service charge onto your
bill. If you are in doubt as to whether
it has been included, it's best to ask
(O serviço está incluído?) or check
the bill carefully. Give the waiter a
bigger tip if you feel the service was
special. Although many waiters will
don a sour face if you don't tip
above the 10 percent included in
the bill, you have no obligation to do
so. Hotels will add a 15 percent
service charge to your bill.

MEDICAL SERVICES

Should you need a doctor while in
Brazil, the hotel you are staying at
will be able to recommend reliable
professionals who often speak
several languages. Many of the
better hotels even have a doctor on
duty. Your consulate will also be
able to supply you with a list of
physicians who speak your
language. In Rio de Janeiro, the Rio
Health Collective (English-speaking)
runs a 24-hour referral service, tel:
294-0282, 325-3327; 239-7401
for the Rio area only.

Check with your insurance
company before traveling – some
plans cover any medical service
that you may require while abroad.

ETIQUETTE

While handshaking is a common
practice, here it is customary to
greet not only friends and relatives
but also complete strangers to
whom you are being introduced with
hugs and kisses. The "social" form
of kissing consists usually of a kiss
on each cheek.

Getting Around

FROM THE AIRPORT

Until you get your bearings and if
you have not pre-arranged airport
transfers through a tour operator or
ground handler, you are
recommended to take a special
airport taxi, for which you pay in
advance at the airport at a fixed rate
set according to your destination.
There will be less of a communication problem, no misunderstanding
about the fare and even if the driver
should take you around by the
"scenic route", you won't be
charged extra for it. However, if you
should decide to take a regular taxi,
check out the fares posted for the
official taxis so that you will have an
idea of what is a normal rate.

PRIVATE TRANSPORT

By Taxi
Taxis are the best way for visitors to
get around cities. It's easy to be
"taken for a ride" in a strange city.
Whenever possible, take a taxi from
your hotel, where someone can tell
the driver where you want to go.

PUBLIC TRANSPORT

By Bus
A comfortable, on-schedule bus
service is available between all major
cities, and to several other South
American countries. Remember that

distances are great and rides can last a few days, so it may be preferable to break up the journey or book one of the luxury (leito) services that offer reclining seats, blankets and on-board services.

DOMESTIC AIR TRAVEL

Within Brazil, the major airlines are Transbrasil, Varig, TAM and Vasp, with several other regional carriers which serve the smaller cities.

Different lines have similar prices for the same routes. For the best value get several quotes. There is a 30 percent discount for night flights (vôo econômico or vôo noturno) with departures between midnight and 6am.

Transbrasil, VASP, Varig and TAM also offer air passes which must be bought outside Brazil. Costing around $490, they are valid for 21 days and limited to 5 or 6 coupons; they offer excellent value for money if you plan to travel extensively within Brazil. Flights need to be pre-booked so contact the airline or your travel agent for more details.

The large airlines also cooperate in a shuttle service between Rio and São Paulo (with flights every half hour), Rio and Brasilia (flights every hour) and Rio and Belo Horizonte (usually about 10 flights per day). Although you may be lucky, a reservation is a good idea.

Where to Stay

HOTELS

Rio de Janeiro (City)
Arpoador Inn
Rua Fransisco Otaviario 177, Ipanema
Tel: (021) 523-006, fax 511-5094
Caesar Park
Avenida Vieira Souto 460, Ipanema
Tel: (021) 525-2525, fax: 521-6000
Email: hotel@caesarpark-rio.com
Copacabana Palace
Avenida Atlântica 1702, Copacabana
Tel: (021) 255-7070, fax: 235-7330
www.orient-express.com
Leme Palace
Avenida Atlântica 656, Leme

Lodges near Manaus

Acajatuba Jungle lodge
Acajatuba River, 60 km (37 miles) from Manaus
Tel/fax: (092) 232-9492
Amazon Ecopark Lodge
Igrape do Taruma-Acu, 20 km (12 miles) from Manaus (close to the Amazon Monkey Jungle where monkey species are rehabilitated)
Tel/fax: (092) 234-0939
Amazon Lodge
Naturis Safaris, Rua Santa Quitéria, 15 – Presidente Vargas
Tel: (092) 622-4144/4146, fax: 622-1420
Amazon Village
Naturis Safaris: details as Amazon Lodge.
Ariau Jungle Tower
Rio Amazonas Tours, Rua Silva Ramos 42
Tel: (092) 234-7308, fax: (092) 233-5615
Email: treetop@internext.com.br
Janauaca Lodge
Anaconda Turismo, Rua Dr. Almino, 36 – Centro
Tel/fax: (092) 233-7642/232-9492
Lago Salvador Lodge
Fontur, Est. da Ponta Negra, s/n –

Km 18 (Tropical Hotel)
Tel: (092) 658-5000/656-2807, fax: 656-2167
Pousada dos Guanavenas
Silves Country (300 km/186 miles from Manaus)
Guanavenas Turismo
Tel: (092) 656-1500, fax: (092) 238-1211
Tapiri Pousada na Selva
s/n – Centr
Tel: (092) 622-2577/ 2578/1173, fax: 622-2177
Mamiraua Sustainable Development Reserve
www.cnpq.br/mamiraua/mamiraua2.htm
Located close to Tefé, one hour's flight from Manaus, at the confluence of the rivers Solimões, Japurá, and Auati-Parani, this little-known reserve now allows a small number of visitors to stay, and is one of the few places where it is possible to experience living in a flooded forest. The reserve has about 400 species of birds and 45 species of mammals, including the white and red uakari monkey and river dolphins.

Tel: (021) 275-8080, fax: 275-8080
Email: leme@othon.com.br
Luxor Copacabana
Avenida Atlântica 2554, Copacabana
Tel: (021) 235-2245, fax: 255-1858
Meridien-Rio
Avenida Atlântica 1020, Leme
Tel: (021) 275-9922, fax: 541-6447
Praia Ipanema
Avenida Vieira Souto 706, Ipanema
Tel: (021) 239-9932, fax: 239-6889

São Paulo (City)
Caesar Park
Rua Augusta 1508/20, Cerqueira Cesar
Tel: (011) 253-6622, fax: 288-6146
Email: mktsales@caesarpark.com.br
Maksoud Plaza
Alameda Campinas 50, Bela Vista
Tel: (011) 253-4411, fax: 253-4544
Email: maksoud@maksoud.com.br
Mofarrej Sheraton
Alameda Santos 1437, Cerqueira

Cesar
Tel: (011) 253-5544, fax: 280-8670
São Paulo Hilton
Avenida Ipiranga 165 Centro
Tel: (011) 256-0033, fax: 257-3033

Belém
Note: Belém, the capital of Pará, is a major river and ocean port, the main commercial center for the lower Amazon. Its mercantile monopoly has been broken by the new highways to Manaus, Santarém and Porto Velho, but old habits (like 100% profit margins) die hard. If you are outfitting an Amazon expedition in Belém, shop around. The same holds true for hotels and services.

Hilton International Belém
Avenida Pres. Vargas 882 (Praça da República)
Tel: (091) 223-6500, fax: 225-2942

Tourist Information

Brazil's national tourism board Embratur, headquartered in Rio de Janeiro, will send information abroad. Write to: Embratur, Rua Mariz e Barros, 13, 9° andar, Praça da Bandeira, 20000 Rio de Janeiro, RJ, Brazil or log onto www.embratur.gov.br Each state has its own tourism secretary and there are many useful websites to help plan your visit to the country. Check www.brazil.org.uk run by the Brazilian Tourist Office in London, which is extremely helpful and will also send out information and give up-to-date advice. The address is 32 Green Street, London W1Y 4AT, tel: (020) 7399-9000.

Cambará
Rua 16 de Novembro 300 (Cidade Velha)
Tel: 224-2422, fax: 224-2422
Novotel
Avenida Bernardo Sayão 4804 (Guamá)
Tel: 229-8011, fax: 229-8709
Plaza
Praça da Bandeira 130 (Cidade Velha) Tel: 224-2800
Regente
Avenida Gov. José Malcher 485 (Center)
Tel: 241-1222, fax: 224-0343
Vanja
Rua Benjamin Constante 1164
Tel: 222-6457, fax: 222-6709
Ver-O-Peso
Avenida Boulevard Castilhos França 208/214 (Center)
Tel: 224-2267, 241-4236
With restaurant, overlooking port.
Verde Oliva
Rua Boaventura da Silva 1179 (São Brás)
Tel: 224-7682
Vidonho's
Rua ó de Almeida 476 (Center)
Tel: 225-1444

Belo Horizonte
Belo Horizonte Othon Palace
Avenida Afonso Pena 1050, Centro
Tel: (031) 273-3844,
fax: 212-2318

Brasília
Brasília Carlton
Setor Hoteleiro Sul, Quadra 5, Bloco G
Tel: (061) 224-8819, fax: 226-8109
Eron Brasília
Setor Hoteleiro Norte, Quadra 5, Lote A
Tel: (061) 321-1777, fax: 226-2698

Mato Grosso/Pantanal Corumbá
Nacional Palace
Rua America 936
Tel: (067) 231-6868,
fax: (067) 231-6202
Pousada do Cachimbo
Rua Alan Kardec 4
Dom Bosco
Tel/fax: (067) 231-3910
Santa Monica
Rua Antonio Maria Coelho 345
Tel: (067) 231-3001,
fax: (067) 231-7880

Cuiabá
Aurea Palace
Avenida General Mello 63
Centro
Tel/fax: (065) 322-3377
Fazenda Mato Grosso
Rua Antonio Dorielio 1200
Coxipó
Tel: (065) 611-1200

Pantanal
The following establishments offer all-inclusive packages, which cost about $120 per person per day.
Botel Amazonas/Botel Corumbá
(boat hotels)
Information in Corumbá
Tel: (067) 231-2871,
fax: (067) 2231-1025
Cabana do Lontra
Estrada Miranda-Corumbá
Information in Aquidauna
Tel: (067) 384-4898/383-4532
Specializes in fishing tours.
Cabanas do Pantanal
Rio Piraim
Information in Cuiabá
Tel: (065) 321-4142
On the northern edge of the Pantanal. Chalet rooms with meals included. Boat trips, horse-riding, fishing are also available.
Caiman Ecological Reserve
Rua Campos Bicudo, 98
36 km (22 miles) from Miranda, 236

km (147 miles) from Campo Grande
Tel: (11) 883-6622, fax: 883-6037
Four comfortable lodges with excellent wildlife-watching and facilities. Highly recommended.

Olinda
Marolinda
Avenida Beira-Mar 1615
Tel: (081) 429-1699, fax: 326-6934
Pousada dos Quatro Cantos
Rua Prudente de Morais 441
Tel: (081) 429-3333, fax: 429-1845

Recife
Recife Palace
Avenida Boa Viagem 4070,
Boa Viagem
Tel: (081) 465-6688, fax: 465-6767

Salvador
Bahia Othon Palace
Avenida Presidente Vargas 2456,
Ondina
Tel: (071) 247-1044, fax: 245-4877
Email: bahia@othon.com.br
Club Mediterranee
Estrada Itaparica-Nazaré, km 13
Itaparica
Tel: (071) 833-1141, fax: 241-0100

Santarém
Santarém Palace
Avenida Rui Barbosa 726
Tel: 522-5285
Fax: 523-2820
Tropical Santarém
Avenida Mendonça, Furtado 4120
Tel: 523-2800, fax: 522-2631

What to Eat

A country as large and diverse as Brazil naturally has regional food specialties. Immigrants, too, influence Brazilian cuisine. In some parts of the south, the cuisine reflects a German influence; Italian and Japanese immigrants brought their cooking skills to São Paulo. Some of the most traditional Brazilian dishes are adaptations of Portuguese or African foods. But the staples for many Brazilians are rice, beans and manioc.

Lunch is the heaviest meal of the day and you might find it very heavy indeed for the hot climate.

Breakfast is most commonly *café com leite* (hot milk with coffee) with bread and sometimes fruit. Supper is often taken quite late.

Although not a great variety of herbs are used, Brazilian food is tastily seasoned, not usually peppery – with the exception of some very spicy dishes from Bahia. Many Brazilians do enjoy hot pepper *(pimenta)* and the local *malagueta* chilis can be infernally fiery or pleasantly nippy, depending on how they're prepared. But the pepper sauce (most restaurants prepare their own, sometimes jealously guarding the recipe) is almost always served separately so the option is yours.

Considered Brazil's national dish (although not found in all parts of the country), *feijoada* consists of black beans simmered with a variety of dried, salted and smoked meats. Originally made out of odds and ends to feed the slaves, nowadays the tail, ears, feet, etc. of a pig are thrown in. *Feijoada* for lunch on Saturday has become somewhat of an institution in Rio de Janeiro, where it is served completa with white rice, finely shredded kale *(couve)*, *farofa* (manioc root meal toasted with butter) and sliced oranges.

The most unusual Brazilian food is found in Bahia, where a distinct African influence can be tasted in the *dendê* palm oil and coconut milk. The Bahianos are fond of pepper and many dishes call for ground raw peanuts or cashew nuts and dried shrimp. Some of the most famous Bahian dishes are *vatapá* (fresh and dried shrimp, fish, ground raw peanuts, coconut milk, *dendê* oil and seasonings thickened with bread into a creamy mush); *moqueca* (fish, shrimp, crab or a mixture of seafood in a *dendê* oil and coconut milk sauce); *xinxim de galinha* (a chicken fricassé with *dendê* oil, dried shrimp and ground raw peanuts); *caruru* (a shrimp-okra gumbo with *dendê* oil); *bobó de camarão* (cooked and mashed manioc root with shrimp, *dendê* oil and coconut milk); and *acarajé* (a patty made of ground beans fried in *dendê* oil and filled with *vatapá*, dried shrimp and pimenta).

Although it is delicious, note that the palm oil as well as the coconut milk can be too rich for some delicate digestive tracts.

Seafood is plentiful all along the coast, but the northeast is particularly famed for its fish, shrimp, crabs and lobster. Sometimes cooked with coconut milk, other ingredients that add a nice touch to Brazilian seafood dishes are coriander, lemon juice and garlic. Try *peixe a Brasileira*, a fish stew served with *pirão* (manioc root meal cooked with broth from the stew to the consistency of porridge) and a traditional dish served along the coast. One of the tastiest varieties of fish is *badejo*, a sea bass with firm white meat.

A favorite with foreign visitors and very popular all over Brazil is the *churrasco* or barbecue, which originated with the southern gaucho cowboys who roasted meat over an open fire. Some of the finest *churrasco* can be eaten in the south. Most *churrascarias* offer a *rodizio* option: for a set price diners eat all they can of a variety of meats.

Drinking Notes

Brazilians are great social drinkers and love to sit for hours talking and often singing with friends over drinks. During the hottest months, this will usually be in open air restaurants where most of the people will be ordering *chope*, cold draft beer, perfect for the hot weather. Brazilian beers are really very good. Take note that although *cerveja* means beer, it is usually used to refer to bottled beer only. Brazil's own unique brew is *cachaça*, a strong liquor distilled from sugar cane; it's a type of rum, but with its own distinct flavor. Although usually colorless, it can also be amber. Each region boasts of its locally produced *cachaça*, also called *pinga*, *cana* or *aguardente*, but traditional producers include the states of Minas Gerais, Rio de Janeiro, São Paulo and in the northeastern states where the sugar cane has long been a cash crop.

Out of *cachaça*, some of the most delightful mixed drinks are concocted. Tops is the popular *caipirinha*, also considered the national drink. It's really a simple concoction of crushed lime – peel included – and sugar topped with plenty of ice. Variations on this drink are made using vodka or rum, but you should try the real thing. Some bars and restaurants mix their *caipirinhas* sweeter than you may want – order yours com *pouco açúcar* (with a small amount of sugar) or even *sem açúcar* (without sugar). *Batidas* are beaten in the blender or shaken and come in as many varieties as there are types of fruit in the tropics. Basically fruit juice with *cachaça*, some are also prepared with sweetened condensed milk. Favorites are *batida de maracujá* (passion fruit) and *batida de coco* (coconut milk), exotic flavors for visitors from cooler climates. When sipping *batidas*, don't forget that the cachaça makes them a potent drink, even though they taste like fruit juice.

Finally there is wonderful Brazilian coffee. Café is roasted dark, ground fine, prepared strong and taken with plenty of sugar. Coffee mixed with hot milk *(café com leite)* is the traditional breakfast beverage throughout Brazil. Other than at breakfast, it is served black in tiny demitasse cups, never with a meal. (And decaffeinated is not in the Brazilian vocabulary.) These *cafezinhos* or "little coffees", offered to visitors at any home or office, are served piping hot at *botequim* (there are even little stand-up bars that serve only *cafezinho*). However you like it, Brazilian coffee makes the perfect ending to every meal.

A few exotic dishes found in the Amazon include those prepared with *tucupi* (made from manioc leaves and having a slightly numbing effect on the tongue), especially *pato no tucupi* (duck) and *tacacá* broth with manioc starch. There are also many varieties of fruit that are found nowhere else.

The rivers produce a great variety of fish, including piranha. River fish is also the staple in the Pantanal.

Two Portuguese dishes that are popular in Brazil are *bacalhau* (imported dried salted codfish) and *cosido*, a glorified "boiled dinner" of meats and vegetables (usually several root vegetables, squash and cabbage and/or kale) served with *pirão* made out of broth. Also try delicate palmito palm heart, served as a salad, soup or pastry filling.

Salgadinhos are a Brazilian style of finger food, served as appetizers, canapés, ordered with

Boat Tours

Local boat tours and excursions are available in coastal and riverside cities as well as local ferries across rivers and bays and to islands. Sometimes it is possible to hire a schooner and yacht complete with crew for a special outing. Boat trips can be taken on the Rio São Francisco in the northeast and in the Pantanal marshlands of Mato Grosso, popular for wildlife safaris and angling. The Blue Star Line will take passengers on its freighters which call at several Atlantic coast ports. Passenger trips with cargo operators can be booked through Strand cruise and Travel Centre, London, tel: (020) 7836-6363. There are now many cruises which stop along the Brazilian coast on the way down to Buenos Aires or up to the Caribbean: contact the Brazilian Tourist office in London on (020) 7399-9000 or www. Brazil.org.uk for an up-to-date list of cruise operators.

Although it is not normally possible to see much wildlife on a river boat trip in Amazonia, it is an ideal way to gain an insight into the daily life of the river and its inhabitants, with the opportunity to meet some of the indigenous people. Manaus is the hub of Brazilian Amazonia, with services west to Tabatinga on the border of Colombia and Peru, south to Porto Velho on the Rio Madeira, northwest to São Gabriel da Cachoeira on the Rio Negro and west to Santarém and Belém at the mouth of the Amazon. There are Amazon River boat trips lasting a day or two to up to a week or more. These range from luxury floating hotels to more rustic accommodations.

The government-owned ENASA service for tourists ply the Belém–Manaus–Santarém–Belém run over 12 days. The modern boats have air-conditioned cabins, swimming pool, etc. Cost is US$540 per person in a 4-person cabin. Offices in Avenida Pres Vargas 41, Belém, and Rua Marechal Deoforo, Manaus. Hammock space on the somewhat irregular ENASA passenger boats costs US$54 from Belém to Manaus, and vice versa.

Other services make the run, often in better conditions than ENASA boats. They take five to six days from Manaus to Belém, as well as working along the more remote tributaries to Colombia and Peru. Arranging a passage is a matter of patience and luck in most cases as there are no fixed schedules for bookings.

Lower Amazon

The *Catamarã Pará*, an ENASA boat, travels between Belém and Manaus. A double cabin costs US$540 or a quadruple cabin US$496.

Book through: ENASA, Rua Marechal Deodoro 61, tel: (092) 633-3280; or BR Online Travel to Brazil, 1110 Brickell Avenue Suite 404 – Miami FL 33131, USA. Tel (toll-free): (1-888) 527-2745 or (305) 3790005, fax: (305)-379-9397
Email: brol@brol.com
www.brol.com

Tour Boats out of Manaus

Where addresses are not given, see *Travel Agents, page 342.*
Viverdi Tours:
Amazon Clipper and Selly Clipper – Three days and two nights on the Amazon River, US$395; four days and three nights on the Negro River, US$550.
Amazon Nut Safaris:
Rua 5 de Setembro, 43 Sao Reimundo, Manaus, tel: (092) 234-5860, fax: (092) 671-1415.
Cassiquira – 24 persons, double cabins with air conditioning, US$135 per day.
Iguana, comfortable accommodation for up to eight passengers in four double cabins, US$135 per day.
Bumerangue – an open boat without cabins; space for six passengers in stockbeds.
Canoros: *Tuna*
Rua Papa João 23, No. 325 – Centro Cívico, Curitiba – PR – Cep: 80530-020, tel/fax: (041) 254-8024, or (041) 352-3876.
Email: canoros@buziosonline. com.br
Swallows and Amazon Tours
Rua Quintino Bocaiuva, 189 Sala 13, Manaus
Tel/fax: (092) 622-1246
www.swallowsandamazonstours. com
Highly recommended small company with wide range of riverboat tours and new 7 cabin air-conditioned houseboat.
Amazon Tours & Cruises:
Head Office
Requena 336, Iquitos, Peru
Tel: (51-94) 233-931, fax: (51-94) 231-265
Email: amazon@amazoncruises. com.pe
Based in Peru, the company offers a wide range of Amazon Tours including the M/V Marcelita which has 26 air-conditioned cabins and sails from Manaus via Leticia in Colombia to Iquitos in Peru on a monthly scheduled basis.

a round of beer or as a quick snack at a lunch counter – a native alternative to US-style fast food chains. *Salgadinhos* are usually small pastries stuffed with cheese, ham, shrimp, chicken, ground beef, palmito, etc. There are also fish balls and meat croquettes, breaded shrimp and miniature quiches. Some of the bakeries have excellent *salgadinhos* which you can either take home or eat at the counter with a fruit juice or soft drink. Other tasty snack foods include *pão de queijo* (a cheesy quick bread), and *pastel* (two layers of a thinly rolled pasta-like dough with a filling sealed between, deep-fried). Instead of French-fried potatoes, try *aipim frito* (deep-fried manioc root).

Culture

MUSEUMS

Rio de Janeiro
Folk Art Museum (Museu do Folclore Edison Carneiro), Rua do Catete, tel: (021) 285-0891.
Tues–Fri 11am–6pm; Sat, Sun and holidays 3–6pm.
Indian Museum (Museu do Indio) Rua das Palmeiras 55, Botafogo, tel: (021) 286-8799.
Tues–Fri 10am–5pm;
Sat and Sun 1–5pm.
Museum of Modern Art (Museu de Arte Moderna), Avenida Infante D. Henrique, 85, Parque do Flamengo, tel: (021) 210-2188.
Tues–Sun noon–6pm.
National History Museum (Museu Histórico Nacional), Praça Marechal Ancora (near Praça 15 de Novembro), Centro, tel: (021) 240-9529.
Tues–Fri 10am–5.30pm; Sat, Sun and holidays 2.30–5.30pm.
National Museum (Museu Nacional) Quinta da Boa Vista, São Cristóvão, tel: (021) 567-6316.
Tues–Sun 10am–4pm.
National Museum of Fine Arts (Museu Nacional de Belas Artes) Avenida Rio Branco 199, Centro, tel: (021) 240-9869.
Tues–Fri 10am–6pm; Sat, Sun and holidays 2–6pm.

São Paulo
Museum of Brazilian Art (Museu de Arte Brasileira), Rua Alagoas 903, Higienópolis, tel: (011) 826-4233.
Tues–Fri 2–10pm; Sat, Sun and holidays 1–6pm.
Museum of Contemporary Art (Museu de Arte Contemporanea) Parque do Ibirapuera, Pavilhão da Bienal, 3° andar, tel: (011) 571-9610.
Tues, Wed, Fri noon–5.30 pm, Thurs noon–8pm, Sat–Sun 10am–5.30pm, closed holidays.
Museum of Image and Sound – Cinema (Museu da Imagem e do Som), Avenida Europa 158, Jardim Europa, tel: (011) 852-9197.
Tues–Fri 2–6pm.
Museum of Modern Art (Museu de Arte Moderna), Parque do

Music & Dance

Musical forms have developed in different parts of the country, many with accompanying forms of dance. While the Brazilian influence (especially in jazz) is heard around the world, what little is known of Brazilian music outside the country is just the tip of the iceberg. Take in a concert by a popular singer or ask your hotel to recommend a nightclub with live Brazilian music: *bossa nova, samba, choro* and *seresta* are popular in Rio and São Paulo.

During Carnival, you'll see and hear plenty of music and dancing in the streets, mostly *samba* in Rio and *frevo* in the northeast. There are also all-year shows for tourists. If you like what you hear, bring back some tapes and CDs.

The classical music and dance season runs from Carnival in February through mid-December. Besides local presentations, major cities (mainly Rio, São Paulo and Brasília) are included in tours by many international artists. One of the most important classical music festivals in South America takes place in July each year in Campos do Jordão in the state of São Paulo.

Ibirapuera, Grande Marquise, tel: (011) 549-9688.
Tues–Fri 1–7pm; Sat and Sun 11am–7pm.

Manaus
Indian Museum (Museu do Indio) Rua Duque de Caxias, Avenida 7 de Setembro, tel: (092) 234-1422 Mon–Fri 8am–noon and 2–5pm, Sat 8.30–1130am, closed Sun.
Museum of the Port of Manaus (Museu do Porto de Manaus) Boulevard Vivaldo Lima, Centro, tel: (092) 633-3433.
Tues–Sun 7–11am and 1–5pm.

Ouro Preto
Inconfidencia Historial Museum (Museu da Inconfidência), Praça Tiradentes.
Tues–Sun noon–5.30pm.

Salvador
Afro-Brazilian Museum (Museu Afro-Brasileiro).
Old medical school/Faculdade de Medicina building, Terreiro de Jesus, tel: (071) 243-0384.
Tues–Sat 9am–5pm.

Shopping

Brazil has a tremendous variety of **gemstones** not found elsewhere. It produces amethysts, aquamarines, opals, topazes, the many-colored tourmaline, as well as diamonds, emeralds, rubies and sapphires. It's wiser to buy from a reliable jeweler, where you will get what you pay for and can trust their advice. The three leading jewelers operating nationwide are H. Stern, Amsterdam Sauer and Roditi, but there are other reliable smaller chains.

Another good buy in Brazil are **leather** goods, especially shoes, sandals, bags, wallets and belts. Although found everywhere, some of the finest leather comes from the south. Shoes are plentiful and handmade leather items can be found at handicraft street fairs.

GLOSSARY

A list of the common and scientific names of Amazon mammals and birds mentioned in this book.

This is obviously not a complete listing but includes some of the most frequently seen species.

Mammals

COMMON NAME	SCIENTIFIC NAME
Acouchi, Red	*Myoprocta acouchy (exilis)*
Acouchi, Green	*Myoprocta pratti*
Agouti, Red-rumped	*Dasyprocta agouti*
Alpaca	*Lama pacos*
Anteater, Giant	*Myrmecophaga tridactyla*
Anteater, Lesser =	
Collared Anteater =	
Southern Tamandua	*Tamandua tetradactyla*
Anteater, Silky =	
Anteater, Pigmy	*Cyclopes didactylus*
Armadillo, Giant	*Priodontes maximus=*
	Priodontes giganteus
Armadillo, Nine-banded	*Dasypus novemcinctus*
Armadillo, Three-banded	*Tolypeutes matacus*
Armadillo, Yellow =	
Armadillo, Six-banded	*Euphractes sexcinctus*
Bat, Bulldog = Bat, Fishing	*Noctilio leporinus*
Bat, Common Vampire	*Desmodus rotundus*
Bat, Proboscis = Bat, Long-nosed	*Rhynchonycteris naso*
Bear, Spectacled = Bear, Andean	*Tremarctos ornatus*
Capuchin monkey, Brown	*Cebus apella*
Capuchin monkey, Weeping =	
Wedge-capped	*Cebus olivaceus*
Capuchin monkey, White fronted	*Cebus albifrons*
Capybara	*Hydrochaeris hydrochaeris*
Coati, South American	*Nasua nasua*
Cuy, Sacha	*Cavia spec.*
Deer, Gray Brocket	*Mazama gouazoubira*
Deer, Marsh	*Blastocerus dichotomus*
Deer, Red Brocket	*Mazama americana*
Deer, White-tailed	*Odocoileus virginianus*
Dolphin, Amazonian Pink River =	
Bouto/Boto	*Inia geoffrensis*
Dolphin, Gray = Tucuxi	*Sotalia fluviatilis*
Dog, Bush	*Speothos venaticus*
Dog, Small-eared = Dog, Short-eared	*Atelocynus microtis*
Fox, Andean	*Dusicyon griseus*
Fox, Crab-eating	*Cerdocyon thous*
Grison	*Galictis vittata*
Jaguar	*Panthera onca*
Jaguarundi	*Felis jagouarundi =*
	Herpailurus yaguarondi
Kinkajou	*Potos flavus*
Manatee, Amazonian	*Trichechus inunguis*
Margay	*Felis/Leopardus wiedii*
Marmoset, Bare-eared =	
Marmoset, silvery	*Callithrix argentata*
Marmoset, Buffy Tufted-ear	*Callithrix jacchus aurita*
Marmoset, Pigmy	*Cebuella pygmaea*
Marmoset, Tassel-eared	*Callithrix humeralifera*
Marmoset, White Tufted-ear =	
Common	*Callithrix jacchus* (over five sub spec.)
Marmoset, Yellow-handed	*Saguinus midas*
Monkey, Black Howler	*Alouatta caraya*
Monkey, Brown Howler	*Alouatta fusca*
Monkey, Common Squirrel	*Saimiri sciureus*
Monkey, Common Woolly =	
Monkey, Common Humboldt's	*Lagothrix lagothricha*
Monkey, Goeldi's	*Callimico goeldii*
Monkey, Mantled Howler	*Alouatta palliata*
Monkey, Night	*Aotus spec.*
Monkey, Red Howler	*Alouatta seniculus*
Monkey, Red-handed Howler	*Alouatta belzebul/belcebul*
Monkey, Spider =	
Monkey, Black Spider	*Ateles paniscus*
Monkey, Yellow-tailed Woolly	*Lagothrix flavicauda*
Mouse, House	*Mus musculus*
Muriqui =	
Monkey, Woolly Spider	*Brachyteles arachnoides*
Ocelot	*Felis/Leopardus pardalis*
Olingo	*Bassaricyon gabbii*
Oncilla	*Felis tigrina = Leopardus tigrinus*
Opossum, Bare-tailed Woolly	*Caluromys philander*
Opossum, Common	*Didelphis marsupialis*
Opossum, Common Gray four-eyed	*Philander opossum*
Opossum, Water	*Chironectes minimus*
Opossum, Western Woolly	*Caluromys lanatus*
Otter, Giant	*Pteronura brasiliensis*
Otter, Southern River =	
Otter, Neotropical	*Lutra longicaudis*
Paca	*Agouti paca*
Peccary, Collared	*Tayassu tajacu*
Peccary, White-lipped	*Tayassu pecari*
Porcupine, Prehensile-tailed/	
Brazilian	*Coendou prehensilis*
Pudu	*Pudu pudu*
Puma	*Felis concolor*
Raccoon, Crab-eating	*Procyon cancrivorous*
Rat, Amazon Bamboo	*Dactylomus dactylinus*
Rat, Black	*Rattus rattus*
Rat, Spiny	*Proechymis spec.*
Rat, Tree	various spec., including *Echymis,*
	Makalata, Nelomys

Saki, Monk	*Pithecia monachus*
Saki, Red-backed/Brown-bearded	*Chiropotes satanas*
Saki, White-faced/Guianan	*Pithecia pithecia*
Skunk, Striped Hog-nosed	*Conepatus semistriatus*
Sloth, Brown-throated Three-toed	*Bradypus variegatus*
Sloth, Hoffmann's Two-toed	*Choloepus hoffmanni*
Sloth, Maned	*Bradypus torquatus*
Sloth, Pale-throated Three-toed	*Bradypus tridactylus*
Sloth, Southern Two-toed	*Choloepus didactylus*
Squirrel, Fire-vented Tree/Northern Amazon Red	*Sciurus igniventris*
Squirrel, Gray Tree	*Sciurus aestuans*
Squirrel, Southern Amazon Red	*Sciurus spadiceus*
Tamarin, Black-mantle	*Saguinus nigricollis*
Tamarin, Cotton-top	*Saguinus oedipus*
Tamarin, Emperor	*Saguinus imperator*
Tamarin, Golden Lion	*Leontopithecus rosalia*
Tamarin, Golden-handed/Red-handed	*Saguinus midas*
Tamarin, Saddle-back	*Saguinus fuscicollis*
Tapir, Baird's	*Tapirus bairdii*
Tapir, Brazilian/Lowland	*Tapirus terrestris*
Tapir, Mountain	*Tapirus pinchaque*
Tayra	*Eira barbara*
Titi, Dusky	*Callicebus moloch*
Titi, Masked	*Callicebus personatus*
Uakari, Black	*Cacajao melanocephalus*
Uakari, Red	*Cacajao calvus*
Vicuna	*Vicugna vicugna*
Wolf, Maned	*Chrysocyon brachyurus*

Birds

Amazon/Parrot, Festive	*Amazona festiva*
Amazon/Parrot, Mealy	*Amazona farinosa*
Amazon/Parrot, Orange-winged	*Amazona amazonica*
Amazon/Parrot, Red-spectacled	*Amazona pretrei*
Amazon/Parrot, Vinaceous = Amazon, Vinaceous-breasted	*Amazona vinacea*
Anhinga	*Anhinga anhinga*
Ani, Greater	*Crotophaga major*
Antbird, White-bellied	*Myrmeciza longipes*
Antpitta, Fulvous-bellied	*Hylopezus fulviventris*
Antshrike, Variable	*Thamnophilus caerulescens*
Ant-Tanager	*Habia* spec.
Bare-eye, Black-spotted	*Phlegopsis nigromaculata*
Booby, Brown	*Sula leucogaster*
Booby, Masked	*Sula dactylatra*
Buzzard-Eagle, Black-chested	*Geranoaetus melanoleucus*
Cacique, Red-rumped	*Cacicus haemorrhous*
Cacique, Yellow-rumped	*Cacicus cela*
Canastero, Many-striped	*Asthenes flammulata*
Caracara, Carunculated	*Phalcoboenus caruncul atus*

Caracara, Crested	*Polyborus plancus*
Caracara, Mountain	*Phalcoboenus megalopterus*
Caracara, Red-throated	*Daptrius americanus*
Caracara, Yellow-headed	*Milvago chimachima*
Cardinal, Red-capped	*Paroaria gularis*
Chachalaca	*Ortalis* spec.
Chlorophonia, Blue-naped	*Chlorophonia cyanea*
Cock-of-the-Rock, Andean	*Rupicola peruviana*
Cock-of-the-Rock, Guianan	*Rupicola rupicola*
Condor, Andean	*Vultur gryphus*
Conebill, Rufous-browed	*Conirostrum rufum*
Conure, Crimson-bellied	*Pyrrhura perlata*
Conure/Parakeet, Dusky-headed	*Aratinga weddellii*
Conure, Golden	*Guaruba guarouba* = (*Aratinga guarouba*)
Conure/Parrot/Parakeet, Golden-plumed	*Leptosittaca branickii*
Conure/Parakeet, Mitred	*Aratinga mitrata*
Conure/Parakeet, Pearly	*Pyrrhura lepida*
Cormorant, Neotropical = Cormorant, Olivaceous	*Phalacrocorax olivaceus*
Cotinga, Black-faced	*Conioptilon mcilhenny*
Cotinga, Black-necked Red	*Phoenicircus nigricollis*
Cotinga, Plum-throated	*Cotinga maynana*
Cuckoo, Black-bellied	*Piaya melanogaster*
Cuckoo, Black-billed	*Coccyzus erythropthalmus*
Cuckoo, Guira	*Guira guira*
Cuckoo, Squirrel	*Piaya cayana*
Cuckoo, Yellow-billed	*Coccyzus americanus*
Curassow, Black	*Crax alector*
Curassow, Nocturnal	*Nothocrax urumutum*
Curassow, Razor-billed	*Mitu mitu*
Curassow, Southern Horned	*Pauxi unicornis*
Dacnis, Blue	*Dacnis cayana*
Dipper, White-capped	*Cinclus leucocephalus*
Donacobius, Black-capped = Mocking-Thrush, Black-capped	*Donacobius atricapillus*
Duck, Muscovy	*Cairina moschata*
Duck, Torrent	*Merganetta armata*
Eagle, Crested	*Morphnus guianensis*
Eagle, Harpy	*Harpia harpyja*
Egret, Great	*Casmerodius albus* = *Egretta alba*
Emerald, Glittering-throated	*Amazilia fimbriata*
Fairy, Black-eared	*Heliothrix aurita*
Falcon, Aplomado	*Falco femoralis*
Falcon, Bat	*Falco rufigularis*
Flowerpiercer, Masked	*Diglossa cyanea*
Flycatcher, Boat-billed	*Megarhynchus pitangua*
Flycatcher, Fork-tailed	*Tyrannus savana* = *Muscivora tyrannus*
Foliage-gleaner	*Philydor* etc. spec.
Frigatebird, Magnificent	*Fregata magnificens*
Fruitcrow, Bare-necked	*Gymnoderus foetidus*
Gallinule, Azure	*Porphyrula flavirostris*

Gallinule, Purple	*Porphyrula martinica*
Gnateater, Ash-throated	*Conopophaga peruviana*
Gnateater, Slaty	*Conopophaga ardesiaca*
Goose, Orinoco	*Neochen jubata*
Greenlet, Lesser	*Hylophilus minor*
Grosbeak, Slate-colored	*Pitylus grossus*
Ground-Cuckoo, Banded	*Neomorphus radiolosus*
Ground-Cuckoo, Red-billed	*Neomorphus pucheranii*
Ground-Dove, Ruddy	*Columbina talpacoti*
Guan, Andean	*Penelope montagnii*
Guan, Black-fronted Piping	*Pipile jacutinga =*
	Aburria jacutinga
Guan, Common Piping	*Aburria pipile*
Guan, Marail	*Penelope marail*
Guan, Rusty-margined	*Penelope superciliaris*
Hawk, Puna	*Buteo poecilochrous*
Hawk, Red-backed	*Buteo polyosoma*
Hawk, Roadside	*Buteo magnirostris*
Helmetcrest, Bearded	*Oxypogon guerinii*
Hermit, Straight-billed	*Phaethornis bourcieri*
Heron, Agami = Heron,	
Chestnut-bellied	*Agamia agami*
Heron, Boat-billed	*Cochlearius cochlearius*
Heron, Cocoi =	
Heron, White-necked	*Ardea cocoi*
Heron, Great Blue	*Ardea herodias*
Heron, Little Blue	*Egretta caerulea = Florida caerulea*
Heron, Zigzag	*Zebrilus undulatus*
Hillstar, Andean	*Oreotrochilus estella*
Hoatzin	*Opisthocomus hoazin*
Honeycreeper, Green	*Chlorophanes spiza*
Honeycreeper, Red-legged	*Cyanerpes cyaneus*
Hornero, Pale-legged	*Furnarius leucopus*
Hummingbird, Giant	*Patagona gigas*
Hummingbird, Swallow-tailed	*Eupetomena macroura*
Hummingbird, Sword-billed	*Ensifera ensifera*
Ibis, Green	*Mesembrinibis cayennensis*
Ibis, Scarlet	*Eudocimus ruber*
Ibis, Wood =	
Wood-Stork, American	*Mycteria americana*
Inca, Collared	*Coeligena torquata*
Jabirú	*Jabiru mycteria*
Jacamar	*Galbula* spec.
Jacana, Wattled	*Jacana jacana*
Jay, Azure	*Cyanocorax caeruleus*
Jay, Green	*Cyanocorax yncas*
Jewelfront = Jewelfront, Gould's	*Polyplancta aurescens*
Kingfisher, Amazon	*Chloroceryle amazona*
Kingfisher, Green	*Chloroceryle americana*
Kingfisher, Green-and-Rufous	*Chloroceryle inda*
Kingfisher, Pigmy	*Chloroceryle aenea*
Kingfisher, Ringed	*Ceryle torquata*
Kiskadee, Great	*Pitangus sulfuratus*
Kite, Snail = Kite, Everglade	*Rostrhamus sociabilis*
Lapwing, Andean	*Vanellus resplendens*

Lapwing, Southern	*Vanellus chilensis*
Leaftosser, Short-billed	*Sclerurus rufigularis*
Macaw, Blue-and-Yellow	*Ara ararauna*
Macaw, Glaucous	*Anodorhynchus glaucus*
Macaw, Hyacinth	*Anodorhynchus hyacinthinus*
Macaw, Red-and-Green	*Ara chloroptera*
Macaw, Red-fronted	*Ara rubrogenys*
Macaw, Scarlet	*Ara macao*
Macaw, Spix's	*Cyanopsitta spixii*
Macaw, Yellow-collared =	
Macaw, Golden-collared =	
Macaw, Golden-naped	*Propyrrhura auricollis =*
	Ara auricollis
Manakin, Blue =	
Manakin, Swallow-tailed	*Chiroxiphia caudata*
Manakin, Blue-backed	*Chiroxiphia pareola*
Manakin, Golden-headed	*Pipra erythrocephala*
Manakin, Pin-tailed	*Ilicura militaris*
Manakin, White-bearded	*Manacus manacus*
Manakin, Wire-tailed	*Pipra filicauda =*
	Teleonema filicauda
Mango, Black-throated	*Anthracocorax nigricollis*
Martin, Bank = Swallow, Bank	*Riparia riparia*
Martin, Southern	*Procne modesta*
Merganser, Brazilian	*Mergus octosetaceus*
Metaltail, Tyrian	*Metallura tyrianthina*
Motmot, Highland	*Momotus aequatorialis*
Mountain-Tanager,	
Scarlet-bellied	*Anisognathus igniventris*
Mountaineer, Bearded	*Oreonympha nobilis*
Nighthawk, Band-tailed	*Nyctiprogne leucopyga*
Nighthawk, Short-tailed	*Lurocalis semitorquatus*
Nightjar, Blackish	*Caprimulgus nigrescens*
Nothura, Spotted =	
Tinamou, Spotted	*Nothura maculosa*
Nunbird, Black-fronted	*Monasa nigrifrons*
Nunlet, Rusty-breasted	*Nonnula rubecula*
Oropendola, Russet-backed	*Psarocolius angustifrons*
Osprey	*Pandion haliaetus*
Owl, Tropical Screech	*Otus choliba*
Palmcreeper, Point-tailed	*Berlepschia* spec.
[Parakeet, see also Conure]	
Parakeet, Andean	*Bolborhynchus orbygnesius*
Parakeet, Canary-winged	*Brotogeris versicolurus*
Parakeet, Cobalt-winged	*Brotogeris cyanoptera*
Parakeet, Golden-winged	*Brotogeris chrysopterus*
Parakeet, Tui	*Brotogeris sanctithomae*
[Parrot, see also Amazon &	
Conure]	
Parrot, Hawk-headed =	
Parrot, Red Fan	*Deroptyus accipitrinus*
Parrot, Purple-bellied =	
Parrot, Blue-bellied	*Trichlaria malachitacea*
Parrot, Red-capped =	
Parrot, Pileated	*Pionopsitta pileata*

Parrot, Scaly-headed =	
Parrot, Red-tailed	*Pionus maximiliani*
Pauraque	*Nyctidromus albicollis*
Pelican, Brown	*Pelecanus occidentalis*
Pigeon, Ruddy	*Columba subvinacea*
Pigmy-Tyrant	*Euscarthmus* spec.
Piha, Screaming	*Lipaugus vociferans*
Plover, Collared	*Charadrius collaris*
Potoo, Common	*Nyctibius griseus*
Potoo, Great	*Nyctibius grandis*
Potoo, Long-tailed	*Nyctibius aethereus*
Puffbird, Black-breasted	*Notharchus pectoralis*
Puffbird, Rufous-necked	*Malacoptila rufa*
Puffbird, Spotted	*Bucco tamiata*
Quetzal, Golden-headed	*Pharomachrus auriceps*
Quetzal, Pavonine	*Pharomachrus pavoninus*
Redstart, Golden-fronted	*Myioborus ornatus*
Rhea, Greater	*Rhea americana*
Sapphirewing, Great	*Pterophanes cyanopterus*
Screamer, Horned	*Anhima cornuta*
Scythebill, Red-billed	*Campylorhamphus trochilirostris*
Seedeater,	
Chestnut-bellied	*Sporophila castaneiventris*
Seriema, Red-legged	*Cariama cristata*
Shrike-Vireo, Slaty-capped	*Smaragdolanius leucotis*
Sierra-Finch, Plumbeous	*Phrygilus unicolor*
Skimmer, Black	*Rynchops niger* (or *nigra*)
Snipe, Andean	*Gallinago jamesoni*
Snipe, Noble	*Gallinago nobilis*
Spadebill	*Platyrinchus* spec.
Spinetail	*Synallaxis* spec.
Spoonbill, Roseate	*Ajaia ajaja* = (*Platalea ajaja*)
Stork, Jabiru *see* Jabirú	
Stork, Wood =	
Wood-Ibis, American	*Mycteria americana*
Sunbittern	*Eurypyga helias*
Sungrebe = Finfoot	*Heliornis fulicula*
Swallow, Barn	*Hirundo rustica*
Swallow-Wing	*Chelidoptera tenebrosa*
Swift, Andean	*Aeronautes andecolus*
Swift, Short-tailed	*Chaetura brachyura*
Swift, White-collared	*Streptoprocne zonaris*
Tanager, Blue-gray	*Thraupis episcopus*
Tanager, Blue-whiskered	*Tangara johannae*
Tanager, Dusky-faced	*Mitrospingus cassinii*
Tanager, Flame-crested	*Tachyphonus cristatus*
Tanager, Golden-crowned	*Iridosornis rufivertex*
Tanager, Golden-hooded	*Tangara larvata*
Tanager, Grass-green	*Chlorornis riefferii*
Tanager, Gray-and-Gold	*Tangara palmeri*
Tanager, Green-and-gold	*Tangara schrankii*
Tanager, Green-headed	*Tangara seledon*
Tanager, Opal-rumped	*Tangara velia*
Tanager, Palm	*Thraupis palmarum*
Tanager, Paradise	*Tangara chilensis*

Tanager, Red-necked	*Tangara cyanocephala*
Tanager, Scarlet-and-White	*Erythrothlypis salmoni*
Tanager, Silver-beaked	*Ramphocelus carbo*
Tanager, Tawny-crested	*Tachyphonus delatrii*
Teal, Puna	*Anas puna*
Tern, Large-billed	*Phaetusa simplex*
Tern, Yellow-billed	*Sterna superciliaris*
Thistletail, White-chinned	*Schizoeaca fuliginosa*
Thornbill, Rainbow-bearded	*Chalcostigma herrani*
Thrush, Cocoa	*Turdus fumigatus*
Thrush, Glossy Black	*Turdus serranus*
Thrush, Great	*Turdus fuscater*
Thrush, Lawrence's	*Turdus lawrencii*
Thrush, Pale-breasted	*Turdus leucomelas*
Thrush, Rufous-bellied	*Turdus rufiventris*
Tiger-Heron, Rufescent	*Tigrisoma lineatum*
Tinamou, Brown	*Crypturellus obsoletus*
Tinamou, Great	*Tinamus major*
Tinamou, Red-winged	*Rhynchotus rufescens*
Tinamou, Solitary	*Tinamus solitarius*
Toucan, Gray-breasted	
Mountain	*Andigena hypoleuca/hypoglauca*
Toucan, Red-breasted	*Ramphastos dicolorus*
Toucan, Toco	*Ramphastos toco*
Toucan, Yellow-ridged	*Ramphastos culminatus*
Toucanet, Spot-billed	*Selenidera maculirostris*
Trainbearer, Black-tailed	*Lesbia victoriae*
Troupial	*Icterus jamacaii*
Trumpeter, Gray-winged	*Psophia crepitans*
Trumpeter, Pale-winged	*Psophia leucoptera*
Tyrannulet, White-throated	*Mecocerculus leucophrys*
Tyrant, White-winged Black	*Knipolegus aterrimus*
Umbrellabird, Amazonian	*Cephalopterus ornatus*
Umbrellabird, Long-wattled	*Cephalopterus penduliger*
Vulture, Black	*Coragyps atratus*
Vulture, Greater Yellow-headed	*Cathartes melambrotus*
Vulture, Lesser Yellow-headed	*Cathartes burrovianus*
Vulture, King	*Sarcoramphus papa*
Vulture, Turkey	*Cathartes aura*
Warbler, Black-and-White	*Mniotilta varia*
Warbler, Blackburnian	*Dendroica fusca*
Water-Tyrant, Masked	*Fluvicola nengeta*
Woodcreeper	various genera like
	Xiphorhynchus,
	Lepidocolaptes etc.
Woodpecker, Cream-colored	*Celeus flavus*
Woodpecker,	
Crimson-crested	*Campephilus melanoleucos*
Woodpecker, Red-necked	*Campephilus rubricollis*
Woodpecker, Yellow-tufted	*Melanerpes cruentatus*
Wood-Quail, Spot-winged	*Odontophorus capueira*
Wood-Rail, Gray-necked	*Aramides cajanea*
Wren, Musician	*Cyphorhinus arada*
Wren, Nightingale	*Microcerculus marginatus*
Wren,	*Campylorhynchus*
Thrush-like	*turdinus*

Further Reading

General Natural History and the Environment

Anhalzer, J.J. 1990. **National Parks of Ecuador**. Imprenta Mariscal, Quito.

Bright, M. 2000. **Andes to Amazon**. BBC Books, London. A lavishly illustrated guide to Latin American wildlife, covering habitats from the high Andes to the surrounding oceans.

Campbell, D., and Hammond, H.D. (eds) 1989. **Floristic Inventory of Tropical Countries**. New York Botanical Garden.

Diamond, A.W., and Lovejoy, T.E. 1984. **Conservation of Tropical Forest Birds**. International Council for Bird Preservation Technical, Cambridge.

Dixon, J.R., and Soini, P. 1986. **The Reptiles of the Upper Amazon Basil, Iquitos Region, Peru**. Milwaukee Public Museum, Milwaukee.

Dunning, J.S. 1982. **South American Land Birds, A Photographic Aid to Identification**. Harrowood Books, Newton Square, Pennsylvania.

Eisenberg, J.F. 1989. **Mammals of the Neotropics. Vol.I: The Northern Neotropics**. University of Chicago Press.

Emmons, L.H., and Feer, F. 1990. **Neotropical Rainforest Animals, A Field Guide**. University of Chicago Press.

French, R. 1976. **A Guide to the Birds of Trinidad and Tobago**. Harrowood, Valley Forge, Pennsylvania.

Fjeldsa, J. and Krabbe, N. 1989. **Birds of the High Andes**. Apollo Books, Svendborg.

Hilty, S.L. and Brown, W.L. 1986. **A Guide to the Birds of Colombia**. Princeton University Press, Princeton.

Jacobs, M. 1988. **The tropical rain forest, A First Encounter**. Springer Verlag.

Kricher, J. 1989. **A Neotropical Companion: A Guide to the Animals, Plants**. Princeton.

Meyer de Schauensee, R. and Phelps, W.H. 1978. **A Guide to the Birds of Venezuela**. Princeton University Press.

Murphy, W.L. 1987. **A Birder's Guide to Trinidad and Tobago**. Peregrine Enterprises, Maryland.

Orlog, C.C. 1984. **Las Aves Argentinas**. Administracion de Parques Nacionales.

Ridgely, R.S., and Tudor, G. 1989. **The Birds of South America**. Vol. I. *The Oscine Passerines*. University of Texas Press.

Smith, A. 1990. **Explorers of the Amazon**. Viking, London.

Smith, N.J.H. 1982. **Rainforest Corridors: The Trans-amazon Colonization Scheme**. University of California Press, Berkeley.

Tierney, P 2000. **Darkness in El Dorado**. W.W. Norton & Company, New York. The book that sparked off heated debate about the ethics of anthropologists and journalists working in the Amazon.

Brazil

Bradbury, A. 1990 . **Backcountry Brazil: the Pantanal, Amazon, and the North-east Coast**. Hunter Publications, Edson.

Cockburn, A., and Hecht, S.B. 1989. **The Fate of the Forest: Developers, Destroyers, and Defenders of the Amazon**. Verso, London and New York.

Fleming, Peter. 1999. **Brazilian Adventure**. Northwestern University Press, Evanston.

Goulding, M. 1989. **Amazon, the Flooded Forest**. BBC Books.

Hemming, J. 1987. **Amazon Frontier, The Defeat of the Brazilian Indians**. Macmillan, London.

Mee, M. and Mayo, S. **Amazon**. Royal Botanical Gardens, Kew, 1988. A beautifully illustrated guide to the flora of the Amazon by the famous botanical artist, Margaret Mee, with text by Simon Mayo.

Padua, M.T.J., and Filho, A.F.C. **Parques Nacionais do Brasil**. Instituto de Cooperacao Iberoamerica.

Penny, N.D., and Arias, J.R. 1982. **Insects of an Amazon Forest**. Columbia University Press, New York.

Smith, N.J.H, 1999. **The Amazon River Forest: A Natural History of Plants, Animals and People**. Oxford University Press, Oxford.

Trupp, F. 1983. **Amazonas**. Verlag Schroll, Wien.

Colombia

Castano U.C. 1990. **Guia del Sistema de Parques Nacionales de Colombia**. INDERENA, Bogotá.

Dix, R.H. 1983. **The Politics of Colombia**. Praeger, NY.

Harding, C. 1996. **Colombia in Focus**. Latin American Bureau. Authoritative and up-to-date overview of the people, politics, economy and culture.

Hemming, J. 1984. **The Search for El Dorado**. Bogotá. An account of the Spanish conquest of Colombia.

Hilty. S.L., and Brown W.L. 1986. **A guide to the Birds of Colombia**. Princeton University Press. Princeton.

MMA **National Parks Guide**.

Tailly, F. de 1981. **Colombia**. Delachaux and Niestle. With photographs and text in Spanish and English.

Any traveler going to Colombia should read the works of the Nobel prize-winning author Gabriel García Márquez. His short stories and classic work **One Hundred Years of Solitude** (many editions) give an invaluable insight into Colombia's past and society.

Venezuela

Caufield, C. 1985. **In the Rainforest**. Picador.

Craighead George, J. and Allen, G. 1990. **One day in the Tropical Rain Forest**. Ty Crowell, New York.

O'Hanlon, R. 1990. **In Trouble Again: A Journey Between the Orinoco and the Amazon**. Vintage, New York.

Ecuador

Darwin, C. **The Voyage of the "Beagle", 1835–36.**
Hassaurek, F. **Four Years Among The Ecuadorians, 1861–65.**
Michaux, H. 1928. *Ecuador – A Travel Journal.*
Von Hagen, V.W 1949. **Ecuador and the Galápagos Islands: A History.**
Whymper, E. 1891. **Travel Among The Great Andes of the Ecuador.**
Canaday, C. and Jost, L. **Common Birds of Amazonian Ecuador.**
Cuvi, P. 1985. **In The Eyes Of My People.**
Hurtado, O. 1980. **Political Power in Ecuador.**
Jackson, M. 1987. **Galápagos: A Natural History Market.**

Meisch, L. 1987. *Otavalo –* **Weaving, Costume and the Market.**
Rachowiecki, R., Thurber, M. and Wagenhauser, B. 1997. **Climbing and Hiking in Ecuador.**
Ridgeley, R. and Greenfield P. 2001. **The Birds of Ecuador.**

Peru

Frost, P. **Exploring Cuzco.** Available in all Cuzco bookshops.
Muscutt, K. 1988. **Warriors of the Clouds: A Lost Civilization in the Upper Amazon of Peru,** University of New Mexico Press, Albuquerque.
Parris, M. 1993. **Inka Kola,** Phoenix Press, London.

Other Insight Guides

Companion titles to *Amazon Wildlife* cover South America comprehensively. *Insight Guide: South America* provides a 430-page overview of 11 countries, combining incisive text with stunning photography. In addition, there are individual Insight Guides to *Argentina, Buenos Aires, Brazil, Rio de Janeiro, Chile, Ecuador, Peru* and *Venezuela.* There is an **Insight Pocket Guide** to *Peru,* and **Insight Compact Guides** to *Rio de Janeiro* and *Chile.* **Insight Fleximaps,** whose laminated waterproof finish makes them ideal for use in a rainforest, cover *Argentina, Buenos Aires, Sao Paulo, Ecuador* and *Peru.*

Feedback

We do our best to ensure the information in our books is as accurate and up-to-date as possible. The books are updated on a regular basis, using local contacts, who painstakingly add, amend and correct as required. However, some mistakes and omissions are inevitable and we are ultimately reliant on our readers to put us in the picture.

We would welcome your feedback on any details related to your experiences using the book "on the road". Maybe we recommended a hotel that you liked (or another that you didn't), as well as interesting new attractions, or facts and figures you have found out about the country itself. The more details you can give us (particularly with regard to addresses, e-mails and telephone numbers), the better.

We will acknowledge all contributions, and we'll offer an Insight Guide to the best letters received.

Please write to us at:
Insight Guides
PO Box 7910
London SE1 1WE
United Kingdom
Or send e-mail to:
insight@apaguide.demon.co.uk

ART & PHOTO CREDITS

INSIGHT GUIDE
amazon WILDLIFE

Cartographic Editor **Zoë Goodwin**
Production **Linton Donaldson**
Art Direction **Klaus Geisler,
Derrick Lim. Tanvir Virdee**
Picture Research **Hilary Genin**

Index

Numbers in italics refer to photographs

✻ INSIGHT GUIDES

The world's largest collection of visual travel guides

Insight Guides – the Classic Series that puts you in the picture

Alaska	China	Hungary	Munich	South Africa
Alsace	Cologne			South America
Amazon Wildlife	Continental Europe	Iceland	Namibia	South Tyrol
American Southwest	Corsica	India	Native America	Southeast Asia
Amsterdam	Costa Rica	India's Western	Nepal	Wildlife
Argentina	Crete	Himalaya	Netherlands	Spain
Asia, East	Cuba	India, South	New England	Spain, Northern
Asia, South	Cyprus	Indian Wildlife	New Orleans	Spain, Southern
Asia, Southeast	Czech & Slovak	Indonesia	New York City	Sri Lanka
Athens	Republics	Ireland	New York State	Sweden
Atlanta		Israel	New Zealand	Switzerland
Australia	Delhi, Jaipur & Agra	Istanbul	Nile	Sydney
Austria	Denmark	Italy	Normandy	Syria & Lebanon
	Dominican Republic	Italy, Northern	Norway	
Bahamas	Dresden	Italy, Southern		Taiwan
Bali	Dublin		Old South	Tenerife
Baltic States	Düsseldorf	Jamaica	Oman & The UAE	Texas
Bangkok		Japan	Oxford	Thailand
Barbados	East African Wildlife	Java		Tokyo
Barcelona	Eastern Europe	Jerusalem	Pacific Northwest	Trinidad & Tobago
Bay of Naples	Ecuador	Jordan	Pakistan	Tunisia
Beijing	Edinburgh		Paris	Turkey
Belgium	Egypt	Kathmandu	Peru	Turkish Coast
Belize	England	Kenya	Philadelphia	Tuscany
Berlin		Korea	Philippines	
Bermuda	Finland		Poland	Umbria
Boston	Florence	Laos & Cambodia	Portugal	USA: On The Road
Brazil	Florida	Lisbon	Prague	USA: Western States
Brittany	France	Loire Valley	Provence	US National Parks: East
Brussels	France, Southwest	London	Puerto Rico	US National Parks: West
Budapest	Frankfurt	Los Angeles		
Buenos Aires	French Riviera		Rajasthan	Vancouver
Burgundy		Madeira	Rhine	Venezuela
Burma (Myanmar)	Gambia & Senegal	Madrid	Rio de Janeiro	Venice
	Germany	Malaysia	Rockies	Vienna
Cairo	Glasgow	Mallorca & Ibiza	Rome	Vietnam
Calcutta	Gran Canaria	Malta	Russia	
California	Great Britain	Mauritius, Réunion		Wales
California, Northern	Greece	& Seychelles	St Petersburg	Washington DC
California, Southern	Greek Islands	Melbourne	San Francisco	Waterways of Europe
Canada	Guatemala, Belize &	Mexico City	Sardinia	Wild West
Caribbean	Yucatán	Mexico	Scandinavia	
Catalonia		Miami	Scotland	Yemen
Channel Islands	Hamburg	Montreal	Seattle	
Chicago	Hawaii	Morocco	Sicily	
Chile	Hong Kong	Moscow	Singapore	

Complementing the above titles are 120 easy-to-carry Insight Compact Guides, 120 Insight Pocket Guides with full-size pull-out maps and more than 100 laminated easy-fold Insight Maps